Lecture Notes in Biomathematics 99

T.K. Stokes J.M. McGlade R. Law (Eds.)

The Exploitation of Evolving Resources

Proceedings of an International Conference, held at Jülich, Germany, September 3-5, 1991

Springer-Verlag
Berlin Heidelberg New York
London Paris Tokyo
Hong Kong Barcelona
Budapest

Editors

Kevin Stokes
MAFF, Fisheries Laboratory
Pakefield Road
Lowestoft, NR33, OHT, U. K.

Jacqueline M. McGlade
Department of Biological Sciences
University of Warwick
Coventry, CV4 7AL, U. K.

Richard Law
Department of Biology
University of York
York, Y01 5DD, U. K.

Mathematics Subject Classification (1991): 92Bxx

ISBN 3-540-57242-2 Springer-Verlag Berlin Heidelberg New York
ISBN 0-387-57242-2 Springer-Verlag New York Berlin Heidelberg

Printed in Germany

46/3140-543210 - Printed on acid-free paper

PREFACE

The impact of man on the biosphere is profound. Quite apart from our capacity to destroy natural ecosystems and to drive species to extinction, we mould the evolution of the survivors by the selection pressures we apply to them. This has implications for the continued health of our natural biological resources and for the way in which we seek to optimise yield from those resources.

Of these biological resources, fish stocks are particularly important to mankind as a source of protein. On a global basis, fish stocks provide the major source of protein for human consumption from natural ecosystems, amounting to some seventy million tonnes in 1970. Although fisheries management has been extensively developed over the last century, it has not hitherto considered the evolutionary consequences of fishing activity. While this omission may not have been serious in the past, the ever increasing intensity of exploitation and the deteriorating health of fish stocks has generated an urgent need for a better understanding of evolution driven by harvesting and the implications of this for fish stock management.

The foundations for this understanding for the most part come from recent developments in evolutionary biology and are not generally available to fisheries scientists. The purpose of this book is to provide this basis in a form that is both accessible and relevant to fisheries biology.

The papers in the book stem from a meeting at the Theoretical Ecology Working Group, Forschungszentrum, Jülich GmbH, FRG, in September 1991 held to investigate the consequences of selective exploitation of evolving biological resources. The meeting brought together fisheries scientists, biologists and mathematicians involved in research in the fields of quantitative and population genetics, life history evolution and fisheries management, in order:

1) To provide an international forum for fisheries biologists and evolutionary biologists to exchange ideas and knowledge on the management of evolving biological resources.

2) To facilitate the synthesis of knowledge from these hitherto disparate subjects into a single discipline.

3) To identify research priorities in this new discipline.

4) To produce a written record of the proceedings as a spur to future developments in the subject.

The meeting focused on several major research topics relevant to the management of evolving biological resources, namely: Selection Differentials, Reaction Norms, Selection Responses and Management and Evolution. These subjects have been retained as organising themes for the papers in this book.

Clearly, the strength of selection caused by fishing is vital for arguments about the evolutionary consequences of exploitation. As discussed in the papers in the first section on Selection Differentials, we can get some insights into this using information from major exploited fish stocks. It is evident from these studies that the selection differentials can be far from negligible.

Unfortunately, the phenotypic traits most obviously under selection (e.g. body size and life-history traits) are notoriously labile, and likely to change with the changing environment. Shifts in the environment may come about through fishing itself (e.g. a reduction in the population density, leading to greater availability of food), and this creates major difficulties in interpreting the phenotypic changes observed in fish stocks. These issues are considered in the section on Reaction Norms.

The amount and kind of genetic change depends on the genetic architecture of the traits under selection. The section on Selection Responses discusses the genetic covariance functions and matrices that can be used to predict genetic changes in the short term, and how these matrices may themselves undergo evolution in the longer term.

Given that exploitation causes evolution, the question arises as to what would be responsible and effective ways of managing exploited populations. The final, wide-ranging, section on Management and Evolution tackles this question, and considers issues such as ESOHS (Evolutionary stable Optimal Harvesting Strategies), the implications of size-selective harvesting for managing evolving resources and the importance of evolutionary changes relative to other management issues.

The success of the meeting that provided the stimulus for this book owed much to the financial and organisational support of the Forschungszentrum, Jülich GmbH; we would like to put on record our gratitude for all the help provided.

TABLE OF CONTENTS

PART I
SELECTION DIFFERENTIALS

FISHING AS A CAUSE OF EVOLUTION IN FISHES

David Policansky

Board on Environmental Studies and Toxicology, National Research Council, Washington, D.C., 20418, U.S.A.

ABSTRACT

This paper discusses the degree to which fishing causes evolution in fishes and how to detect and measure it. Fishing mortality is often very high and nonrandom with respect to several life-history traits that are at least partly heritable. Therefore, it seems likely that fishing causes evolution in fishes. Selective harvest of experimental populations has produced genetic changes in them, and natural selective agents, especially selective predation, have also caused genetic changes in populations of various species. However, the action of many other factors makes the detection and measurement of evolution difficult, so many observations that show changes in life-history traits of exploited fish populations are not sufficient by themselves to establish the occurrence of evolution. The difficulty of detecting and measuring evolution by observation alone should not be interpreted as evidence that evolution is not occurring; instead, it provides a opportunity for experimental research that has theoretical and practical importance. Some experimental approaches, which should be accompanied by simulations, are discussed.

INTRODUCTION

Fishing is a very old human activity. As fish production by fishing increases in response to increasing demand for fish, the rate of harvest of many fish stocks is increasing also. One clear result of this pressure has been the decline of a number of fish stocks (e.g. Pacific mackerel, Pacific sardine, several Georges Bank groundfish species). This result is well known and has been the subject of a very large literature. But another result also seems likely, although often difficult to demonstrate, namely evolution of exploited populations as a result of harvesting.

Recently, the possibility that fish populations can undergo genetically based changes in response to fishing has become more widely recognized. Several taxonomic groups, notably insects and bacteria, have evolved rapidly in response to unintended selection resulting from the use of pesticides and antibiotics (e.g. National Research Council 1986a, 1986b). There is evidence from experimental fish populations that variation in life-history traits such as growth rate, fecundity, behavior, and age and size at sexual maturation has at least some genetic basis; thus, those traits have the potential to be altered by selection.

Natural processes analogous to fishing pressure, mainly selective predation, have resulted in evolution of a variety of traits, including growth rates, color, morphology, and mating patterns of freshwater crustaceans (Strong 1973), of fishes (H. Gross 1978, Endler 1980, Bell and Richkind 1981, Reznick 1982, Reznick and Endler 1982), plants (Marquis 1984), and of many other organisms as well (Endler 1986). Given the nonrandomness of almost all fishing gear and the high fishing mortality that occurs in many fish stocks, it seems that fish populations must have evolved in response to the long periods of intense exploitation they have been subjected to. Yet, in most cases that have been considered, the jury is still out: it remains difficult to show that evolution has occurred. In this paper, I discuss evidence for evolution as a result of fishing, its probable manifestations, and how they might be detected.

THE PREREQUISITES FOR EVOLUTION

For a fish population to evolve[1] in response to fishing, two factors must be present: heritable variations in the population (genetically based phenotypic variation) and fishing that causes differential reproduction of the different genotypes.

HERITABLE POPULATION VARIATION

Many phenotypic traits have been shown to be influenced by genotypic variation in fishes (Kirpichnikov 1981, Doyle 1983, Gall 1983, Wohlfarth 1983). Some of these are related to size, and hence likely to be affected differentially by fishing gear (see Table 1 for a small sample, with heritability estimates). Other traits with heritable variation include color patterns in guppies (Endler 1980), laterality in flounders (Policansky 1982), egg size in rainbow trout(Gall 1975), number of lateral plates in sticklebacks (Hagen 1973) and age and size at maturation in Pacific and Atlantic salmon (Ricker 1981, Gjerde 1984, Thorpe et al. 1983, M. Gross 1991) and in male platyfish, *Xiphophorus maculatus* (McKenzie et al. 1983). The platyfish example is discussed in more detail below.

DIFFERENTIAL REPRODUCTION OF GENOTYPES CAUSED BY FISHING

The most obvious way that fishing might cause differential reproduction of genotypes is through gear selectivity. Commercial and recreational fishing remove large proportions of fishes from many natural populations, and the removal is often nonrandom. Size-selectivity of fishing gear has been known for centuries. As early as the 14th century, people in England were concerned about the effects of fishing with a 'wondyrchoum', a type of trawl with a fine mesh. They feared that small fish were being killed in enormous numbers, to the 'great damage of the whole commons of the kingdom' (March 1970). By 1716, minimum

---

[1]I use the term 'evolution' in this paper only to mean 'genetically based change in phenotype', recognising that the broad sense includes nongenetic change.

Table 1. Heritabilities (± SE) of some quantitative traits in fishes.

Trait	Heritability	Species	Source
Age at maturity	48 (± .20)	*Salmo salar*	Gjerde, 1984
Body weight	50 (± .13)	*Ictalurus punctatus*	Dunham & Smitherman, 1983
Total length	.35	*Ictalurus punctatus*	Bondari, 1983
Growth rate	26 (± .12)	*Oncorhynchus mykiss**	Kinghorn, 1983
Growth rate	08 (- .15)	*Salmo salar*	Refstie & Steine, 1978
Rapid growth	-.10 (± .02)	*Tilapia nilotica*	Teichert-Coddington & Smitherman, 1988
Slow growth	.36		
Lateral plate number	.50(- .84)	*Gasterosteus aculeatus*	Hagen, 1973

*Rainbow trout; previously known as *Salmo gairdneri*.

mesh sizes *and minimum size limits* for various fish species were in effect, with potentially severe penalties. The clearly stated goal of those regulations was to ensure the selective capture of larger fish.

The literature on gear selectivity, most of which focuses on size-selectivity, is far too large to begin to summarize here, so only a few examples are given. Rollefsen (1953) compared the selectivity of various fishing gears on cod in Lofoten; Hamley (1975) reviewed gillnet selectivity; and Myhre (1969) compared the selectivity of various fishing gears on Pacific halibut (*Hippoglossus stenolepis*). Myhre's results (Figure 1) are illustrative. Many gears, including longlines (Myhre 1969), electrofishing (Stewart 1975), and trap nets (Laarman and Ryckman 1982) selectively capture larger fish; gillnets tend to capture fish within a particular size range depending on their mesh size (Hamley 1975), while trawls typically capture smaller fish (Myhre 1969; Rollefson 1953). Some of these results are intuitively obvious. Large fish can swim fast enough to escape slow-moving trawls, while very small fish slip through the mesh; with gillnets, a fish that is too small passes through the mesh while one that is too large cannot get far enough into it to get caught. Several factors affect longline catches, especially the choice by fishermen of the time and place they fish and the baits and hooks they use.

A recent study by L. Miller (1992) illustrates how complex the effects of gear selection can be. Miller found that the selectivity of gillnets for walleye (*Stizostedion vitreum*) in Red Lakes, Minnesota varied as a function of age, year class, gender, size, and from one year to another. As a result of his findings, Miller pointed out that to make accurate predictions about the selective effect of any fishing regime, it is necessary to obtain information about population characteristics such as year-class strength, growth, and the environmental factors that affect them.

Although size-selectivity is the most obvious selective aspect of nonrandom harvesting, other traits can be selected. Some are related to size (e.g., swimming speed, as described by Rudstam et al. 1984); others, such as behavior or morphology, may not be.

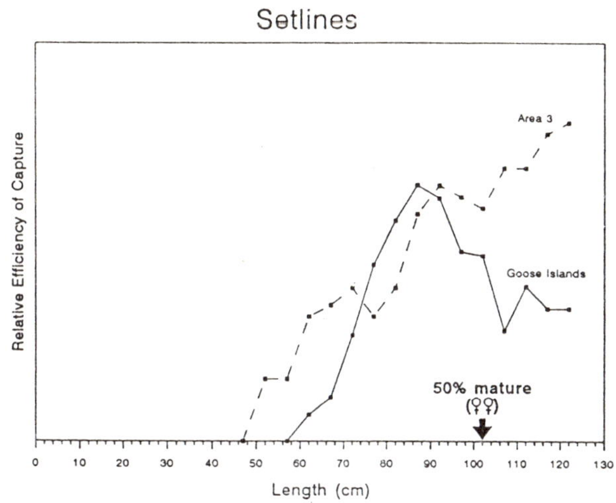

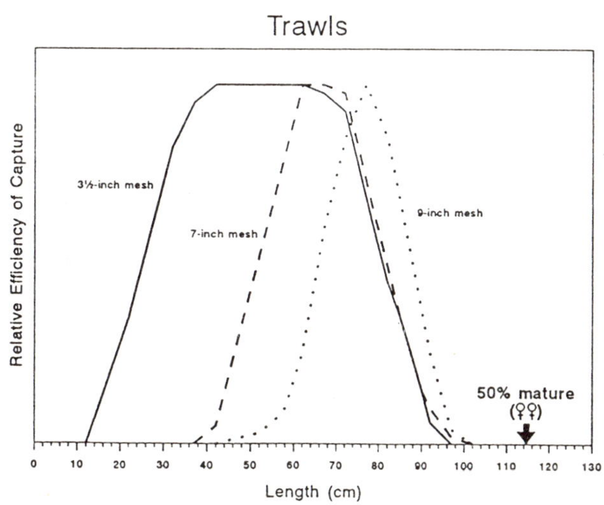

Figure 1. Efficiency of setlines (top) and trawls (bottom) in capturing Pacific halibut (*Hippoglossus stenolepis*) as a function of their length. The Y-axes are not numbered, because the absolute capture efficiency varied in different tests. Redrawn from Myhre 1969.

The nonrandom (i.e. selective) removal of individuals from a population is one of the mechanisms of natural and artificial selection. For many species, such as Atlantic cod (*Gadus morhua*), herring (*Clupea harengus*), plaice (*Pleuronectes platessa*), and sole (*Solea solea*), the period of fishing on some stocks has probably exceeded 40 generations. That span would count as a long experiment in fruitfly (*Drosophila*) or mouse genetics and is more than enough time for selection to produce substantial genetic change on almost every quantitative character that has been examined (Falconer 1989, Gall 1983).

In addition to the effects of gear selectivity as commonly understood, fishing can cause evolution by increasing overall mortality of certain life stages. Fishing mortality rate can exceed 0.7 per year in some marine fisheries (Cushing 1968) and reaches 0.8 for North Sea cod (Rijnsdorp et al. 1991); for Pacific salmon returning to rivers to spawn, it can reach 0.9 (e.g. Heard 1991). Fishing mortality can be high enough effectively to change a fish population with many year classes into one with only one or two year classes, because the probability of survival beyond one or two years becomes vanishingly small (Murphy 1968). If there is genetic variability in life-history patterns, populations or species subject to high adult mortality should evolve life histories characterized by early maturity and high fecundity as compared with populations experiencing lower mortality (see, e.g., Pianka 1978, Rowell this volume).

DETECTION OF EVOLUTION

The major difficulty in the detection of evolution in natural fish populations is the disentangling of environmentally based changes from genetically based ones. The difficulty is encountered even when fishing is not a factor. For example, Alm (1959) used 80 years of fishery statistics and more than 20 years of experimentation in the attempt to understand the relationship between size, age, and maturation in fishes, but he could not separate genetic factors from environmental ones. This difficulty is probably responsible for the general lack of attention paid by fishery managers to the genetic effects of selective fishing, despite the existence of considerable discussion in the scientific literature (Beverton et al. 1984). A few of the major sources of confusion are discussed briefly below.

COMPENSATORY GROWTH

Perhaps the most important factor is compensatory growth. In addition to possibly applying selective pressure, fishing can reduce the population size of a fished stock. The result might be reduced competition for food and other resources, allowing faster growth rates, greater fecundity, and even larger maximum sizes. Fishes are quite plastic in their response to environmental conditions (e.g. McKenzie et al. 1983) and evolutionary responses could easily be masked by plastic responses to improved environmental conditions, as suggested by many authors for cod (see Borisov 1978 for review).

BARANOV'S 'FISHING-UP EFFECT'

Seemingly related to compensatory growth, but actually not related, is the 'fishing-up effect' of Baranov (1918), which applies in particular to long-lived species. Baranov pointed out that in an area that is fished each year, old fish have been subjected to repeated fishing for longer than younger ones. Thus, a greater proportion of old (= slow growing) fish have been removed from the population than younger (= fast growing) fish, and the *average* of the growth rates of the remaining fish is higher than for the original population as a whole. Although repeated fishing can select for earlier-maturing fish (Borisov 1978), it will increase the mean growth rate in the population even if the heritability for growth rate is zero.

MIGRATION

Another difficulty in detecting evolution results from the difficulty in identifying the limits of a population. Fish often migrate. If a population that is fished appears to change over time with respect to some character such as growth rate or size at maturity, it is possible that the reason is related to migration into the area of a different stock. Mechanical and genetic markers can help to reduce confusion caused by migration.

ENVIRONMENTAL CHANGES

Yet another confounding factor is the possibility of a change in the environment not directly related to fishing. Such changes could include changes in ocean currents, water temperature, or productivity, environmental pollution, introduction of exotic species, and others. Those changes can affect the distribution, abundance, age structure, growth rates, and many other characteristics of fish populations. As an example, Ricker (1981) reported a significant negative correlation between ocean temperature and size of Fraser River sockeye salmon (*Oncorhynchus nerka*).

THE PLATYFISH

To separate the genetic and environmental factors influencing age, weight, and maturation, McKenzie et al. (1983) raised two genotypes of male platyfish - early and late maturers - in a wide variety of environmental conditions (feeding regimes and temperatures) in three different laboratories. The genotypes, which occur naturally, appear to differ only by an allelic substitution at a single, sex-linked locus (Kallman and Borkoski 1978). The anal fin of male platyfish turns into an intromittent organ (gonopodium) as they mature, so their state of maturation can be unambiguously determined by external observation. We weighed the fish and examined them for signs of sexual maturation once a week or more often. The results made clear that both genetic and environmental variation had profound effects on the age and weight of fish at sexual maturation. Figure 2 shows that the environmental variations caused early maturing (P^1P^3) fish to mature as small as 52 mg and as large as 450 mg; they matured as early as 9 and as late as 30 weeks. The late-maturing (P^1P^4) fish matured as small as 100 mg (one, not shown, matured at 52 mg) and as large as 1,800 mg; the time to maturation ranged from 13 to 45 weeks. Yet the genotypic effects are also dramatic: there is almost no overlap between the age-weight envelopes of the two genotypes. The size difference might have been somewhat exaggerated by the environmental conditions; the best-fed late maturers received somewhat more food than the best-fed early maturers. Nonetheless, the tendency of the late maturers to mature at larger sizes is clear.

The results of this experiment show clearly two things that are important to the present discussion. First, a single allelic substitution at a single genetic locus can result in large differences in the age and weight of maturation of fish in identical environmental conditions. It is obvious from Figure 2 that a single generation of selection on a mixed population could substantially alter the genetically determined ages and weights of maturation of males in the population. Second, there is an enormous effect of the environment on traits such as maturation and growth rate in fishes. This makes the disentanglement of genetic and environmental effects of fishing on fish very complicated.

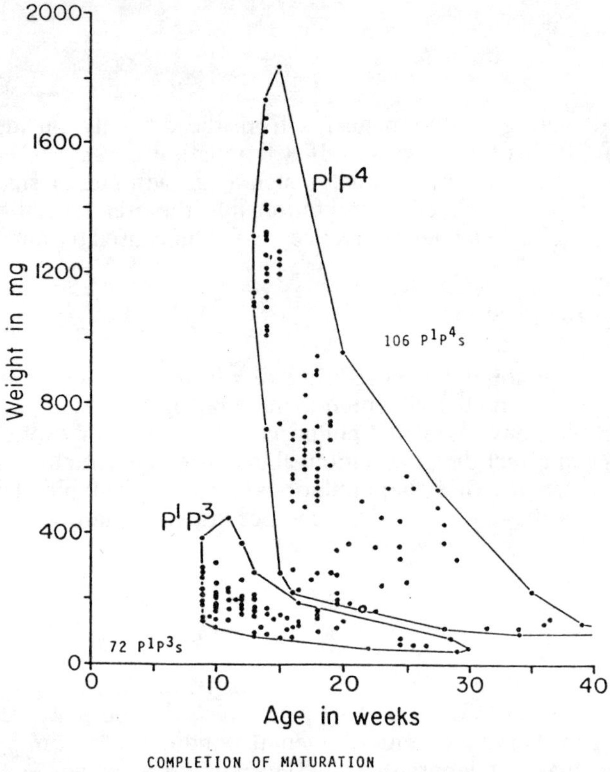

Figure 2. Age and weight at completion of maturation of 178 male platyfish, showing the effects of genotype and of environmental variation. For details, see text. From McKenzie et al. 1983.

Although I believe the experiments on male platyfish provide a useful model for understanding the potential for evolution in commercial species, some cautions are in order. First, the study was done on males only, although Kallman and Borkoski (1978) reported a similar relationship between size and age at maturity in female platyfish of a different genotype. Second, it is possible that the inheritance of age and weight at maturation is more complex in many other fishes than in platyfish, although those traits are usually heritable (e.g. Kirpichnikov 1981). Finally, the interactions between environment and genotype in natural populations are undoubtedly more complicated than they were in our fish, which were raised in individual containers. None of the above cautions, however, seems to invalidate the general applicability of the platyfish model or its ability to provide insight as to how quickly selective fishing might change gene frequencies in exploited populations.

THEORY, EXPERIMENTS, AND OBSERVATIONS

The theoretical basis for expecting exploited populations to evolve is clear and has been discussed in several papers. Results of some experiments confirm the theoretical expectation. However, few cases of observed changes in exploited natural populations can be attributed with certainty to evolution caused by selective fishing, although many cases are suggestive.

THEORY

The most convincing arguments for evolution in specific cases appear to have been made for pandalid shrimp and salmon. Some pandalid shrimp mature first as males and then change into females. Charnov (1981) argued that sex change in the shrimp is independent of size. Instead, it depends most strongly on mortality rates. Indeed, if the mortality rates are high enough, the optimal age for sex change could be less than one year, so some shrimp should mature first as females. Charnov cites data from North Sea shrimp (Jensen 1965) as supporting his model. Similar arguments have been made concerning mixed reproductive strategies in male salmon (M. Gross 1984, 1985, 1991, Myers 1983, 1986). Some salmon mature at young ages and small sizes (jacks) and engage in 'sneaky' mating, fertilizing eggs when refuges are available. 'Normal', hooknose males instead rely on fighting other males to ensure access to females. There appears to be a genetic component to the different life-history patterns, although the decision mechanism is not clear (Gross 1984, 1991). The sneaky strategy is risky; some jacks are forced to fight, which increases their mortality. There is an evolutionary balance between the sneaky and the fighting strategy. As in the case of the shrimp, the theory was developed to analyze a naturally occurring polymorphism not involving fishing, but fishing mortality would affect the hooknose form disproportionately. If the mortality associated with one strategy increases, the other becomes relatively favored. Fishing, which increases the mortality of adult salmon, but not immatures, would favor the sneaky strategy according to the model, so jacks should be more frequent in heavily fished populations (Myers 1983, 1986, Gross 1991). Unfortunately, good data on this question are difficult to find (e.g. Myers et al. 1986), although Ricker (1972) reported a gradual increase in the proportion of sockeye salmon jacks in the Columbia River, which he attributed to the effect of fishing selectively for larger fish.

The general argument that selective fishing can cause evolution has been made many times, the first to my knowledge being that of R. Miller (1957). Miller suggested four possible effects of selective fishing; the two most plausible were selection for slow growth rate by differential removal of faster-growing fish, and selection for increased intelligence through differential removal by angling of less intelligent fish. He concluded that there was no evidence for any genetically based change in growth rates, but that brown trout (*Salmo trutta*), well known among anglers for being hard to catch, were probably more intelligent than native U.S. trouts because of their much longer exposure to angling (they were introduced to the United States from Europe). Moav et al. (1978), Borisov (1978), Kirpichnikov (1981), and Nelson and Soulé (1987) are among those who have discussed the probable effects of selective fishing on evolutionary change. Nelson and Soulé discussed many of the difficulties of showing that evolution has occurred. Law (1979) carefully analyzed the effects of age-specific predation and pointed out that his results should apply equally to the effects of age-specific harvesting, i.e., fishing. Law noted that despite 'the importance of understanding our influence on exploited populations, our role as an agent of selection has rarely been considered.' Law and Grey (1989) analyzed a model of age-specific cropping of a population and concluded that age-specific cropping does cause evolution of life histories and yields; the nature of the evolution is related to the distribution of the ages that are cropped.

OBSERVATIONS

A variety of changes have been observed in populations that have been exploited for long periods. Schmitt and Skud (1978) reported reduced abundance, increased growth rate, increased fecundity at age, reduced age at maturity, and increased size at maturity in Pacific halibut, which are fished with longlines. De Veen (1976) found increased growth rate, increased size at maturity, and no change in age at maturity in the European sole, typically fished with trawls. However, Rijnsdorp et al. (1991), in a much more extensive survey, found no consistent trends in size at age, age at maturity, or growth rate in European sole, plaice, or cod. Marked changes did occur over several years, but they often were reversed in later years. Handford et al. (1977) reported *decreased* growth rate, size at maturity, and condition and increased age at maturity in lake whitefish (*Coregonus clupeaformis*) after they had been fished with gillnets for 34 years. Gwahaba (1973) reported decreased size at maturity in exploited Nile perch (*Tilapia nilotica*) in Lake George, Uganda. In general, European cod, fished by a variety of gear but mainly trawls, have shown increased growth rate, decreased age at maturity, and decreased mean size over the years (e.g., Borisov 1978, Rollefsen 1938), but such trends are not apparent in the data of Rijnsdorp et al. (1991). Clark and Alexander (1985) discussed the controversy over the cause of the observed decrease in size of brown trout in Michigan's Au Sable River, which has been attributed to fishing pressure and to reduced sewage enrichment over time. They concluded that insufficient information was available to decide which factor was more important. Pacific salmon, in which adults are fished with various gears but juveniles are generally not taken, have shown decreased size at maturity and decreased mean size (Ricker 1981). Pandalid shrimp reproduce first as males and then as females; in heavily fished populations, the time of sex change is earlier than in unexploited populations (Charnov 1981). A genetic basis for the change, has not been demonstrated in these cases, although most of the changes are consistent with expectations based on the selectivity of the gear. The cases for genetic change in shrimp (Charnov 1981), pink salmon (*Oncorhynchus gorbuscha*; Ricker 1981), and whitefish (Handford et al. 1977) are convincing.

In addition to changes in harvested populations, many observations have strongly suggested that selective natural predation has caused evolution of prey species. The best known is probably Kettlewell's work on the peppered moth, *Biston betularia* (Kettlewell 1973). I will mention only a few other examples that seem analogous to fishing here. Strong (1973) showed that in lakes with predatory fish that selected large prey, freshwater amphipods (*Hyalella azteca*) spent less time in amplexus, when they would appear large to a predator, than they did in lakes without fish. He reared the animals in the laboratory and found that the variation in times of amplexus was heritable. Reznick and Endler studied guppies (*Poecilia reticulata*) from various streams with various predators that focused on different life-history stages. They found heritable differences in life histories that corresponded to the predicted effects of the various predators (Reznick 1982, Reznick and Endler 1982). Bell and Richkind (1981) are among authors who have shown a relationship between predation on threespine sticklebacks (*Gasterosteus aculeatus*) and the number of lateral plates they have. The variation in plate number and morphology is heritable (Hagen 1973).

EXPERIMENTS

Silliman (1975) harvested two experimental populations of the cichlid fish *Tilapia mossambica*. In the population from which large fish were selectively taken, growth rates of males decreased and the difference from the unselectively harvested population was heritable. Law (1979) described the experimental harvest of adult pike (*Esox lucius*) in Lake Windermere by the Freshwater Biological Association coupled with careful demographic analysis. Three-year old females were producing approximately 80,000 eggs in 1975, up from around 50,000 in 1963. The increase was not shown to be genetic, but is consistent with the expected evolutionary response to increased mortality on reproductive-age fish. Edley and Law (1988) culled two genetically similar populations of *Daphnia magna*, removing small individuals from one population and large individuals from the other. The former evolved to grow rapidly through the vulnerable size, while the latter evolved reduced growth, lengthening the time they spent below the vulnerable size. Reznick et al. (1990) transplanted guppies from streams in which their natural predators preyed on adults to streams where a different predator preyed only on juveniles. After 11 years, heritable changes in life histories had occurred in the direction that they predicted.

THE FUTURE: NEW APPROACHES TO THE PROBLEM

Despite arguments and evidence that evolution might be occurring in exploited fish populations, fishery managers (and others) have not paid much attention to the evolution of their charges as a serious problem. And one cannot blame them: too often, it is impossible to show that evolution has occurred or to provide an accurate, plausible prediction of how evolution might occur in the future. In the meantime, the fishery manager is worried about overharvesting, allocation of fishing effort, poaching, bycatch, and other important matters about which better information is available. But questions concerning the intensity of the selective harvest, the nature and speed of potential response (e.g. Law 1991), and the desirability and feasibility of managing with evolution in mind are important nonetheless. How should they be addressed? Two general approaches suggest themselves. The first is an intensified and improved program to detect and measure the effects of evolution in exploited populations; the second, more promising approach is experimental, and it includes the need for simulations.

INTENSIFIED ATTEMPTS AT DETECTION AND MEASUREMENT

This approach would involve more detailed and careful assessments of gear selectivity and of life-history traits such as growth rates, age and size at maturity, fecundity, maximum size, and so on. I am not optimistic about the probability of success of this approach in general. It is not easy to imagine a more intensive, careful, detailed analysis of life-history traits than that of Rijnsdorp et al. (1991), yet even their analysis does not allow a firm conclusion as to whether or not evolution has occurred. Their long-term data set shows that trends can be observed, but if the investigation continues for long enough, the trends sometimes disappear or reverse direction. Even consistent trends, however, would not have been sufficient to prove that evolution was occurring. Miller's study of walleye reveals the surprising complexity of the selective effects of gillnets.

Many of the fish populations that have been studied were being harvested long before the investigations began, so any evolution might have occurred too early to be detected. Therefore, while long-term data sets like those reported by Rijnsdorp et al. (1991) and those collected by the International Pacific Halibut Commission (e.g. Bell 1981) are enormously valuable for many reasons, it seems unrealistic to expect that by themselves they will permit unambiguous detection of evolution caused by fishing. Of course, if a new species or stock is exploited, then the collection of such data becomes crucial; fishing-caused evolution is most likely in the early stages of a fishery.

EXPERIMENTS ON COMMERCIALLY EXPLOITED POPULATIONS AND SIMULATIONS

Larkin (1972) urged fishery managers to 'boldly *experiment* with new ways of managing resources for the public benefit.' For detecting and measuring evolution in exploited populations, experiments could include fishing with various kinds of gears and at varying intensities, at different times of year, in different places, and so on. Such experiments would be most feasible on closed systems, such as rivers or lakes, or on stocks with restricted distributions. Possibly, the techniques of genetic engineering would allow the use of readily identifiable markers for genes affecting one or more life-history traits, which would make data collection easier. Nelson and Soulé (1987) discuss the use of electrophoretic markers linked to genes affecting growth rates in natural populations.

Because experimentation with a public resource or with potential profits is not undertaken lightly, true experimentation in fishery management is still uncommon, even 20 years after Larkin's advice, and after several others have offered similar advice (McAllister and Peterman 1992a). However, recent work by McAllister and colleagues shows promise. McAllister et al. (1992) presented an experimental harvest design that permits the escapement of large pink salmon as opposed to the current fishery, which permits the escapement of smaller fish. They argued that the design would allow managers to test for the effects of size-selective fishing and to increase mean body size and annual catch biomass of pink salmon if such effects occurred. The obvious question arises as to whether the potential profit from the experiment and resulting changes in fishing practices would justify the cost of doing the work. McAllister and Peterman (1992b) addressed the question by means of simulations and decision analysis of the economic performance of the proposed experiment and reported that, under a variety of discount rates, the expected economic return was greater than that of not doing the experiment in most cases they looked at. Miller (1992) recommended a similar approach. Although predicting responses to selection might be intractable, he suggested that an accurate measurement of the selection differentials of fishing gear might be sufficient to guide management in many cases.

Simulation is crucial. Because many iterations using different parameters for selection differentials and possible responses can be run quickly and inexpensively, areas of concern or interest can be identified, and the costs and potential benefits (if any) of various experiments can be assessed.

CONCLUSIONS

For evolution of life-history traits to occur as a result of fishing, there must be heritable variation and fishing must cause differential reproduction of genotypes. Evidence reviewed here shows that those two prerequisites are present in many (perhaps most) harvested populations. Theoretical arguments make it seem likely that evolution of exploited populations does occur, and studies show that evolution occurs in selectively harvested experimental populations and that natural selective predation causes evolution of prey. Law and Grey (1989) describe the point of view that because the genetic architecture of life-history variation is unknown, it is unprofitable to consider evolutionary responses to selective harvesting. They sensibly comment that ignorance of the architecture will not affect the progress of evolution, and argue that it should not be overlooked. Finally, the data of McKenzie et al. (1983) on the genetic and environmental factors affecting age and size at maturation in platyfish make abundantly clear how evolution might occur. In a fish with the genetic architecture of male platyfish, substantial evolution could occur in a single generation.

Yet the detection and measurement of evolution caused by fishing remains difficult in most cases. In some cases, this might mean that evolution has not occurred; in others, it might reflect only the complexity of the effects of various factors. But the difficulties should not dissuade researchers and managers. Experiments and simulations hold particular promise, and the evolution of exploited populations should be regarded as a fertile ground for experimental and theoretical research as well as being of practical concern to managers.

ACKNOWLEDGEMENTS

This work grew out of conversations I had with R. Doyle and others at a Dahlem Conference in 1984. I am most grateful to R. Doyle, especially, and to R. Beverton, C. Bublitz, R. Deriso, M. Gross, A. Kapuscinski, R. Kendall, R. Law, J. Magnuson, R. May, R. Myers, K. Nelson, G. Orians, R. Peterman, D. Reznick, K. Sherman, M. Sissenwine, J. Sohn, K. Stokes, and C. Sullivan for data, discussions, and manuscripts. The critical comments of two anonymous reviewers were most helpful. The writing of this paper was supported by a grant from the Kobelt Fund of the National Research Council.

REFERENCES

Alm, G. 1959. Connection between maturity, size and age in fishes. Institute of Freshwater Research Drottningholm Reports **40**: 5-145.

Baranov, F. I. 1918. On the question of the biological basis of fisheries. Nauchnyi issledovatelskii ikhtiologicheskii Institut Isvestia **1**: 81-128. (In Russian; translated by W. E. Ricker, mimeograph, 1945.)

Bell, F. H. 1981. The Pacific Halibut: The Resource and the Fishery. Alaska Northwest Publishing Company, Anchorage.

Bell, M. A., and K. E. Richkind. 1981. Clinal variation of lateral plates in threespine stickleback fish. American Naturalist **117**: 113-132.

Beverton, R. J., J. G. Cooke, J. B. Csirke, R. W. Doyle, G. Hempel, S. J. Holt, A. D. MacCall, D. Policansky, J. Roughgarden, J. G. Shepherd, M. P. Sissenwine, and P. H. Wiebe. 1984. Dynamics of Single Species: Group Report. Pp. 13-58 in R. M. May, ed., Exploitation of Marine Communities. Dahlem Konferenzen. Springer, Berlin.

Bondari, K. 1983. Response to bidirectional selection for body weight in channel catfish. Aquaculture **33**: 73-81.

Borisov, V. M. 1978. The selective effect of fishing on the population structure of species with a long life cycle. Journal of Ichthyology **18**: 896-904.

Charnov, E. L. 1981. Sex reversal in *Pandalus borealis*: effect of a shrimp fishery? Marine Biology Letters **2**: 53-57.

Clark, R. D., and G. R. Alexander. 1985. Effects of a slotted size limit on the brown trout fishery, Au Sable River, Michigan. Pp. 74-84 in F. Richardson and R. H. Hamre, eds. Wild Trout III: Proceedings of the Symposium. Federation of Fly Fishers and Trout Unlimited, Vienna, Virginia.

Cushing, D. H. 1968. Fisheries Biology: A Study in Population Dynamics. University of Wisconsin Press, Madison.

de Veen, J. F. 1976. On changes in some biological parameters in the North Sea sole (*Solea solea* L). Journal du Conseil International pour l'Exploration de la Mer **37**: 60-90.

Doyle, R. W. 1983. An approach to the quantitative analysis of domestication selection in aquaculture. Aquaculture **33**: 167-185.

Dunham, R. A., and R. O. Smitherman. 1983. Response to selection and realized heritability for body weight in three strains of channel catfish, *Ictalurus punctatus*, grown in earthen ponds. Aquaculture **33**: 89-96.

Edley, M. T., and R. Law. 1988. Evolution of life histories and yields in experimental populations of *Daphnia magna*. Biological Journal of the Linnean Society **34**: 309-326.

Endler, J. A. 1980. Natural selection on color patterns in *Poecilia reticulata*. Evolution **34**: 76-91.

Endler, J. A. 1986. Natural Selection in the Wild. Monographs in Population Biology **21**. Princeton University Press, Princeton, New Jersey.

Falconer, D. S. 1989. Introduction to Quantitative Genetics. 3rd Edition. Longman, Harlow.

Gall, G. E. 1975. Genetics of reproduction in domesticated rainbow trout. Journal of Animal Science **40**: 19-28.

Gall, G. E. 1983. Genetics of fish: A summary of discussions. Aquaculture **33**: 383-394.

Gjerde, B. 1984. Response to individual selection for age at sexual maturity for age at sexual maturity in Atlantic salmon. Aquaculture **38**: 229-240.

Gross, H. P. 1978. Natural selection by predators on the defensive apparatus of the three-spined stickleback, *Gasterosteus aculeatus* L. Canadian Journal of Zoology **56**: 398-413.

Gross, M. R. 1984. Sunfish, salmon, and the evolution of alternative reproductive strategies and tactic in fishes. Pp. 55-75 in G. Potts and R. Wootton, eds. Fish Reproduction: Strategies and Tactics. Academic Press, London.

Gross, M. R. 1985. Disruptive selection for alternative life histories in salmon. Nature **313**: 47-48.

Gross, M. R. 1991. Salmon breeding behavior and life history evolution in changing environments. Ecology **72**: 1180-1186.

Gwahaba, J. J. 1973. Effects of fishing on the *Tilapia nilotica* (Linné 1757) population in Lake George, Uganda over the past 20 years. East African Wildlife Journal **11**: 317-328).

Hagen, D. W. 1973. The inheritance of numbers of lateral plates and gill rakers in *Gasterosteus aculeatus*. Heredity **30**: 303-312.

Hamley, J. M. 1975. Review of gillnet selectivity. Journal of the Fisheries Research Board of Canada **32**: 1943-1969.

Handford, P., G. Bell., and T. Reimchen. 1977. A gillnet fishery considered as an experiment in artificial selection. Journal of the Fisheries Research Board of Canada **34**: 954-961.

Heard, W. R. 1991. Life history of pink salmon (*Oncorhynchus gorbuscha*). Pp. 119-230 in C. Groot and L. Margolis, eds., Pacific Salmon Life Histories. University of British Columbia Press, Vancouver.

Jensen, A. J. C. 1965. *Pandalus borealis* in the Skagerak (length, growth, and changes in the stock and fishery yield). Rapports et Procès-Verbeaux des Réunions, Conseil International pour l'Exploration de la Mer **156**: 109-111.

Kallman, K. D., and V. Borkoski. 1978. A sex-linked gene controlling the onset of sexual maturity in female and male platyfish (*Xiphophorus maculatus*), fecundity in females and adult size in males. Genetics **89**: 78-119.

Kettlewell, H. B. D. 1973. The Evolution of Melanism. Clarendon Press, Oxford.

Kinghorn, B. 1983. Genetic variation in food conversion efficiency and growth in rainbow trout. Aquaculture **32**: 141-155.

Kirpichnikov, V. S. 1981. Genetic Bases of Fish Selection. Translated from Russian by G.G. Gause. Springer, Berlin.

Laarman, P.W., and J.R. Ryckman. 1982. Relative size selectivity of trap nets for eight species of fish. North American Journal of Fisheries Management **2**: 33-37.

Larkin, P. A. 1972. A confidential memorandum on fisheries science. Pp.189-197 in B. Rothschild, ed., World Fisheries Policy: Multidisciplinary Views. University of Washington Press, Seattle.

Law, R. 1979. Optimal life histories under age-specific predation. American Naturalist **114**: 399-417.

Law, R. 1991. On the quantitative genetics of correlated characters under directional selection in age-structured populations. Philosophical Transactions of the Royal Society of London, Series B **331**: 213-223.

Law, R., and D. R. Grey. 1989. Evolution of yields from populations with age-specific cropping. Evolutionary Ecology **1989**: 343-359.

March, E. J. 1970. Sailing Trawlers. David & Charles, Newton Abbot, Devon.

Marquis, R. J. 1984. Leaf herbivores decrease fitness of a tropical plant. Science **226**: 537-539.

McAllister, M. K., and R. M. Peterman. 1992a. Experimental design in the management of fisheries: a review. North American Journal of Fisheries Management **12**: 1-18.

McAllister, M. K., and R. M. Peterman. 1992b. Decision analysis of a large-scale fishing experiment designed to test for a genetic effect of size-selective fishing on British Columbia pink salmon (*Oncorhynchus gorbuscha*). Canadian Journal of Fisheries and Aquatic Sciences **49**: 1305-1314.

McAllister, M. K., R. M. Peterman, and D. M. Gillis. 1992. Statistical evaluation of a large-scale fishing experiment designed to test for a genetic effect of size-selective fishing on British Columbia pink salmon (*Oncorhynchus gorbuscha*). Canadian Journal of Fisheries and Aquatic Sciences **49**: 1294-1304.

McKenzie, W. D., D. Crews, K. D. Kallman, D. Policansky, and J. J. Sohn. 1983. Age, weight and the genetics of sexual maturation in the platyfish, *Xiphophorus maculatus*. Copeia **1983**: 770-774.

Miller, L. M. 1992. Genetic Alteration of Fish Size by a Size-Selective Fishery. Unpublished M.Sc. Thesis, University of Minnesota.

Miller, R. R. 1957. Have the genetic patterns of fishes been altered by introductions or selective fishing? Journal of the Fisheries Research Board of Canada **14**: 797-806.

Moav, R., T. Brody, and G. Hulata. 1978. Genetic improvement of wild fish populations. Science **201**: 1090-1094.

Murphy, G. I. 1968. Pattern in life history and the environment. American Naturalist **102**: 391-403.

Myers, R. A. 1983. Evolutionary change in the proportion of precocious parr and its effect on yield in Atlantic salmon. I.C.E.S. C.M/ **M**:14: 16p.

Myers, R. A. 1986. Game theory and the evolution of Atlantic salmon (*Salmo salar*) age at maturation. Pp. 53-61 in D. J. Meerburg, ed., Salmonid Age at Maturity. Canadian Special Publications in Fisheries and Aquatic Sciences **89**.

Myhre, R. J. 1969. Gear Selection and Pacific Halibut. Report Number **51** of the International Pacific Halibut Commission. Seattle, Washington.

National Research Council. 1986a. Pesticide Resistance: Strategies and Tactics for Management. National Academy Press, Washington, D.C.

National Research Council. 1986b. Ecological Knowledge and Environmental Problem-Solving: Concepts and Case Studies. National Academy Press, Washington, D.C.

Nelson, K., and M. Soulé. 1987. Genetical Conservation of Exploited Fishes. Pp. 345-368 in N. Ryman and F. Utter, eds. Population Genetics and Fishery Management. University of Washington Press, Seattle.

Pianka, E. R. 1978. Evolutionary Ecology. Second Edition. Harper & Row, New York.

Policansky, D. 1982. Flatfishes and the inheritance of asymmetries. The Behavioral and Brain Sciences **5**: 262-266.

Refstie, T., and T. A. Steine. 1978. Selection experiments with salmon. III. Genetic and environmental sources of variation in length and weight of Atlantic salmon in the freshwater phase. Aquaculture **14**: 221-234.

Reznick, D. A. 1982. The impact of predation on life history evolution in Trinadadian guppies: genetic basis of observed life history patterns. Evolution **36**: 1236-1250.

Reznick, D. A., H. Bryga, and J. A. Endler. 1990. Experimentally induced life-history evolution in a natural population. Nature **346**: 357-359.

Reznick, D. A., and J. A. Endler. 1982. The impact of predation on life history evolution in Trinidadian guppies (*Poecilia reticulata*). Evolution **36**: 160-177.

Ricker, W. E. 1972. Hereditary and environmental factors affecting certain salmonid populations. Pp. 27-160 in R. C. Simon and P. A. Larkin, eds. The Stock Concept in Pacific Salmon. H. R. MacMillan Lectures in Fisheries. University of British Columbia, Vancouver.

Ricker, W. E. 1981. Changes in the average size and average age of Pacific salmon. Canadian Journal of Fisheries and Aquatic Sciences **38**: 1636-1656.

Rijnsdorp, A. D., N. Daan, F. A. van Beek, and H. J. L. Heessen. 1991. Reproductive variability in North Sea plaice, sole, and cod. Journal du Conseil International pour l'Exploration de la Mer **47**: 352-375.

Rollefsen, G. 1938. Changes in mean age and growth-rate of the year-classes in the Arcto-Norwegian stock of cod. Rapports et Proces-Verbeaux des Reunions, Conseil International pour l'Exploration de la Mer. **108**: 37-44.

Rollefsen, G. 1953. The selectivity of different fishing gear used in Lofoten. Journal du Conseil International pour l'Exploration de la Mer **19**: 191-194.

Rudstam, L. G., J. J. Magnuson, and W. M. Tonn. 1984. Size selectivity of passive fishing gear: a correction for encounter probability applied to gill nets. Canadian Journal of Fisheries and Aquatic Sciences **41**: 1252-1255.

Schmitt, C. C., and B. E. Skud. 1978. Relation of Fecundity to Long-Term Changes in Growth, Abundance and Recruitment. Scientific Report **66**. International Pacific Halibut Commission, Seattle, Washington.

Silliman, R. 1975. Selective and unselective exploitation of experimental populations of *Tilapia mossambica*. Fishery Bulletin **73**: 495-507.

Stewart, P. A. M. 1975. Catch selectivity by electrical fishing systems. Journal du Conseil International pour l'Exploration de la Mer **36**: 106-109.

Strong, D. R., Jr. 1973. Amphipod amplexus, the significance of ecotypic variation. Ecology **54**: 1383-1388.

Teichert-Coddington, D. R., and R. O. Smitherman. 1988. Lack of response by *Tilapia nilotica* to mass selection for rapid early growth. Transactions of the American Fisheries Society **117**: 297-300.

Thorpe, J. E., R. I. G. Morgan, C. Talbot, and M. S. Miles. 1983. The inheritance of developmental rates in Atlantic salmon, *Salmo salar* L. Aquaculture **33**: 119-128.

Wohlfarth, G. W. 1983. Genetics of fish: applications to warm water fishes. Aquaculture **33**: 373-381.

SELECTION DIFFERENTIALS IN MALE AND FEMALE NORTH SEA PLAICE AND CHANGES IN MATURATION AND FECUNDITY

A.D. Rijnsdorp

Netherlands Institute for Fisheries Research, P.O.Box 68, 1970 AB IJmuiden, The Netherlands

ABSTRACT

This paper presents an analysis of the selective effects of fishing on reproductive parameters of North Sea plaice. This species has been exploited intensively since the late 19th century, and the fishing mortality substantially exceeded the level of natural mortality. Since the start of biological research in the early 1900s, life history parameters such as growth, onset of sexual maturity and fecundity have changed. In order to evaluate possible genetical effects of exploitation, the mortality imposed by fishing is estimated for males and females separately and factors determining the relationship between fishing mortality and age are analysed. Fitness profiles of various reproductive traits are studied under the present regime of exploitation, employing a simple allocation model of surplus production over somatic growth and reproduction. This model describes the cost of reproduction in terms of decelerated somatic growth and reduced future reproduction. Selection differentials are estimated from the fitness profiles and are compared to the selection differentials estimated from the observed changes in reproductive parameters. It is concluded that the direction of the observed changes is in agreement with the expected change due to selective fishing.

INTRODUCTION

The additional mortality induced by fishing may act as an artificial selection pressure on life history parameters, such as growth rate, length and age at first sexual maturity and reproductive investment (Nelson and Soulé 1987). The way in which these parameters may change will depend on their heritability and the fishing intensity and pattern of exploitation. In this paper a method is presented to estimate the selection pressure exerted by the fisheries. The relationship between fishing mortality and age, i.e. the exploitation pattern, is studied and factors that determine the exploitation pattern are analysed. The fitness of various reproductive traits, taken here as combinations of length at first maturity and reproductive investment, are studied under the present regime of exploitation, employing a simple allocation model of surplus production over somatic growth and reproduction that describes the cost of reproduction in terms of decelerated somatic growth and reduced future reproduction.

The method is explored using data on North Sea plaice distinguishing between males and females. This species has been exploited for several centuries and has shown changes in reproductive parameters and somatic growth. Selection differentials are calculated from the observed changes in reproductive parameters and compared with those estimated from the exploitation pattern of the fisheries.

DEVELOPMENTS IN THE FISHERIES

Cushing (1988) gives the following account on the major developments in the demersal fisheries in the North Sea. In the 18th and 19th century, the fishery was mainly conducted with sailing vessels deploying small beam trawls. With the introduction of steam trawlers in the late 19th century the traditional beam trawl was replaced by the otter trawl and the number of vessels expanded substantially. In conjunction the area fished greatly expanded and covered all of the North Sea by the end of the 19th century. In the first half of the 20th century the steam trawlers were gradually replaced by motor trawlers, although the principal gear, the otter trawl, remained largely the same. Finally in the early 1960s, heavy beam trawlers quickly outcompeted the otter trawlers. According to Thursby-Pelham (quoted in Wimpenny 1953), plaice landings from the North Sea were already at a level of 47,000 tonnes in 1892. Landing statistics collected by the International Council for the Exploration of the Sea (ICES) since 1906 show that the plaice landings varied around 50,000 tonnes annually until the mid 1950s (Bannister 1978). Since then they increased to a level of around 150,000 tonnes in the 1980s (Anon. 1991).

GENERAL BIOLOGY OF PLAICE

Plaice is a flatfish species living in temperate waters down to a depth of about 100 m. The distribution range extends from the Bay of Biscay in the south to the Barents Sea and the waters around Iceland in the North. The areas and intensity of spawning in the North Sea and adjacent waters indicate that plaice abundance is highest in the southern and southeastern North Sea (Wimpenny 1953, Harding et al. 1978). Tagging studies have shown that the population in the North Sea is composed of various sub-groups that partially mix on the summer feeding grounds but separate on their spawning grounds in winter (de Veen, 1978).

The different life history stages are spatially segregated (Wimpenny 1953, Rijnsdorp and van Beek 1991). After spawning, the pelagic eggs and larvae drift with the residual current in open sea. At the end of the larval stage, plaice settle in very shallow nursery areas on sandy beaches with a marked preference for estuarine areas. In the following years they leave the inshore areas and disperse over the southern and central North Sea. Beverton and Holt (1957) showed that the off-shore movement of juvenile plaice could be described by a process of diffusion. After maturation plaice show distinct migrations between their feeding areas in summer and spawning grounds in winter (de Veen 1978).

Natural mortality (*M*) of adult age groups was estimated at 0.08 for female plaice and 0.14 for male plaice (Beverton 1964), although in general a value of 0.1 is accepted for both sexes. A recent review of 0-group mortality rates indicated an annual mortality rate from eggs to juveniles of $M_0 = 10.38$ (Beverton and Iles 1992).

The annual cycle of plaice is composed of a distinct spawning period between December and March during which plaice stop feeding, and a growing period between March and November. During the spawning period adult male and female plaice lose respectively 27 per cent and 44 per cent of their total energy content, while juvenile plaice lose only 9 per cent (Rijnsdorp and Ibelings 1989). During the growing period energy is mainly accumulated in the soma. Ovary growth does not start before October and 50 per cent of the energy for ovary growth is reallocated from the soma (Dawson and Grimm 1980, Rijnsdorp 1990). Total energy accumulated during the growing period, here defined as surplus production, is a continuous increasing function of body size, but in male plaice surplus production levels off at a body size above 30 cm (Rijnsdorp and Ibelings 1989).

CHANGES IN POPULATION PARAMETERS

GROWTH

Changes in growth of North Sea plaice have been reported by Bannister (1978) and Rijnsdorp (1991). At present, the juvenile growth, reflected in the length attained at the end of their 1st and 2nd year of life, is higher than those observed between 1900-1950. Changes in growth in the adult stage are less well documented. Comparison of the mean weight at age in the 1st quarter between the 1930's and the 1980s indicates that the rate of growth was higher in the 1930's from about 8 years onwards. The weight of younger age groups in the 1930's however, were lower than at present, reflecting the then lower growth rate of juveniles. The above data have to be interpreted with caution, because they do not take account of growth differences between the sexes. Market sampling data collected for males and females separately showed an increase in the length at age of the younger age groups in both sexes, and a decrease in the older age groups of female plaice in the period 1950 - 1970 (Bannister 1978). Rijnsdorp and van Beek (1991) observed a similar increase in the length at age of 4-year old plaice in the 1960s and concluded that this was probably related to an increase in the availability of benthic food.

Changes in growth rate may (in part) be related to density effects. Beverton and Holt (1957) showed that growth rate of plaice in the southern North Sea was substantially reduced during the second world war coinciding with a threefold increase in the stock size. An analysis of density-dependent aspects in changes in growth between 1958 and 1989, however did not give unequivocal evidence for density-dependent effects. Only the decrease in the length of I- and II-group plaice in the 1980s might be related to the above average level of year class strength in the period (Rijnsdorp and van Beek 1991).

MATURITY

Changes in the onset of sexual maturity, given by Rijnsdorp (1989), indicate that in the early 1900s, depending on the area within the North Sea, females became sexually mature at a length of 31 - 43 cm, and an age of V-VII years. In 1985-1986 the L_{mat} had decreased to 30 - 35 cm (10-16 %) and A_{mat} to IV-V years (30-35 %). In the early 1900s males became sexually mature at 30 - 37 cm and V-VI years, whereas in 1985-1986 L_{mat} had decreased to 20-24 cm (35 per cent) and A_{mat} to II-III years (50-60 %). The observed change in length (L_{mat}) and age at 50 per cent maturity (A_{mat}) in females was larger than the between year variability as observed in the period 1958-1989 (Rijnsdorp et al. 1991). A regression of L_{mat} of 4-year old females against time for the time period 1963-1988 indicated a decrease of 0.026 cm per year, although the decrease did not differ significantly from zero ($r^2 = 0.024$, n = 22).

FECUNDITY

Changes in fecundity between three time periods (I - 1900-1910, II - 1948-1949 and III - 1977-1985) were reviewed by Rijnsdorp (1991). Fecundity of 30 cm females increased between periods II and III by 30 -100 per cent, but not between period I and II. Fecundity of 50 cm females did not change substantially between period I and III (4 per cent), but was drastically reduced in period II by 21 per cent.- 38 per cent. The reduction in fecundity in period II of the larger females coincided with the reduced growth in the preceding years and a threefold increase in stock size during the second world war. However, since no substantial change was observed in size-specific ovary weights, it was concluded that the energy allocation over reproduction and growth did not change in response of a increased stock size, but the trade-off between egg number and egg size. The observed changes in fecundity suggests that the reproductive investment may have increased in smaller, but not in larger females.

SURPLUS ENERGY PRODUCTION

Somatic growth, maturation and reproduction are interacting processes that are related to the allocation of surplus energy (Ware 1980, Roff 1983). Surplus production (SP_x), defined as the energy available in excess of maintenance requirements that is available for somatic growth and reproduction, can be described as a continuously increasing function of body weight W_x (Reiss 1989).

$$SP_x = \alpha W_x^{\beta} \qquad (1)$$

Figure 1 shows that this model gives a reasonable, though not a perfect, description of the observed relationship between *SP* and *W* for both male and female plaice.

The reproductive investment at age *x* (m_x) can be described as

$$m_x = \gamma W_x^{\delta} \qquad (2)$$

In the present paper m_x is expressed in grammes which has the advantage that males and females can be treated in equal terms. Furthermore, it is assumed that $\delta = 1.0$ based on the fact that fecundity shows an almost proportional relationship with body weight (Rijnsdorp et al. 1983, Rijnsdorp 1991). For females $\gamma = 0.16$, corresponding to the observed gonadosomatic index (Rijnsdorp 1990). For males a comparable index was calculated from the ratio of the male over female energy loss during the spawning season times the female gonadosomatic index: $\gamma = (0.27/0.44) \times 0.16 = 0.10$.

The somatic growth can then be described by

$$W_{x+1} = SP_x - m_x + W_x \qquad (3)$$

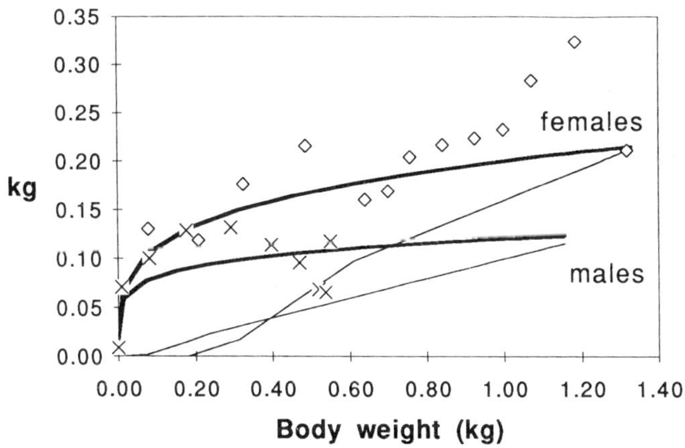

Figure 1. Surplus production (kg) against body weight for male (x) and female (◊) plaice. The thick lines indicate the relationship $SP = \alpha W^{\beta}$, with $\alpha = 0.12$, $\beta = 0.17$ (males) and $\alpha = 0.20$, $\beta = 0.25$ (females). The thin lines indicate the reproductive investment Rep = $\gamma W^{1.0}$, with $\gamma = 0.10$ (males) and $\gamma = 0.16$ (females).

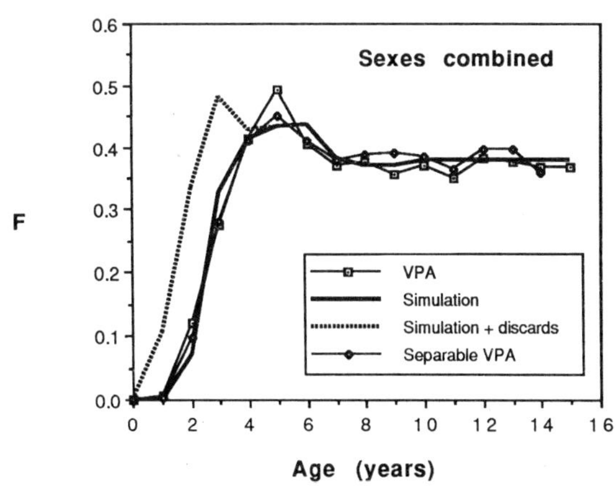

Figure 2. Comparison of the exploitation pattern in the period 1974-1977 as determined by VPA, separable VPA and as calculated from the quarterly distributions of effort and age groups including and excluding discards (respectively simulation + discards and simulation).

With this model, where reproduction clearly causes a reduction in somatic growth, the interaction of surplus production, maturation and reproduction can be modelled to study the effects of changes in maturation or reproduction on somatic growth.

FISHING MORTALITY

The decline of population numbers of a cohort of fish over time period t will be a function of the natural mortality (M) and the fishing mortality (F) according to:

$$N_t = N_0 e^{-(F+M)t} \tag{4}$$

Fishing mortality is age specific and related to fishing effort (E) according to:

$$F_x = q_x E \tag{5}$$

where q_x is the catchability coefficient at age x (Beverton and Holt 1957, Gulland 1983).

The catchability coefficient q is a function of the probability that a fish in the path of a trawl will be captured (catch efficiency), the mesh selection, and the spatial overlap between fishing effort and fish. The array of F by age is called the exploitation pattern.

EXPLOITATION PATTERN

The exploitation pattern is estimated from the annual age compositions of the commercial landings, that are based on market samples for length distributions and otolith samples as reported in Anon (1991). The sampling programme covers all of the main fleets and about 90 per cent of the total international landings. The catch-at-age matrix was analysed by two methods.

Firstly, the separable VPA (Pope and Shepherd 1982) was applied, which gives a least square solution of the exploitation pattern that fits best to the catch-at-age matrix under the assumption of a level of natural mortality (M), fishing mortality on some reference age in the last year and the exploitation factor on the oldest age. Figure 2 shows the exploitation pattern for the period 1974-1977. The exploitation factor on the oldest age was chosen at 0.8 x F_5 because for this value the F 's of the older age groups are constant, which is in line with the high overlap in spatial distributions between these age groups (Rijnsdorp and van Beek 1991).

Secondly, a standard VPA (Gulland 1983) was applied, where the level of F in the last year was 'tuned' on fishing effort, assuming that catchability-at-age is constant over the years and that there is a positive relation between the trend in fishing mortality and effort (Laurec and Shepherd 1983, Pope and Shepherd 1985). It was further assumed that the fishing mortality on the oldest age group XV was equal to the mean of the previous 5 age groups. Effort data used were the number of fishing hours of the Dutch beam trawl fleet (*fh*) corrected for horsepower (*HP*) by $E = fh \times HP$. The exploitation patterns obtained by both methods show a close correspondence (Figure 2).

The fishing mortality in plaice thus characteristically increases with age to a maximum value at age V and decreases slightly thereafter (Figure 2). This pattern corresponds to the observed spatial distribution of the various age groups as described in Rijnsdorp and van Beek (1991). Age group I to IV were shown to have rather distinct spatial distribution patterns with coefficients of overlap between 0.1 - 0.6, whereas age groups V and older showed almost similar spatial patterns with coefficients of overlap between 0.8 and 1.0.

Anon (1987) provides estimates of the exploitation pattern for the period 1974 - 1977 based on spatially disaggregated data on the distribution of age groups and fishing effort under the assumption that the catch efficiency was constant over seasons and age groups. The result is plotted in Figure 2 as 'simulation'. The close agreement with the earlier results indicates that the exploitation pattern is indeed largely determined by the relative distribution of age groups and effort, and that the catch efficiency is approximately constant over the recruited age-groups.

The gradual increase in *F* for age groups I - IV is related to recruitment to the off-shore fishing grounds. In the southeastern North Sea, the age at which plaice leave the nursery areas increases with latitude, from about age group II - III in the eastern part of the Southern Bight to age group II - IV along the Danish coast (Anon 1987). There is some evidence that fast growing fish of a cohort leave the shallow nursery areas at a younger age than slow growing fish (Wimpenny 1953, Beverton and Holt 1957, Rijnsdorp 1989).

MESH SELECTION

At present the major fisheries for flatfish use a cod-end mesh size of 80 mm or less, which corresponds to a 50 per cent retention length of $\leq$ 18 cm. Since the minimum landing size of plaice is 27 cm, large quantities of undersized plaice are discarded at sea (van Beek 1990). The length range of discards is between 10 and 33 cm, representing age groups I - IV. The survival chances of these discards is virtually zero (van Beek et al. 1990). The level of fishing mortality on undersized plaice was estimated by Anon (1987) from the quarterly distribution of fish and effort and is shown together with the *F* on the landed fish in Figure 2. Exploitation rate is apparently high from age II onwards. Fishing mortality on 0-group and I-group may in fact be somewhat higher since discards made by small vessels fishing in the nursery areas were not included. However, at present this fleet is smaller than in the past and the vessels also use special devices to prevent the by-catch of undersized fish (separating panel) or to increase their survival (rotary sieves). Their contribution to the discard mortality will probably be small compared to that of the flatfish fisheries itself. In the past they may have contributed substantially to the fishing mortality of I- and II-group plaice (Boddeke 1989).

DIFFERENCES IN CATCHABILITY OF MALES AND FEMALES

Beverton (1964) showed that males have a higher catchability than females. Figure 3 shows that fishing mortalities on males are indeed higher. In order to trace the origin of this difference, the annual F-values were partitioned over the quarters in proportion to the quarterly landings of the various age-groups by sex. Figure 4 shows that the differences in *F* are restricted to the first quarter when plaice spawns. This suggests that behavioural differences during spawning affect their vulnerability to fishing gear, which may be related to the fact that males stay for a much longer period on the spawning grounds than females (Wimpenny 1953, Beverton 1964, Rijnsdorp 1989), as well as to differences in vertical migrations during spawning time (Arnold et al. in prep).

CONCLUSION

Fishing mortality is mainly determined by the overlap of spatial distributions of the various age groups on the one hand and that of fishing effort on the other, although the increase in vulnerability of males during the spawning period suggests that also a behavioural component is involved.

HISTORIC LEVEL OF FISHING MORTALITY

Although no estimates of the level and pattern of exploitation are available for the late 19th and early 20th century, it may be inferred from the high landings that it probably did not differ much from the 1930 level. The estimated F-values for the latter period range from 0.3 (Bannister 1978) to 0.7 (Beverton and Holt 1957). Obviously these values are not very precise and it seems likely that the average fishing mortality has been in the same order of magnitude (0.5) as in recent years (Anon 1991).

Some inferences on the exploitation pattern for the historic period can be made from the distribution of effort. In the first half of the century, the United Kingdom (UK) was responsible for 35-60 per cent of the total international landings. The distribution maps of the UK effort given by Bannister (1978) indicate that effort was distributed all over the southern and central North Sea. Fishing effort of the other countries e.g. Netherlands, Germany and Denmark, taking account of 40-60 per cent of the international landings, was concentrated in the southern and southeastern parts (Wimpenny 1953). During the 20th century the fishing grounds of the trawler fleets have not changed very much, although the fishing effort in the coastal nursery grounds, which also employed smaller mesh size, may have been higher in the past. The main changes that occurred are in the types of gear used and the power of the fishing vessels. Data on the relationships of catch efficiency and fish size of commercial gears used in the past are not available, but it is likely that catch efficiency may have declined with fish size in the historic period as a result of lower towing speeds and lower weight of the gear. The historic exploitation pattern thus might have shown a more distinct peak on the youngest age-groups. However, without more detailed information, it seems justified to assume that the exploitation pattern as observed in the last 30 years also applied to the previous period.

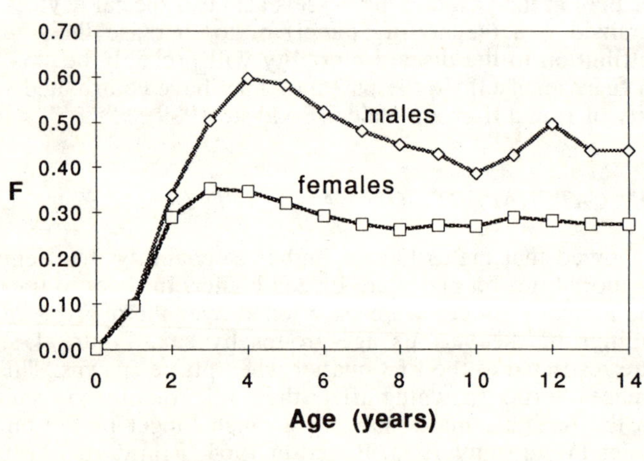

Figure 3. Exploitation pattern of male and female in the period 1958-1986 from VPA excluding discards.

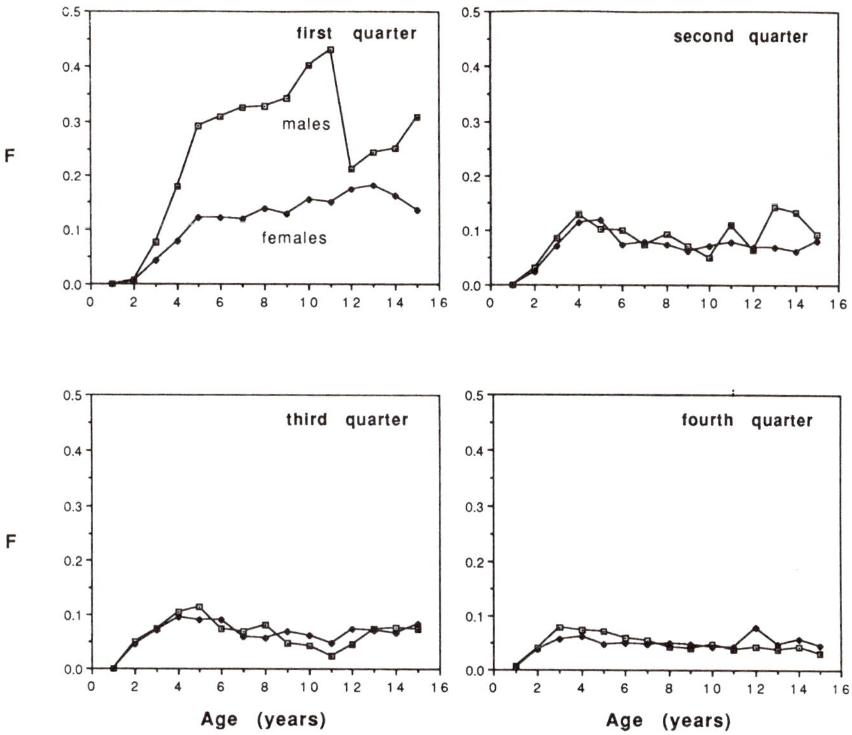

Figure 4. Quarterly exploitation patterns of males and females in the period 1974-1977 excluding discards.

SELECTION DIFFERENTIALS

With the above information we now can evaluate the effect of exploitation on the fitness of various reproductive traits. The first step is to model the interaction between somatic growth, reproductive investment and maturation using equation (3) and parameter values from Table 1. The onset of sexual maturity was simulated by specifying a length at 50 per cent maturity and a range of 10 cm over which the maturity linearly increased from 0 per cent to 100 per cent. Figure 5 shows the effects of the L_{mat} on somatic growth. The dashed line indicates the expected somatic growth without maturation. The somatic growth of a female that matures at a small size lags behind that of a female maturing at a larger size, but the reduction in somatic growth becomes smaller when L_{mat}decreases (Figure 5). For instance, a L_{mat} =20 (A_{mat} = II) does not decrease somatic growth substantially until age V. Obviously the reproductive investment at age III and IV, which is a fixed proportion of the body weight, is still insignificant compared to the total available surplus production.

Table 1. Parameter values used in the simulation of growth and reproductive investment in male and female plaice.

		MALES	FEMALES
Bertalanffy growth parameters:			
$L_t = L_{inf}(1\text{-}e^{-Kt})$			
	L_{inf}	39.7	51.3
	K	0.41	0.24
	t_0	0.31	0.005
Surplus production (kg):			
$SP = \alpha W^{\beta}$			
	α	012	0.20
	β	0.17	0.25
Length at first maturity (cm)	1985	22	33
	1910	32	38
Reproductive investment (kg):			
$m_t = \gamma W_t^{\delta}$	γ	0.10	0.16
	δ	1.00	1.00
Fecundity (eggs.g^{-1})			265

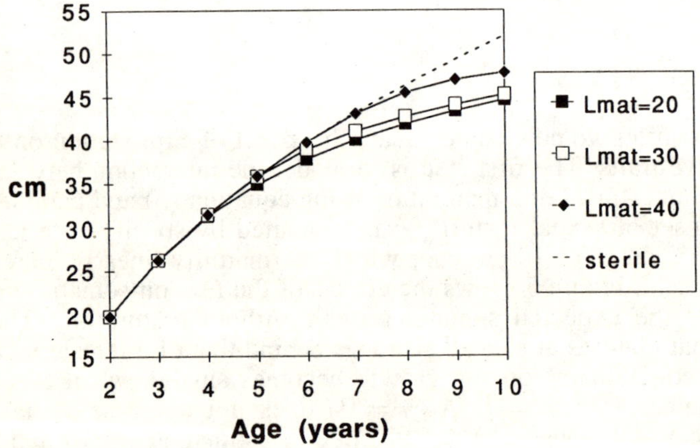

Figure 5. Simulated somatic growth curves of female plaice varying in the length at first maturity (L_{mat}) with a surplus production - body weight relationship according to Table 1. The dashed line indicate the somatic growth of females that do not become sexually mature.

Table 2. Life table of female North Sea plaice giving an estimate of $\lambda = 1.056$. Length is given by the von Bertalanffy growth equation from Table1. Weight (W_x)=10^{-5} x Length^3; P_x is the proportion maturity (Rijnsdorp and Ibelings 1989); M_x is the natural mortality; F_x is the average fishing mortality from VPA for 1958-1986 including discards;. b_x=W_x x P_x x 265 x 0.5, represents the number of female eggs produced; l_x represents the survival of one female offspring.

Age	Length (cm)	W_x (kg)	P_x	M_x	F_x	b_x x10^3	l_x x10^{-6}	$\lambda^{-x} l_x b_x$
0	0.0	0.000	0.00	10.38	0.00	.0	10^6	0.000
1	10.8	0.013	0.00	0.20	0.14	.0	30.99	0.000
2	19.4	0.073	0.01	0.08	0.39	.0	22.06	0.000
3	26.2	0.180	0.06	0.08	0.41	.1	13.74	0.001
4	31.5	0.313	0.29	0.08	0.35	3.5	8.45	0.024
5	35.7	0.456	0.80	0.08	0.32	38.6	5.49	0.162
6	39.0	0.594	0.99	0.08	0.29	77.2	3.68	0.205
7	41.6	0.721	1.00	0.08	0.27	95.5	2.53	0.165
8	43.7	0.833	1.00	0.08	0.26	110.3	1.78	0.127
9	45.3	0.928	1.00	0.08	0.27	123.0	1.26	0.095
10	46.5	1.009	1.00	0.08	0.27	133.6	0.89	0.069
11	47.5	1.075	1.00	0.08	0.29	142.4	0.62	0.049
12	48.3	1.129	1.00	0.08	0.28	149.6	0.43	0.033
13	49.0	1.173	1.00	0.08	0.27	155.5	0.30	0.023
14	49.4	1.209	1.00	0.09	0.27	160.1	0.21	0.016
15	49.8	1.237	1.00	0.09	0.27	163.9	0.14	0.010
16	50.1	1.260	1.00	0.09	0.27	166.9	0.10	0.007
17	50.4	1.278	1.00	0.09	0.27	169.3	0.07	0.005
18	50.6	1.292	1.00	0.10	0.27	171.2	0.05	0.003
19	50.7	1.303	1.00	0.10	0.27	172.7	0.03	0.002
20	50.8	1.312	1.00	0.11	0.27	173.9	0.02	0.001
21	50.9	1.320	1.00	0.13	0.27	174.8	0.02	0.001
22	51.0	1.325	1.00	0.14	0.27	175.6	0.01	0.001
23	51.0	1.330	1.00	0.17	0.27	176.2	0.01	0.000
24	51.1	1.333	1.00	0.21	0.27	176.6	0.00	0.000
25	51.1	1.336	1.00	0.27	0.27	177.0	0.00	0.000
26	51.1	1.338	1.00	0.38	0.27	177.3	0.00	0.000
27	51.2	1.340	1.00	0.60	0.27	177.5	0.00	0.000
28	51.2	1.341	1.00	1.06	0.27	177.7	0.00	0.000
29	51.2	1.342	1.00	2.21	0.27	177.8	0.00	0.000
30	51.2	1.343	1.00	4.42	0.27	177.9	0.00	0.000

The second step is the calculation of the fitness using Fisher's reproductive value :

$$V_0 = \Sigma\lambda^{-x} l_x\, b_x \tag{6}$$

where λ is the intrinsic rate of population increase, l_x is the survival to age x and b_x is the number of females born at age x. The fitness of a reproductive trait will depend on the population growth parameter λ. Therefore, the present value of λ was estimated for the total population of female plaice from the life table given in Table 2, using empirical estimates for all relevant parameters. Table 1 gives the parameter values used for modelling somatic growth and reproductive investment. The l_x was estimated according to equation (4) using estimates for M_x and the average F_x from sexes separated VPA for the period 1958-1986, after correction for discarding. Natural mortality was assumed to increase geometrically with age from a low of 0.08 (females) and 0.14 (males) at age II to reduce the adult life span to a realistic value of 30 years. M_0 = 10.38 and M_1 was taken at 0.20. The calculations gave an estimate of λ = 1.056.

Fitness profiles for L_{mat} and reproductive investment, assuming λ = 1.00, are given in Figure 6. Fitness values are expressed relative to the fitness values of the present reproductive tactic (male: L_{mat}= 22 cm, γ = 0.10; female: L_{mat} = 33 cm, γ = 0.16). In both males and females fitness decreases with increasing L_{mat}. The fitness of the L_{mat} in the early 1900s compared to the present L_{mat} is 0.42 (males) and 0.80 (females). Assuming λ = 1.056, these ratios become 0.37 (males) and 0.76 (females). Figure 6 also shows that fitness has not yet reached the maximum level at the current level of exploitation, suggesting that there is still selection for a reduction in the L_{mat} With regard to reproductive investment, the simulations indicate that there is selection for an increase in reproductive investment.

The selection differential that corresponds to the estimated fitness profile for L_{mat} at λ = 1.00 can be approximated by calculating the average L_{mat} weighted over the fitness profile. The calculation is illustrated in Figure 7. With an average L_{mat} = 35 cm, halfway between the present and historic value, and with a S.D = 2.5 cm (Rijnsdorp 1989), the weighted average L_{mat} was calculated as 34.73 cm, giving a selection differential of 0.27 cm. At λ = 1.056 the selection differentials becomes 0.33 cm. For males (L_{mat} = 27 cm, SD = 5 cm) the selection differential was calculated as 1.89 and 2.09 cm for λ=1.00 and λ=1.056 respectively.

DISCUSSION

The accuracy of the estimated fitness values depends on the relationship between surplus production and body weight, the measure of reproduction, the mortality estimates and the innate rate of population increase (l). The two former parameters are likely to be less certain. The fit between the observed values and the assumed surplus production -body weight relationship is far from perfect (Figure 1), making the fitness estimates necessarily crude. Also the use of reproductive energy as a measure of gene propagation is a rather crude simplification. In females this assumption neglects the possibility of differences in the viability of eggs produced at different ages or maternal body sizes. In males, gene propagation is an even more difficult question, and it may be speculated that in addition to the available reproductive energy, also the competitive power in relation to other males will play an important role. The latter may be a function of body size. If true this would reduce the fitness of precocious males. It was also assumed that there is no mortality cost of reproduction, although the opposite can be expected (e.g. Roff 1983, 1991). Comparison of the fitness ratio for λ = 1.0 and λ = 1.056 indicated only a small influence on the estimated difference in fitness ratio. Although a value close to unity will be likely, the true value during the period 1900 - 1990 will probably have varied around unity. The mortality estimates will be reasonably accurate since they are dominated by the fishing mortality that is well estimated for the recent period. Backwards extrapolation to the historic period however remains speculative. Given the above considerations, the fitness profiles should be taken as crude estimates indicating the likely level and direction of selection.

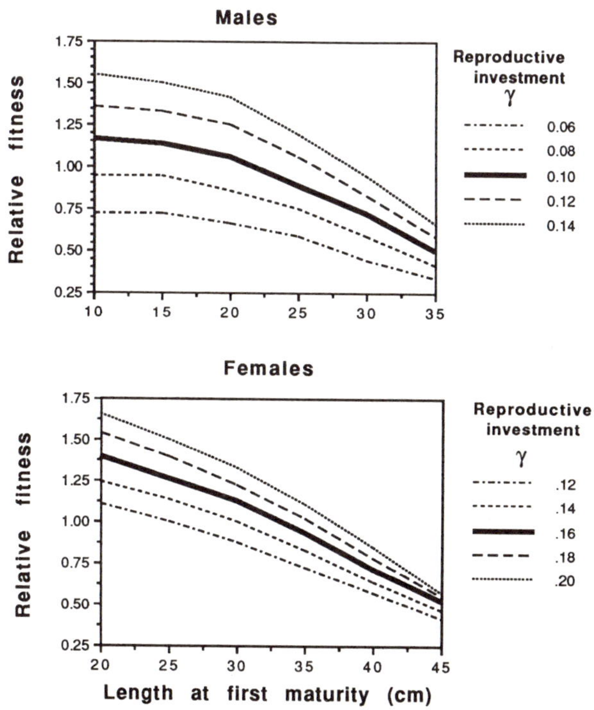

Figure 6. Fitness profiles for length at first maturity (L_{mat}) for various levels of reproductive investment (γ) with $\lambda = 1.0$. The fitness is expressed relative to that at the average observed length at first maturity.and the observed reproductive investment. males: $L_{mat} =$ 22 cm, $\gamma = 0.10$; females: $L_{mat} = 33$ cm, $\gamma = 0.16$.

Between 1910 and 1985 the L_{mat} decreased by 10-13 cm in males and 3-7 cm in females. In order to estimate the change per generation, the age at which an individual has produced 50 per cent of its offspring was estimated as an approximation of the generation time. At the present level and pattern of exploitation, the generation time was 4.8 and 7.7 years, corresponding to 15.6 and 9.7 generations, of males and females, respectively. The change in L_{mat} per generation is then 11.5/15.6 = 0.74 cm in males and 5/9.7 = 0.52 cm in females. Within the period 1963-1988 female L_{mat} decreased by 0.026 cm.year^{-1}, or with 7.7x0.026 = 0.20 cm per generation. Although this value did not differ significantly from zero, it suggests that the change may have been smaller in the recent period.

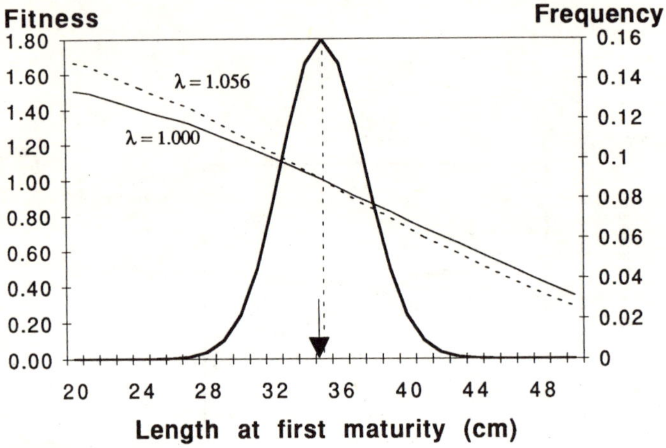

Figure 7. Estimation of the selection differential from the fitness profiles of female plaice with a L_{mat} that is normally distributed with μ = 35 cm and σ = 2.5 cm. The arrow indicates the weighted average length at first maturity (L'_{mat} = 34.73 cm), weighted over the fitness profile at λ = 1.00. The selection differential is the difference between μ and L'_{mat} (S = 0.27 cm). The fitness profile for λ = 1.056 is shown by the dashed line.

An estimate of the 'observed' selection differential (S) can be obtained from $R = h^2 S$, where R is the change in a quantitative trait per generation and h^2 is the heritability coefficient (Falconer 1989). Taking a mean heritability for life history traits of h^2 = 0.24 (Roff 1991), S can be calculated as 0.74/0.24 = 3.1 cm in males and 0.52/0.24 = 2.1 cm for females. The 'predicted' selection differentials, estimated from the fitness profiles for λ = 1.00, were 1.89 and 0.27 cm for males and females respectively. Averaged over males and females the selection differential is 1.08 cm. For λ = 1.056 the selection differentials are 2.09 (males), 0.33 (females) and 1.21 cm (male and female).

The 'observed' selection differential is higher than the 'predicted' value, especially in females. Whether the averaged selection differential for males and females is a correct measure depends on the genetic coding of the onset of sexual maturity. If autosomal genes affect L_{mat}, selection on one sex will influence the other sex as well and different selection pressures on the sexes may result in an average effect. In this case it is essential to study the selection differential of both males and females. However, if maturation is primarily coded by genes on the sex chromosome, the selective effects on male and female will be independent.

The 'observed' and 'predicted' selection differentials were at least qualitatively in agreement. This agreement however, does not necessarily indicate that the observed decrease in L_{mat} is due to a genetical change, it can also be a phenotypic response to improved conditions for growth. Although a discussion of the phenotypic versus genetic changes is beyond the scope of the present paper, a genetic change in L_{mat} is not unlikely since the observed decrease in L_{mat} is in the opposite direction of the expected change due to phenotypic plasticity. Present day variation in L_{mat} suggests that at a higher growth rate in the juvenile phase, as actually observed in North Sea, L_{mat} should increase rather than decrease (Rijnsdorp 1989). The question is even more complicated if we realise that there exists a cline in L_{mat} and A_{mat} within the North Sea, which could be related to differences in growth rate several years before the actual onset of sexual maturity.

The simulations predict that fitness should increase with increasing reproductive investment. The increase in fecundity since the late 1940s of the smaller females (30cm) is unlikely to be due to phenotypic plasticity, because Rijnsdorp (1990) and Millner et al.(1991) obtained evidence that improved conditions for growth will mainly result in a higher somatic growth, but not in a higher size-specific fecundity. It may therefore be tentatively concluded that the increase in fecundity of small fish might be related to a genetical change. In contrast to the prediction however, the fecundity of larger females did not increase, suggesting that the fitness calculation may be deficient in some respect. The model assumed that the cost of reproduction was restricted to a reduced future fecundity, although an additional mortality cost may also occur because body reserves become severely depleted during spawning (Roff 1991). Also a further increase in reproductive investment may be impossible due to morphological constraints.

It is obvious that our ignorance with regard to many of the factors influencing the fitness of reproductive traits in plaice make any conclusion speculative. The best we can achieve is a consistent and concise description of the empirical data in terms of a basic model describing the interaction of growth and reproduction. The at least qualitative agreement between the predicted and expected change in L_{mat} give some support for a genetical change.

In the present study only age-based methods were used to study the mortality imposed by fishing in relation to sex, reproductive status and age. Analysis of the effects of this additional mortality on the fitness of slower and faster growing fish is therefore not possible. Such an influence is not unlikely since it is known that recruitment to the fishing grounds and the onset of sexual maturity is influenced by both size and age. If the catch efficiency of the commercial trawl declines with fish size we further can expect a size-selective component additional to the age-dependent component. These size differentials will mainly affect the fitness of fast- and slow growing genotypes. If the additional mortality is high at an intermediate size groups of plaice we can expect that fast growing genotypes will have an advantage over slower growing genotype because they have a reduced time period of highest vulnerability. If the additional mortality is high beyond a certain size and this size is above the size at first maturity the slow growing genotypes will have an advantage. Future studies of the size-selective aspects of fishing mortality may reveal whether the observed changes in growth rate of juveniles may be in line with the changes which can be expected due to the selective effects of fishing.

REFERENCES

Anon. 1987. Report of the ad hoc meeting of the North Sea Flatfish Working Group, IJmuiden, 2-5 February 1987. I.C.E.S.C.M./**Assess**:14.

Anon. 1991. Report of the North Sea Flatfish Working Group, Copenhagen, 18-24 October 1990. I.C.E.S. C.M./**Assess**:5.

Arnold, G.P., P. Scholes, and F.R. Harden Jones (in prep). The diurnal vertical distribution of plaice (*Pleuronectes platessa* L.) on a spawning ground in the southern North Sea.

Bannister, R.C.A. 1978. Changes in plaice stocks and plaice fisheries in the North Sea. Rapports et Procès-Verbeaux des Réunions, Conseil International pour 1'Exploration de la Mer **172**: 86-101.

Beek, F.A.van. 1990. Discards sampling programme for the North Sea Dutch participation. Internal Report Netherlands Institute for Fisheries Research, D.E.M.V.I.S. **90-303**: 1 - 24.

Beek, F.A.van, P.I. van Leeuwen and A.D. Rijnsdorp 1990. On the survival of plaice and sole discards in the otter-trawl and beam-trawl fisheries in the North Sea. Netherlands Journal of Sea Research. **26**: 151-160.

Beverton, R.J.H. 1964. Differential catchability of male and female plaice in the North Sea and its effect on estimates of stock abundance. Rapports et Procès-Verbeaux des Réunions, Conseil International pour 1'Exploration de la Mer **155**: 103-112.

Beverton, R.J.H. and S.J.Holt 1957. On the dynamics of exploited fish populations. Fish. Investigations, London, Ser.2, Vol. **19**: 533 p.

Beverton, R.J.H. and T.C. Iles 1992. Mortality rates of 0-group plaice (*Pleuronectes platessa* L.), dab (*Limanda limanda* L.) and turbot (*Scophthalmus maximus* L.) in European waters II. Comparison of mortality rates and construction of life-table for 0-group plaice. Netherlands Journal of Sea Research. **28**: 000-000 (in press)

Boddeke, R. 1989. Management of the brown shrimp (*Cragnon cragnon*) stock in Dutch coastal waters. Pp. 35-62 in J.F. Caddy, ed. Marine invertebrate fisheries: their assessment and management. John Wiley & Sons. Chichester.

Cushing, D.H. 1988. The provident sea. Cambridge University Press, Cambridge.

Dawson, A.S. and A.S. Grimm 1980. Quantitative changes in the protein, lipid and energy content of the carcass, ovaries and liver of adult female plaice, *Pleuronectes platessa* Linnean Journal of Fish Biology **16**: 493-504.

Falconer, D.S. 1989. Introduction to quantitative genetics. 3rd edition. Longman, Essex.

Gulland, J.A. 1983. Fish stock assessment: a manual of basic methods. John Wiley & Sons, Chichester.

Harding, D., J.H. Nichols and D.S. Tungate 1978. The spawning of the plaice (*Pleuronectes platessa* L.) in the southern North Sea and English Channel.Rapports et Procès-Verbeaux des Réunions, Conseil International pour l'Exploration de la Mer **172**: 102-113.

Laurec, A. & J.G. Shepherd 1983. On the analysis of catch and effort data. Journal du Conseil International pour l'Exploration de la Mer **41**: 81-84.

Millner, R.S., C.L. Whiting, M. Greer Walker and P.Witthames 1991. Growth increment, condition and fecundity in sole (*Solea solea* L) from the North Sea and eastern English Channel. Netherlands Journal of Sea Research **27**: 433-439.

Nelson, K and M. Soulé 1986. Genetical conservation of exploited fishes. Pp. 345-368 in N. Ryman and F. Utter, eds., Population genetics and fishery management. Washington Sea Grant, University of Washington Press, Seattle.

Pope, J.G. and J.G. Shepherd 1982. A simple method for the consistent interpretation of catch-at-age data. Journal du Conseil International pour l'Exploration de la Mer **40**: 176-184.

Pope, J.G. and J.G. Shepherd 1985. A comparison of the performance of various methods for tuning VPAs using effort data. Journal du Conseil International pour l'Exploration de la Mer **42**: 129-151.

Reiss, M.J. 1989. The allometry of growth and reproduction. Cambridge University Press, Cambridge.

Rijnsdorp, A.D. 1989. Maturation of male and female North Sea plaice (*Pleuronectes platessa* L.). Journal du Conseil International pour l'Exploration de la Mer **46**: 35-51.

Rijnsdorp, A.D. 1990. The mechanism of energy allocation over reproduction and somatic growth in North Sea plaice, *Pleuronectes platessa* L. Netherlands Journal of Sea Research **25**: 279-290.

Rijnsdorp, A.D. 1991. Changes in fecundity of female North Sea plaice, *Pleuronectes platessa* L, between three periods since 1900. I.C.E.S.Journal of Marine Scinece, **48**: 253-280.

Rijnsdorp, A.D., F. van Lent and K. Groeneveld 1983. Fecundity and the energetics of reproduction and growth of North Sea plaice (*Pleuronectes platessa* L.). I.C.E.S. C.M./**G**: 31.

Rijnsdorp, A.D. and B. Ibelings 1989. Sexual dimorphism in the energetics of reproduction and growth of North Sea plaice, *Pleuronectes platessa* Linnean Journal of Fish Biology **35**: 401-415.

Rijnsdorp, A.D., N. Daan, F.A. van Beek and H.J.L Heessen 1991. Reproductive variability in North Sea plaice, sole, and cod. Journal du Conseil International pour l'Exploration de la Mer **47**: 352-375.

Rijnsdorp, A.D. and F.A. Beek 1991. Changes in growth of North Sea plaice (*Pleuronectes platessa* L.) and sole (*Solea solea* L.). Netherlands Journal of Sea Research. **27**: 441-457.

Roff, D.A. 1983. An allocation model of growth and reproduction in fish. Canadian Journal of Fisheries and Aquatic Sciences. **40**: 1395-1404.

Roff, D.A. 1991. The evolution of life-history variation in fishes, with particular reference to flatfishes. Netherlands Journal of Sea Research **27**: 197-207.

Veen, J.F.de 1978. On the selective tidal transport in the migration of North Sea plaice (*Pleuronectes platessa_*) and other flatfish species. Netherlands Journal of Sea Research **12**: 115-147.

Ware, D.M. 1980. Bioenergetics and stock and recruitment. Canadian Journal of Fisheries and Aquatic Sciences **37**: 1012-1024.

Wimpenny, R.S. 1953. The plaice. Arnold, London..

GROWTH AND FECUNDITY CHANGES IN FLATFISH

Joseph Horwood

MAFF, Directorate of Fisheries Research, Fisheries Laboratory, Pakefield Road., Lowestoft, Suffolk NR33 OHT, UK.

ABSTRACT

Data are presented for various stocks of flatfish in the Bristol Channel, Northern Celtic Sea and the North Sea which demonstrate recent changes in growth and fecundity. Although stock biomasses have decreased in recent decades, these reductions are small relative to historical changes. It is argued that a density-dependence is unlikely to be the cause of these recent changes in demographic parameters.

INTRODUCTION

Large temporal changes are observed in demographic parameters of the large populations of temperate, adult, marine fish (e.g. Cushing 1975, 1981, 1982). These changes are frequently associated with density-independent shifts in the ecosystem and with global cycles. It is much more difficult to substantiate the implication of any density-dependent effects in adult fish.

Two important aspects are associated with this difficulty. First, internationally, a huge effort is devoted to the sampling of commercially exploited stocks of fish, with the objective of providing annual management advice, however, this sampling is frequently inadequate to address scientific questions with any rigour. Second, there is reason to believe that many changes observed in adult fish cannot be attributed to intra-specific, density-dependent effects in the adult life. This study addresses the latter point, and draws upon recent work on the population biology of the sole (*Solea solea*) from the Bristol Channel and northern Celtic Sea (Horwood 1993).

Density-dependence in fish populations does of course exist, for populations have been successfully exploited for over a century. The density-dependent modulation occurs somewhere between fertilization of the eggs and the appearance of juvenile fish in the fisheries, at about age two years. Fecundity and mortality are typically high, and in flatfish about one egg in a million will survive to adulthood. Small differences in the annual cumulative mortality of eggs to juveniles give rise to the observed large fluctuations in the size of year-classes or cohorts. Nevertheless, over wide ranges of adult stock size (N), there is no trend in the numbers of subsequently recruiting fish, the *per capita* survival being proportional to $1/N$.

GROWTH CHANGES IN THE BRISTOL CHANNEL SOLE

The sole stock of the Bristol Channel and northern Celtic Sea, generally between 50-52°N can be regarded as a population, with immigration and emigration rates low in comparison to natural and exploitation rate. Weights at age of the sole are given each year, but for both sexes combined (Anon., 1992). The weights over time are illustrated for ages 3, 6, 9 and 12 years, and an increase in weight can be seen over the twenty year period (Figure 1).

Table 1: Weight (kg) by age (years, 2-15) of male and female sole combined, from the Bristol Channel, 1971-1975, and percentage increase in subsequent periods relative to 1971-75 (data: Anon., 1992).

	YEAR			
AGE	1971-75	1976-80	1981-85	1986-90
2	0.092	12	22	67
3	0.151	13	22	25
4	0.203	13	22	26
5	0.283	13	22	24
6	0.335	13	22	25
7	0.399	13	22	23
8	0.398	13	22	41
9	0.441	13	22	32
10	0.450	13	22	50
11	0.573	12	20	12
12	0.567	13	22	17
13	0.542	12	22	15
14	0.487	16	26	66
15+	0.625	14	23	23

The data are based on a combination of direct weighings in Belgium and indirect means in the U.K. Similar patterns are found for all ages; for example all values are relatively low in 1975 and high in 1980. Variation may be real, but more likely reflects vagaries in the sampling procedures. Consequently, Table 1 presents the weights at age averaged over the five year period 1971-1975, and percentage changes in weight at age, by five year period, relative to the initial period (10 year old sole in 1986-90 were 50 per cent heavier than in 1971-75). It can be seen that the weights at age had all increased by about 13 per cent in 1976-80 and increased by a further 9 per cent by 1981-85. In the latest period results are more variable, but most ages exhibited further increases.

Are these changes real or artefacts of sampling? Horwood (1993) gives mean length at age, by sex, of sole landed by the international fleets (mainly Belgium and U.K.) for 1970-74, and the U.K. fleet over 1985-88. For males, the later period showed that lengths were 7 per cent larger for ages 2-10 years and similar thereafter, whereas for females, ages 6-10 years were 9 per cent larger and 6 per cent for older ages. This is consistent with the changes in weight, described above, but the information is not totally independent.

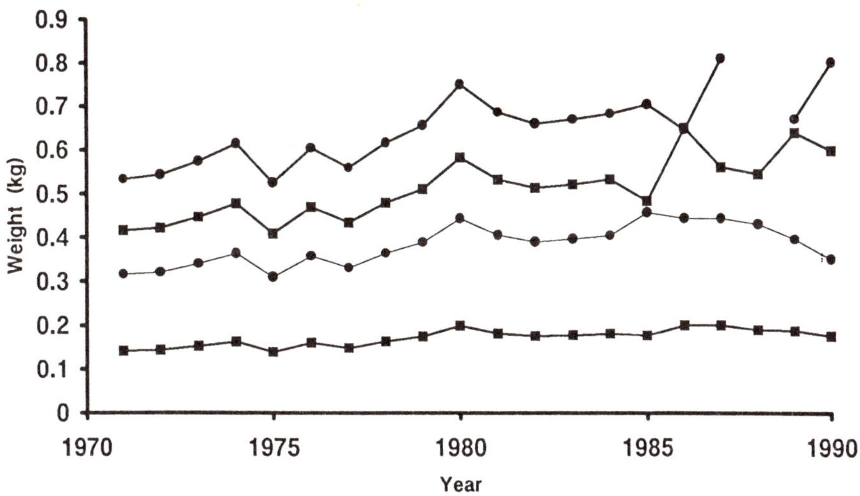

Figure 1. Weight (kg) by year of sole aged 3, 6, 9 and 12 years, males and females combined (data: Anon., 1992).

However, in 1983 there was a change in the legal mesh size used in the Bristol Channel, with an increase from 75 mm to 80mm. With an 80 mm mesh net almost all fish of 30 cm and above are retained. Consequently the increase in the mean length of, at least, female sole over the age of six years cannot be attributed to the change in legal mesh size. The changes in weights and length at age before 1983, and of older females after 1983, do appear to be real.

DENSITY DEPENDENCE?

Can the above increases in size be attributed to intra-specific, density-dependence in the adults? Figure 2 gives two realizations of the trajectory of adult stock size from the beginning of exploitation from about 1820. The solid points, from 1971 to 1990, show the results of the I.C.E.S. assessment (Anon., 1992), and they indicate that the adult stock fell from about 4,000 t to 3,000 t. The solid, back-trajectory is based upon simple, constant assumptions of growth and recruitment. The second series is similar but assumes a value for the natural mortality rate of $M = 0.2y^{-1}$ (Horwood 1993) and indicates a greater decline in the stock from 1970.

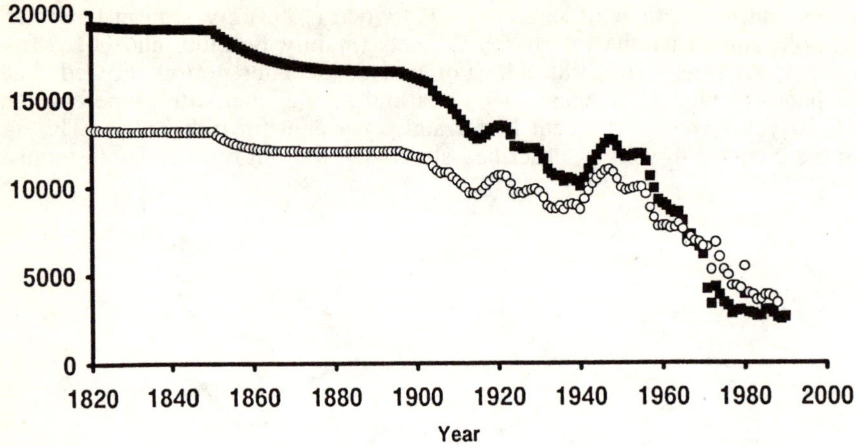

Figure 2: Population trajectories of the Bristol Channel sole (tonnes of adult biomass) for the years 1820-1990. Two graphs are realizations assuming different growth and natural mortality rates (based on Horwood 1993).

These assessments do show a decline in stock size over the period of the increases in growth, although alternative assessments indicate that no decline may have occurred (Horwood 1993). Nevertheless, the main point is that by 1970 the adult stock was already significantly depleted, perhaps by as much as 50-80 per cent (Figure 2).

Figure 3 gives the spawning stock biomass (SSB) per recruit entering the fishery based upon the ICES (Anon. 1992) results. It assumes a constant selection pattern and fishing rate (F). It can be seen that for modest fishing levels of 0.1-0.2y^{-1} the equilibrium stock size is reduced by 60-75 per cent. Such low rates of exploitation are seldom experienced by our commercially important flatfish stocks. Consequently, most of our observations, for at least the last three decades, have been of flatfish stocks that are probably significantly depleted from virgin levels. Figure 3, with the ICES assessment of current fishing mortality on the Bristol Channel sole being 0.44 y^{-1} giving an equilibrium reduction of 90 per cent.

Nicholson (1954) described the abrupt change in demographic parameters associated only with the final stages of resource or population depletion; this he termed scramble competition. It is not easy to relate this concept to the sole, nor to the positive effects of more resource *per capita*. For a contest form of density-dependent competition it might be expected that all changes would already have occurred before the relatively low stock size of 1970 was reached. I am inclined to the opinion that, given that the changes are real, that they are due to factors other than intra-specific, density-dependence in adults.

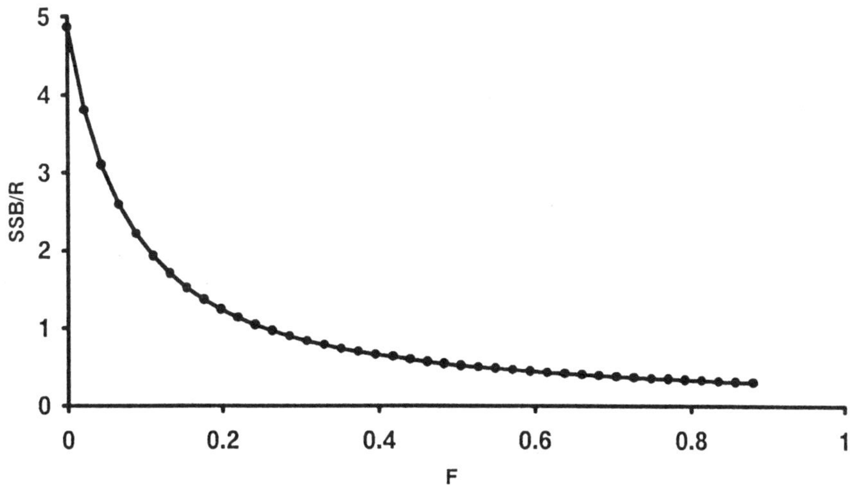

Figure 3: Adult population biomass (kg) per recruit (SSB/R) against annual instantaneous fishing morality rate (F).

GROWTH OF NORTH SEA SOLE

Increases in the weights at age have also been recorded for North Sea sole. Length, weight and maturity of were determined from 1957, and de Veen (1976) found that, from 1963, length at age and condition factor ($W L^{-3}$) increased. From 1957 to 1973 weights of 4-6 year-old females increased by 90-100 per cent and males by about 50 per cent. The size increment of the adults apparently remained constant, the change being due to a larger size of the young sole. Fishing mortality rates have exceeded $0.2y^{-1}$ from 1963 (Anon. 1991). Lengths and weights at age continued to increase to 1986 (van Beek 1988), and the females now mature earlier (van Beek 1985).

De Veen (1976) found no environmental feature correlated with these changes, but noted that they did correlate with the increase in modern beam-trawling. Beam-trawling uses heavy gear specifically designed to disturb the sea bed and the sole which otherwise easily evade the otter-trawlers by burying. He considered that the damage caused by trawling to the sea-bed fauna might make them more available to the sole. It is also possible that the fauna might change to one more suited to sole, such as more productive, soft-bodied forms.

Modern beam-trawling started during the 1960s. In the Bristol Channel and northern Celtic Sea the Belgian beam-trawl fleet fished 28,000 hours y^{-1} and the U.K. fleet 1,000 hours in the early 1970s, but by 1987-89 the beam-trawlers of Belgium fished over 70,000 hours and those of the U.K. 29,000 hours. It is possible that the increase in growth of the Bristol Channel sole has been brought about by disturbances of the sea bed similar to those postulated for the North Sea.

CHANGES IN FECUNDITY

Large, temporal changes in fecundity of North Sea plaice (*Pleuronectes platessa*) have also been detected. The number of eggs, at length, of plaice sampled in 1977-80 had increased by 60 per cent over those sampled in 1947-49, from the Southern Bight, and by 36 per cent from off Flamborough (Horwood et al. 1986). This large increase was not associated with any measurable decline in the population of North Sea plaice, in fact, it occurred after a period when planktonic egg abundances in the eastern North Sea increased, the stock increased and then stabilized. Similar changes were not found in the Irish Sea and Cardigan Bay (Horwood, 1990). The results emphasise that significant density-independent changes can occur in marine fish populations.

CONCLUSIONS

The main point of the study is to illustrate that changes in demographic parameters of the large, marine fish populations do occur, but that it is unlikely that these are associated with any density-dependent effects in the adults. Figures 2 and 3 illustrate the same point, that over the recent two or three decades, fishing mortality rates have been so high that adult stock sizes have long been drastically depleted. It is unlikely that further intra-specific, density-dependent effects will occur, or be detected, in adults.

REFERENCES

Anon. 1991. Report of the North Sea flatfish Working Group. I.C.E.S. C.M./**Assess 5**: 226.

Anon. 1992. Report of the Irish Sea and Bristol Channel Working Group. I.C.E.S. C.M./**Assess 1**: 208.

Beek, F.A. van 1985. On the maturity of North Sea sole in Dutch market samples. I.C.E.S. C.M./**G**57: 20.

Beek, F.A. van 1988. On the growth of sole in the North Sea. I.C.E.S. C.M./**G**24:6.

Cushing, D.H. 1975. Marine Ecology and Fisheries. University Press, Cambridge.

Cushing, D.H. 1981. Fisheries Biology and Study in Population Dynamics. The University of Wisconsin Press.

Cushing, D.H. 1982. Climate and Fisheries. Academic Press, London.

Horwood, J.W. 1990. Fecundity and maturity of plaice (*Pleuronectes platessa*) from Cardigan Bay. Journal of the Marine Biological Association, U.K. **70**: 515-529.

Horwood, J.W. 1993. The Bristol Channel sole (*Solea solea* (L.)). Advances in Marine Biology **29**: 215-367.

Horwood, J.W., R.C.A. Bannister and G.J. Howlett 1986. Comparative fecundity of North Sea plaice (*Pleuronectes platessa* L.). Proceedings of the Royal Society, London, B **228**: 401-431.

Nicholson, A.J. 1954. An outline of the dynamics of animal populations. Australian Journal of Zoology **2**: 9-65.

Veen, J.F. de 1976. On changes in some biological parameters in the North Sea sole (*Solea solea* L.) Journal du Conseil International pour l'Exploration de la Mer **37**: 60-90.

THE EFFECTS OF FISHING ON THE TIMING OF MATURITY IN NORTH SEA COD (*GADUS MORHUA* L.)

Catherine A Rowell

Dept of Biology, University of York, York YO1 5DD

ABSTRACT

Although there is a body of theoretical work on evolutionary responses to harvesting there have been relatively few attempts to apply this to commercially harvested wild populations. In this paper the long term data on North Sea cod are examined for trends in size at maturity. The preliminary work shows that there are differences between areas of the North Sea in the recent data, and so the long term analysis is confirmed to the one area (Round Fish Area 4) for which there was sufficient data. A shift towards maturation at smaller size was found. As there is no evidence for a corresponding change in growth rates it is likely this reflects a change towards maturation at younger ages.

A model of the effects of increased adult mortality on the optimal timing of maturation was also constructed. The model suggested an increasingly large advantage to fish maturing early as the adult mortality increased: a change in the same direction as was found in the data. No record of an environmental trend that could be expected to cause a shift towards maturation at lower ages, and a smaller size was found. It is suggested, therefore, that the observed change may be the result of an evolutionary change resulting from selection imposed by harvesting.

INTRODUCTION

Fishing, whether size selective or not, increases the risk of mortality of the fish at some point in its life cycle, usually after the larval and juvenile stages where most of natural mortality occurs. Interest in responses to externally imposed mortality dates from the early life-history investigations of Fisher (1930) and Cole (1954). Initial studies were theoretical (e.g. Gadgil and Bossert 1970, Schaffer 1974, Law 1979, Michod 1979). Supporting evidence came more slowly, (Reznick and Bryga 1987, Edley and Law 1988,

Reznick et al. 1990) and, with some notable exceptions, (Handford et al. 1977, Turner 1977, Borisov 1978, Ricker 1981, Law and Grey 1989), few attempts have been made to apply this theory to commercially exploited wild populations. In this paper a model of the change in the optimal timing of reproduction caused by a change in the mortality schedule is constructed and applied to an exploited stock of fish.

Much of the work on the optimal timing of, and investment in, reproduction assumes that growth is determinate and that organisms stop growing at maturity. In many fish, however, a large part of growth occurs after maturity, and fecundity is markedly size or age dependent (Bagenal 1966). Consequently there is likely to be a trade-off between maturation and other life history parameters, e.g. growth as has been suggested by Roff (1982, 1984) and Rijnsdorp (1990). In such a case an iteroparous fish might grow rapidly prior to maturation then grow more slowly, as energy is diverted to gonadal tissue. If a late maturing fish continued its fast growth for one or more seasons after the precocious fish started reproducing, then at the onset of maturation it would be substantially heavier than a fish of the same age that had matured earlier. If fecundity were size dependent the late maturing fish would produce correspondingly more gametes. Where the fecundity weight relation is steep the lifetime's egg production of late maturing fish could exceed that of precocious fish, but this would also depend on the rate of mortality. If mortality rates were high in the late juvenile and adult fish, an early maturer would be more likely to spawn successfully since it has a higher chance of reaching maturity. Where chances of late maturers surviving until maturation are particularly low the selective advantage would be with the fish with the lower potential lifetimes fecundity and the faster maturation.

A number of studies have found a genetic component to the age and size of maturation (Gardner 1976, Naevdal et al. 1979, Kirpichnikov 1981, Gjedrem 1983, McKenzie et al. 1983, Policansky 1983, Reznick et al. 1990), although the work of Godø and Moksness (1987) suggests that this may not always be the case. If the timing of mortality can affect the optimal timing of reproduction, and if non-zero heritabilities are postulated, changes in the timing of maturation might be expected in commercially fished stocks as exploitation changes the pattern of mortality. There are a number of cases where changes in age or size of maturity have been reported for commercially exploited stocks. These include Northern stocks of plaice (Rijnsdorp 1990) and sole, (De Veen 1976), Arcto-Norwegian cod (Jørgensen 1988, 1990), Newfoundland cod (Kohler 1964, Beacham 1983), and North Sea cod (Graham 1923, Oosthuizen and Daan 1974). In most cases these have been attributed to changes in density, or temperature (Kohler 1964, Beacham 1983, Jørgensen 1988). Oosthuizen and Daan (1974) noted in their case that there was no obvious trend in potentially explanatory variables, leaving open the possibility of some other cause.

This chapter is concerned with the role of exploitation on the maturation of North Sea cod (*Gadus morhua* L.). The choice of this species and geographical area was made because there were recent data readily available from the International Commission for the Exploration of the Seas and reports of size at maturity going back almost one hundred years.

The background to the investigation is as follows. In 1893 Holt reported the stage of maturity of fish landed at Grimsby from coastal fishing vessels (Holt 1893). Graham (1923) reported results from a similar study in the same area, and remarked that the size at maturation was less than Holt's estimate. Oosthuizen and Daan (1974) writing on the reproductive biology of cod point out a trend towards reduced sizes at maturation with time (Their Figure 7 is reproduced as Figure 1). They note: ‘From inspection of figure

7 there can be little doubt that the both male and female cod mature at significantly smaller length than during the first half of the century' (Oosthuizen and Daan 1974). They suggest that this might have been due to fishing pressures since they could find no significant trend in environmental factors such as temperature, or stock density. However Hislop (1984) pointed out that Poulsen (1934) found ages at maturity much more similar to present ones. Moreover Rijnsdorp et al. (1991) found that the variation between years in the length at which 50 per cent of the population is mature is enough to swamp any trend on the scale that Oosthuizen and Daan (1974) reported.

None of these studies considered the different areas of the North Sea separately. Since some include spawning grounds, it seemed possible that there could be differences in the proportion of fish mature in each area. Small mature fish are more likely to be found in a spawning area and large immature ones in a non-spawning area. This could bias the findings of both Oosthuizen and Daan (1974), most of whose data came from a different area from that sampled in the earlier studies, and Rijnsdorp et al. (1991) who could have been misled by different proportions of their sample coming from each area in different years. It would also explain the discrepancy between the findings of Graham and those of Poulsen, since the studies were carried out in different areas of the North Sea.

In the following investigation a model of the reproductive success of fish maturing at different ages was used to predict whether increased fishing mortality on a fish with the North Sea cod life history

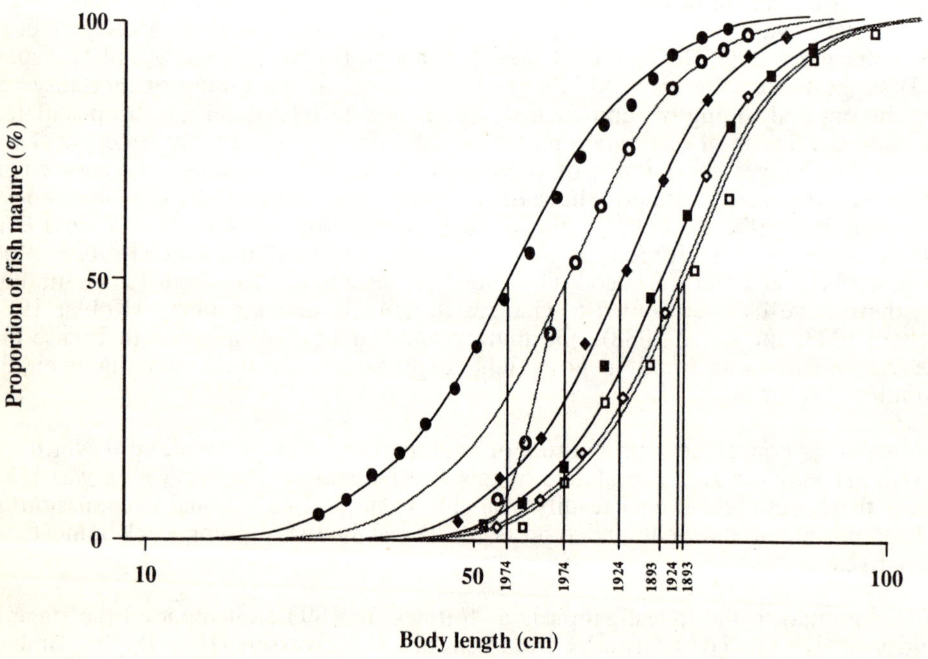

Figure 1. Length-maturity ogives (estimated from logit transformation of North Sea cod for males (solid symbols) and females (open symbols) according to data by Holt (1893) (■ □), Graham (◆ ◇) and the present research (● ○): L_{50} values indicated by date. Redrawn from Oosthuizen and Daan (1974).

parameters would change optimal age at maturation and if so what the direction of these changes might be. The reported changes in size at maturity in this stock were then reassessed making allowances for the potential effects of between area variation in size at maturity. Lastly the two results were compared to see how the observations correspond to the general predictions of the model.

MATERIALS AND METHODS

MODEL

The first step was to construct a model to examine the relative fitness of fish maturing at various ages. The average lifetime's fecundity, R_o , was taken as a measure of fitness, instead of the more commonly used *r*, the instantaneous rate of increase. R_o can be used if it is assumed that the survival of the larval fish is strictly density dependent (Law and Grey 1989). In the interests of simplicity only the females are considered in this model, and it is assumed that the differences between the sexes is not sufficient to render this invalid. It is assumed that the growth rate changes at the onset of maturity. A direct weight/fecundity relationship is also assumed. These two assumptions mean that delaying maturity will result in a higher weight at maturation and so a higher initial fecundity. The resultant model bears a similarity to the models of Roff (1984), and Jones and Johnston (1977) in that it shares the assumption that growth is slower after maturation, because of a trade-off between growth and reproduction. However the model developed in this paper is considerably simpler, while being adequate for the present purpose.

The reasoning behind the model is as follows. The fecundity of an individual at any given age at maturity is related to the weight of the fish according to equation 1:

$$R_i = W_i f \, , \qquad (1)$$

where W_i is the weight at age *i*, and *f* is the fecundity/weight relationship. The weight at age of the fish will depend on when it matures, with a different suite of weights for each age at maturity.

Equation 1 gives the actual fecundity of an individual of age *i*. The average expected fecundity of a fish at age *i*, however, will be influenced by the probability of surviving to age *i*. The average lifetimes fecundity is the sum of the fecundity at each age, multiplied by the probability of reaching that age. This is expressed in Equation 2,

$$R_0 = \sum_{i=a}^{n} W_i f(\Pi_{j=0}^{i-1} p_j) \qquad (2)$$

where R_o is the lifetimes reproductive output, *a* is the age at maturity, W_a is the weight at maturity, p_j is the probability of surviving year *j*, and *n* is the maximum age.

The values for weight at age, given maturation at various ages are calculated using Law and Grey's (1989) method. Essentially they divide the growth into two parts: pre-maturation and post-maturation. The pre-maturation curve is described by :

$$W_i = W_\infty(1-B(1-\alpha)^i)^3 \, . \qquad (3)$$

After maturity the growth is slower and is described by:

$$W_i = \overline{W}_\infty(1\text{-}A\ [(1\text{-}\ g)^{1/3}(1\text{-}\ \alpha)]^i)^3$$

A and *B* are constants defined by comparison to the von Bertalanffy growth curve, *g* is the gonadosomatic index, and *a* is the third constant defined by the gonadosomatic index and the von Bertalanffy growth parameter *K*. For fish that mature at the average age of maturity, *v*, the curve made from these two equations is defined as being equal to the von Bertalanffy curve. This relationship is used to set the values for the parameters *A* and *B* . These equations can then be used to calculate the growth curves for fish maturing at ages other than *v*.

By using Law and Grey's (1989) method the suite of post-maturation weights for fish maturing at different ages can be calculated and used to calculate the lifetime's fecundity for each maturation age. The optimal age at maturation, (i.e.that giving the highest value for R_o) can be assessed under different mortality schedules.

Parameters for the North Sea cod model

In order to apply this model to the North Sea cod a number of parameters have to be supplied. These are given below with the information used to determine the appropriate value of each.

The von Bertalanffy parameters $\overline{W}_\infty$ =17.393 kg, K = 0.269, t_o = 0.87 y, were taken from the estimates given by Daan (1974). Their values for the northern North Sea were used since my intended comparisons were among fish from Round Fish Area (RFA) 4. The average age at maturation found in the North Sea as a whole was taken from Hislop's (1984) paper which showed a table of the average proportion mature at a given age. The conversion constant from weight to eggs is given as $f = 569\ g^{-1}$ (Oosthuizen and Daan 1974, Daan 1974).

The gonadosomatic index (weight of eggs/total fish weight) was taken as $g = 0.2$. This was calculated from the following information. The number of eggs produced by cod of a given weight was taken from the weight/fecundity relationship *(f)* given above. The weight of individual eggs was calculated from the dry weights of cod eggs given as 0.764 mg in Hislop (1984), and the overall relation of dry to wet fish weight of 0.2 given in Daan (1974). This last may be too large a ratio but better data were lacking. Combining these gives an estimate for the weight of the eggs to total body weight of 0.212. Using Hislop's (1984) estimate of the fecundity/weight ratio results in a lower value; g =0.177. As a compromise between the two estimates the intermediate value of 0.2 was used.

I restricted the age at maturity to lie between two and eight years, as there are very few records of mature 1 year olds, whilst a value over eight years would give negative values for growth, having exceeded the maximum weight allowable in the model, and then in following years shrinking towards the asymptote. The maximum age is set to 15 yrs, which is compatible with the present pattern of mortality, but may be too low for an unfished stock.

The survival probabilities $(p_0...p_n)$ were taken as one minus the mortality rates given by Macer and Easey (1988) values for proportion. They are shown in Table 1. The probability of surviving natural causes of mortality and fishing mortality are abbreviated to p_{Nat} and p_{Fish} respectively. The total probability of survival p_{Tot} is the product of the two. It should be noted that the estimate of fishing mortality in the first year is likely to be too low as it does not include discards, and that the 0-group (fish aged zero to one year) survival refers to the period after settling only. No estimate of 0-group fishing mortality was available.

Effects of changing the fishing mortality.

The effects of changing fishing mortality on the optimal age at maturation were investigated by running the model with the fishing mortality ($1\text{-}p_{Fish}$) set at the values given by Macer and Easey (1988) and then at 90 per cent, 80 per cent, ... etc. of their fishing mortality down to no fishing mortality. The total mortalities used in each case were the product of the reduced fishing mortality and the unchanged natural mortality. These are shown in Table 2.

Table 1. Probabilities of cod surviving different ages. p_{Nat}, p_{Fish} and p_{Tot} are the probability of surviving natural mortality, fishing mortality and the overall probability of survival respectively.

	Age (yr)					
	0.5-1	1	2	3	4	5+
p_{Nat}	0.20	0.50	0.80	0.85	0.90	0.90
p_{Fish}	?	0.90	0.40	0.40	0.45	0.45
p_{Tot}	0.2	0.45	0.32	0.34	0.40	0.40

Table 2. Probabilities of survival (p_{Tot}) at different levels of fishing intensity.

	Probability of Survival by age				
% 1988 fishing	0 year	1 year	2 year	3 year	4+ year
100%	.20	.45	.32	.34	.40
90%	.20	.45	.37	.39	.45
80%	.20	.46	.42	.44	50
70%	.20	.46	.48	.51	.52
60%	.20	.47	.51	.54	.60
50%	.20	.47	.56	.59	.62
40%	.20	.48	.61	.65	.70
30%	.20	.48	.66	.70	.75
20%	.20	.49	.70	.75	.80
10%	.20	.49	.75	.80	.85
0%	.20	.50	.80	.85	.90

I.C.E.S. DATA

This subsection investigates the possible change in size at maturation of North Sea cod over the last century. The following two questions were addressed:1) Is there any significant difference in the timing of maturity between the areas? I used ICES data on proportion mature at length for 1988 and 1989 classified by area, age, and sex. Sample sizes are shown in Table 3. 2) If RFA 4 is looked at separately, is there an apparent trend in the timing of maturity or is this masked by annual variation? ICES length at maturity data from RFA 4 over the years 1980-1990 (excluding 1981 for which the data were incomplete) were used for this. The rate at which the proportion mature increased with size was compared to that reported in Holt (1893), and Graham (1923). The sizes of the samples used are shown in Table 4.

Table 3. Sample sizes for each Round Fish Area in 1988, 1989.

	Number sampled	
Round Fish Area	1988	1989
1	964	784
2	413	760
3	380	218
4	243	464
5	282	148
6	699	597
7	0	158

Table 4. Sample sizes of data used for the comparison of the pattern of maturity at length from 1890 - 1990.

Sample Size for RFA 4 by year			
Year	male	female	total
1890	201	186	387
1923	2186	1990	4176
1980	94	114	208
1982	55	62	117
1983	85	78	163
1984	192	108	300
1985	277	271	548
1986	145	139	284
1987	315	297	612
1988	116	127	243
1989	342	122	464
1990	134	94	228

The data used to assess the change in size at maturation came from Holt (1893), Graham (1923), and Oosthuizen and Daan (1974). RFA 4 data were collected on the ICES Young Fish Surveys from 1980 to 1990.

The statistical package, GLIM, was used to carry out a logit transformation (logit p = log $(p/(1-p))$ on the proportion of individuals mature, to make the followed by regression analysis based on generalised linear models (See Aitkin et al. (1990) for explanation of how this can be done.)

RESULTS

RESULTS FROM THE MODEL

The lifetime's fecundity for different ages at maturity under different fishing mortalities is shown in Figure 2. The optimal age at maturation under each set of mortalities is that which gives the highest value for the lifetime's fecundity R_o. When there is no fishing mortality there is surprisingly little difference between the average lifetime's fecundity of fish maturing at different ages. As fishing activity increases an optimal age at maturation appears and the differences in fecundity increase. Taking R_o as a measure of fitness we see an increase in the difference in the relative fitness of early and late maturing phenotypes, with increased fishing mortality markedly favouring fish that mature early.

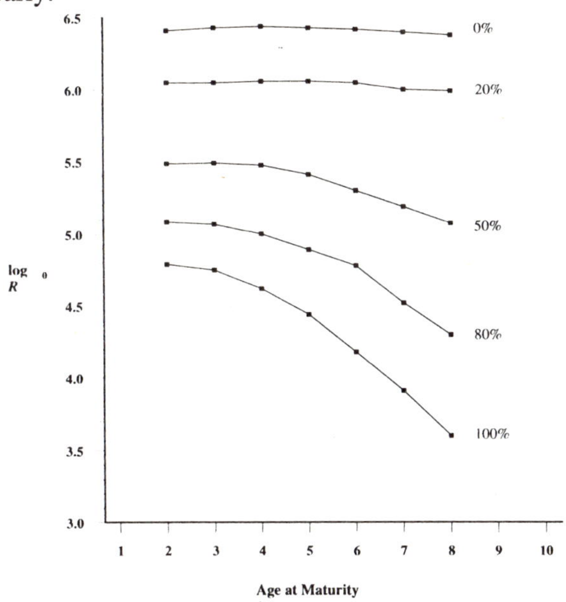

Figure 2. Lifetime's fecundity for fish maturing at different ages. Fishing mortality is varied from its 1988 level (marked 100%) to zero.

RESULTS FROM THE DATA

Differences over areas

The GLIM statistical modelling program was used to fit slopes to the data on proportion mature at a given size. Explanatory variables investigated included length, sex, age (when available), and year of capture. Statistical models containing progressively more explanatory variables were compared to the null model, which assumes these variables have no effect. The scaled deviances given in Table 5 are indicators of the fit of the model. To compare statistical models the deviance of the new model is subtracted from that of the previous one . The differences follow a χ^2 distribution with degrees of freedom equal to the difference between the models' degrees of freedom. The reductions in their deviances, their degrees of freedom, and the significance of these are shown in Table 6.

Table 5. Deviances and degrees of freedom for each model applied to 1988, 1989 data by fishing area.

	Model	Deviance	d.f.
0	null	4472.6	748
1	+ length	1537.1	747
2a	+sex	1503.1	746
2b	+sex.length	1500.1	745
3a	+area	1096.4	740
3b	-area +area.length	1136.9	740
3c	+area +area.length	1047	734
4	+age	1046	733
5a	+year	1039.7	733
5b	-year +year.length	1043.2	733
5c	year +year.length	1035.6	732
6	+year.area	979.9	727

The statistical models are as follows (the models are cumulative).(0) the null model fits a single horizontal line to the data, passing through the mean. (1) Model 1 allows for the effect of length, letting the line have a slope with respect to length. (2) Model 2a allows the intercept of the line to differ between the two sexes, while holding the slope constant. 2b allows the slopes and the intercepts to differ between the sexes. (3) Model 3a allows the intercept of the line to differ between the areas while holding the slope constant. 3b allows the slopes and the intercepts to differ between the sexes. (4) Model 4 added the effect of age, fitting separate lines to each age, within sex, and area groupings. The effect was not significant, so this variable was excluded from further models. (5) Model 3 adds the effect of year, allowing first the intercept, then the slope, and then both to vary between years. (6) Model 6 allows the slopes fitted for area to differ between years.

A significant effect of length, and of sex on maturity was found, indicating that the proportion of fish mature depends on the length of the fish, and that this relationship differs between the sexes. No effect of age was found in the data, in contrast to the data used in Oosthuizen and Daan which shows an effect of age on the proportion mature at a given length. However, the effect of age is significant when it is taken before the area effect is removed. The difference in the significance when the age is removed before and after removing the effect of area implies that there is a large area component to the effect of age. (The area effect is removed first because it has a higher significance level.)

There is a significant difference between areas (effect 3c), in both the slope and the y-intercept. There was also a significant difference in y-intercept and slope between 1989 and 1990 even after the effect of between area effects had been accounted for (effect 5c).

Differences over time

The data from Holt (1893) and Graham (1923) and the ICES RFA 4 data (1980-90) were compared by fitting a number of models. An effect of length and sex was assumed for all data. The deviances and degrees of freedom, and the χ^2 values of the effects are shown in Tables 7 and 8. The models were as follows. (1) The null model - no difference between years. (2) Differences between data from this century (ICES and Graham 1923) and last (Holt 1893). (3) Differences between the ICES data and the 2 earlier sets. (4) Differences between all 3 groups of data. (5) Differences between all years. The slopes and intercepts were allowed to vary both separately and together for each model.

Table 6. χ^2 values, and degrees of freedom for the reduction in deviance gained by fitting each of the models to the 1988 and 1989 data.

Model	Effect	χ^2	d.f.	p
1	length	2936	1	<.001
2a	sex(intercept)	34	1	<.001
2b	sex(slope)	3	1	>.5
3a	area(intercept)	406.7	6	<.001
3b	area(slope)	366.2	6	<.001
3c	area(both)	456.1	12	<.001
4	age(slope)	1	1	>.1
5a	year(intercept)	7.3	1	<.01
5b	year(slope)	3.8	1	<.05
5c	year(both)	11.4	1	<.005
6	year by area	55.6	5	<.001

Table 7. Deviances and degrees of freedom for the models of differences in length at maturity over the last century.

Model		Deviance	d.f.
1	null (length +.sex)	1630.9	602
2a	ICES vs old intercept	1403.1	601
2b	+slope	1036.3	600
3a	Holt vs Graham and ICES:intercept	1572.4	601
3b	+slope	1537.2	600
4a	Holt :vs Graham vs ICES: intercept	1367	600
4b	+slope	985	598
5a	Separate years intercept	968	591
5b	+slope	660.4	580

The slope and *y*-intercept of the 1893 and the 1924 data are significantly different from the combined 1980-1990 data, (effect 1), and from each other as can be seen from the decrease of deviance between models 2 and 4. Figures 3 and 4 show that the trend is toward a shallower slope and higher intercept with time, corresponding to a smaller minimum size at maturity and a reduced slope of the maturation ogive. The reduction in length of the smallest mature individuals is particularly striking (Table 9). There is also a change in the length at which 50 per cent of the fish mature, the $L_{50,}$ as shown in Table 9.

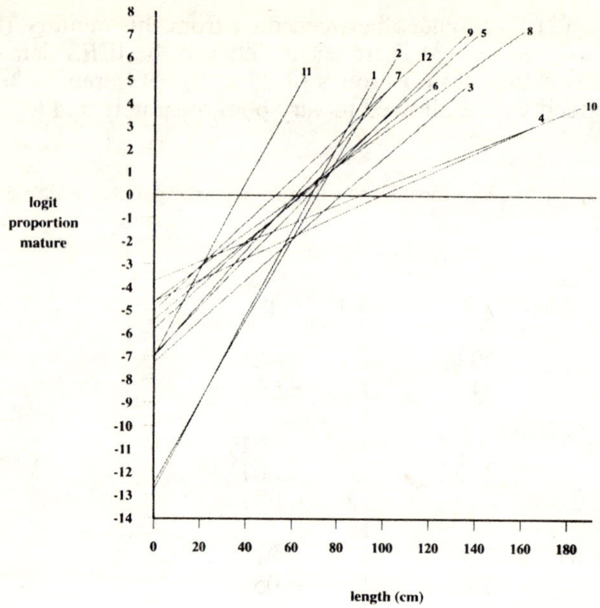

Figure 3. Logit transformed proportion of females mature against fish length in each year. The x-intercept of any line is the point where 50% are mature. Numbers label the years: 1 = 1893 (Holt's) data, 2 = 1923 (Graham's) data, 3 - 12 are 1980, 1982 - 1990 (I.C.E.S.) data.

Table 8. χ^2 values and degrees of freedom for the reduction in deviance gained by fitting each of the models to the data on change over the last century.

Model	**Effect**	χ^2	**d.f.**	**p**
2a	ICES vs Old: intercept	227.8	1	< 0.001
2b	+slope	366.82	1	< 0.001
3a	Holt vs Graham+ICES: intercept	58.5	1	< 0.001
3b	+slope	35.2	1	< 0.001
4a	Holt vs Graham.vs.ICES: intercept	263.7	2	< 0.001
4b	+slope	381	2	< 0.001
5a	Separate years intercept	662	11	< 0.001
5c	+slope	307.59	11	< 0.001

Table 9. Changes of length at 50 per cent maturity, and smallest size at maturity, over the last century.

	L_{50} (cm)		**smallest mature**	
Year	male	female	male	female
1890	70.76	75.85	58	58
1923	64.78	70.50	54	29
1980's	55.70	69.89	15	15

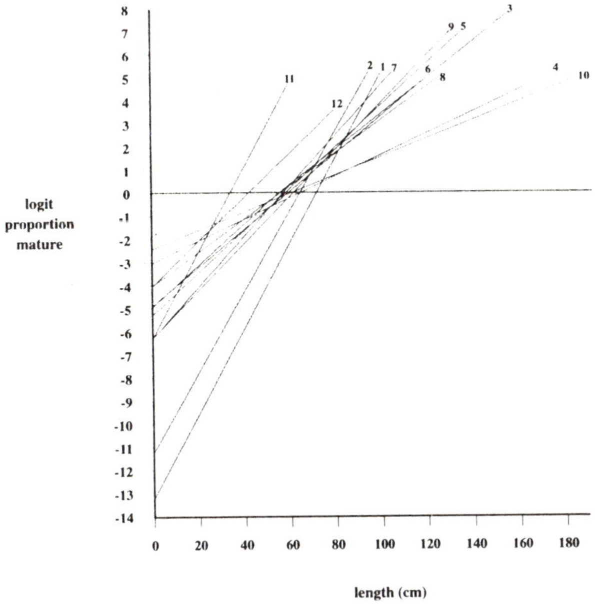

Figure 4. Logit transformed proportion of males mature against fish length in each year. Labelling as in figure 3.

DISCUSSION

DISCUSSION OF THE MODEL

The model contains a number of possible oversimplifications:

1) The assumption that the main trade-off in maturation is with growth repeats the perspective of existing literature (Jones and Johnston 1977, Roff 1984, Law and Grey 1989). However, other trade-offs include, for example, maturity and reproductive lifespan (Roff 1981, 1982) There could also be an increased mortality from predation in fish that mature at very young ages if they remain in the prey size-range of predators for longer. This effect will decrease as fishing mortality increases and the number of large, potentially cannibalistic cod decrease. The possible complications are therefore considerable.

2) The fecundity/weight relationship was assumed to be the same for both sexes. There are no data of any sort for the virility of male cod as a function of size, but it seems reasonable to assume that the increase in sperm production will be in step with the female egg production.

3) The assumption that larval survival is strictly density dependent is almost certainly an oversimplification. However the lack of correlation between stock numbers and recruitment is indirect evidence that some compensatory pre-recruit mortality occurs.

4) Of the parameters used, the values for the gonadosomatic index and the maximum weight are the most uncertain. Calculating the gonadosomatic index involved calculating the wet weight of fish given the dry weight but, as there is no published information of the ratio of dry to wet egg weight, the whole fish dry to wet weight ratio of 0.2 was used instead. As Hislop (1984) comments that an egg may be up to 95 per cent water. I could be overestimating the ratio considerably. If the dry weight were only 5 per cent of the wet weight the gonadosomatic index would be closer to 0.8 than 0.2.

5) The value for W_∞, 17.393, was taken from Daan (1974), but weights of up to twice this value have been reported (Yorrell 1859). Although the maximum size of North Sea cod may therefore be larger than the chosen value, it was retained since I was using an entire set of von Bertalanffy parameters calculated by Daan (1974) and changing the value for W_∞ would have changed the entire curve. If the weights of older fish are underestimated, the estimate of the optimal age at maturity will be biased in favour of younger maturers.

INVESTIGATIONS OF AREA EFFECTS USING I.C.E.S. DATA

The ICES data for 1989-90 showed a definite increase in the proportion mature with length, which differed between the sexes. In general males mature at slightly smaller sizes than females, as observed by Graham (1923), and Oosthuizen and Daan (1974). Holt reported a lack of difference in size at maturity between males and females but the present reanalysis does show such a difference in his data.

In contrast to Oosthuizen and Daan's (1974) results, no difference was found between fish of different ages. They attributed their result to the lack of any mature 2 year old females, but the age effect is significant for their data even when two year olds are excluded from the analysis. Since an age effect was observed in the ICES data before the area effect was taken into account the age effect noticed by Oosthuizen and Daan (1974) may have been largely due to differences between areas.

Comparing the different areas between the years revealed a large effect of area on the pattern of maturation. It is possible that this is due to migration to spawning grounds. It is likely that smaller mature fish would be near the spawning grounds, and large immature ones may be less likely to migrate and so would be disproportionately common in areas that did not hold a spawning ground. This could generate a less extreme form of the pattern found in the Arcto-Norwegian cod where only mature fish are found on the spawning ground (Jørgensen 1988). Unfortunately there is not enough information to be able to tell whether the differences between areas are due to migration.

Alternatively there may be spatial differences between areas in the pattern of maturation, which could be genetic or environmental. The heritability of size at maturation in cod is unknown however, and it is unclear whether there is sufficient genetic isolation between stocks to create a genetic difference. There is sufficient variation in growth and food availability between regions of the North Sea (Daan 1973, 1974) to suggest that these factors could contribute to the variation in length at maturity between areas.

Whatever the cause of variation between areas in length at maturity, the effect is pronounced and makes lumping data across areas inadvisable. In this study, areas were only compared for two years, and so it is not possible to say whether the differences between areas would remain consistent over longer periods.

TRENDS OVER TIME IN MATURITY AT LENGTH

Comparing data for 1893, 1923 and the 1980s from the same geographical area, shows a clear trend towards a smaller minimum size at maturation, and maturation over a wider range of sizes as time progresses. There are a number of possible explanations for this change, of which the possibility of an evolutionary response to the change in mortality, or density dependent effects, or a release from social inhibition of early maturity are the most interesting. Changes in temperature could be a possible cause, but there is no evidence for a long term trend in water temperature (Hill and Dickson 1975) and this is not discussed further.

The model results suggest that, if there is a genetic component to the variation in age at maturity there could be a decrease in the the average age at maturity as fishing pressure increases. The optimal age of maturity decreases from four to two in the model and we see from the historical data that there is some evidence for a similar change in the population over the last century. Since age and length are closely correlated rapid maturation should lead to a smaller size at maturity such as has been shown. However ignorance of the heritability of size at maturation, and lack of detailed information on the fishing mortality prior to the 1960s makes it impossible to judge whether selection on age at maturity is sufficient explanation for the observed change in length at maturation over time.

Density dependent effects due to fishing have been used to explain decreases in the age at maturity for other stocks (Kohler 1964, Jørgensen 1988). This assumes that food availability limits growth at high densities, and as the stock decreases this limitation is removed allowing increased feeding and growth and, where maturity is correlated with size, younger maturation. The population of North Sea cod, however has almost certainly increased since the 1920s (Daan 1978, Macer and Easey 1983), while size at maturation has decreased. If density dependent growth was affecting maturation it could only result in smaller size at maturation if the point of maturity was age dependent. If maturity is length dependent then age at maturity should increase. However there is no

Table 10. Predicted mean length at age of fish maturing at 2 and 4.

Maturing at 2 years		**Maturing at 4 years**	
age (yrs)	length (cm)	age (yrs)	length (cm)
2	24.37	2	24.37
3	48.05	3	46.62
4	67.55	4	63.62
5	79.62	5	76.60
6	88.84	6	86.50
7	91.66	7	94.11
8	101.26	8	99.92

conclusive evidence for a decrease in growth rate (Daan 1974). Therefore fish that mature at smaller sizes are maturing at lower ages, not greater ones as would be the case if the observed changes were due to density dependent growth, which makes it unlikely that density dependent effects are the cause of the observed change in size at maturity.

The lack of any decrease in growth rates would superficially seem to be an argument against the model used here as this predicts a decrease in growth with earlier maturity. However, as Table 10 shows the decrease in size that would be predicted if the whole population matured early would only be of the order of 3 cm.

CONCLUSION

The current conclusions extend those of Oosthuizen and Daan (1974) in suggesting a change towards smaller size at maturation in North Sea cod over the last century. The modelled effects of increasing fishing mortality on the optimal age at maturation suggest that, given heritability of size at maturation, the observed pattern would be consistent with selection favouring early maturation. The most likely alternative explanation, that of density dependent growth was also considered and seems less likely. However, until a genetic basis of age and size at maturation is determined, the issue must remain open to further study and observation.

ACKNOWLEDGEMENTS

Thanks are due to the staff at the MAFF Lowestoft Laboratories, and at ICES for giving me access to the data. Thanks also to Richard Law, Kevin Stokes, and an anonymous reviewer for comments on the text.

REFERENCES

Aitkin, M., D. Anderson, B. Francis and J. Hinde. 1990. Statistical modelling in GLIM. Clarendon Press, Oxford.

Bagenal, T.B. 1966. The ecological and geographical aspects of the fecundity of Plaice. Journal of the Marine Biological Association of the U.K. **46**: 161-186.

Beacham, T.D. 1983. Variability in median size at age at sexual maturity of the Atlantic cod, *Gadus morhua*, on the Scotian shelf in the Northwest Atlantic Ocean. Fishery Bulletin **81**: 303-321.

Borisov, V. M. 1978. The selective effects of fishing on the population structure of species with a long lifecycle. Journal of Ichthyology **18**: 986-904.

Brawn, V. M. 1961a. Aggressive behaviour in the cod (*Gadus callarius*) Behaviour **18**: 108-147.

Brawn, V. M. 1961b. Reproductive behaviour in the cod (*Gadus callarius*) Behaviour **18**: 177-198.

Cole L.C. 1954. The population consequences of lifehistory phenomena. Quarterly Review of Biology **21**: 103-137.

Daan, N. 1973. A quantitative analysis of the food intake of the North Sea cod, *Gadus morhua*. Netherlands Journal of Sea Research **6**: 479-517.

Daan, N. 1974. Growth of North Sea cod, *Gadus morhua.* Netherlands Journal of Sea Research **8**: 27-48.

Daan, N. 1978. Changes in cod stocks and cod fisheries in the North Sea. Rapports et Procès-Verbeaux des Réunions, Conseil International pour l'Exploration de la Mer **172**: 39-57.

De Veen, J. F., 1976. On changes in some biological parameters of North Sea sole (*Solea solea* L.) Journal du Conseil International pour l'Exploration de la Mer **37**: 60-90.

Edley, M. T. and R. Law. 1988. Evolution of life histories and yields in experimental populations of *Daphnia magna.* Biological Journal of the Linnean. Society **34**: 309-326.

Fisher R. A., 1930. The genetical theory of natural selection. Clarendon Press, Oxford..

Gadgil, M. and W. H. Bossert. 1970. Life historical consequences of natural selection. American Naturalist **104**: 1-24.

Gardner,M.L. 1976. A review of factors which may influence the sea age of Atlantic Salmon (*Salmo salar* L.). Journal of Fish Biology **9**: 289-327.

Gjedrem, T. 1983. Genetic variation in quantitative traits and selective breeding in fish and shellfish. Aquaculture **33**: 51-72.

Godø O. R., and E. Moksness 1987. Growth and maturation of Norwegian coastal cod and Northeast Arctic cod under different conditions. Fisheries Research **5**: 235-242.

Graham, M. 1923. The annual cycle in the life of mature cod in the North Sea. Fisheries Investigations Series II vol **VI** no 6.

Handford, P. G. Bell and T. Reimchem. 1977. A gillnet fishery considered as an experiment in artificial selection. Journal of the Fisheries Board of Canada **34**: 954-961.

Hill, H. W. and R. R. Dickson. 1975. Long-term changes in North Sea hydrography. ICES Symposium on the changes in North Sea fish Stocks and their causes. No **1**. 9-12/7/1975.

Hislop, J. R. G, 1984. A comparison of the reproductive tactics and strategies of cod, haddock, whiting and Norway pout in the North Sea. Pp311-329, in G.W. Potts and R.J. Wooton eds., Fish Reproduction Strategies and Tactics.Academic Press, London.

Holt, E.W.L. 1893. North Sea investigations. Journal of the Marine Biological Association of the U.K. **3**: 78-122.

Jones, R. and C. Johnston. 1977. Growth, reproduction and mortality in gadoid fish species. Pp 37 - 62 in J.H. Steele ed., Fisheries Mathematics. Academic Press, London.

Jørgensen T. 1988. Changes in age at sexual maturity of the Northeast Arctic cod, (*Gadus morhua.* L) I.C.E.S. C.M./**G**:42.

Jørgensen T. 1990. Long-term changes in age at sexual maturity of the Northeast arctic cod, (Gadus morhua. L) Journal du Conseil International pour l'Exploration de la Mer **46**: 235-248.

Kirpichnikov, V.S. 1981. Genetic Basis of Fish Selection. Translated by G. G. Gause. Springer-Verlag, Berlin.

Kohler, A. C., 1964. Variation in the growth of Atlantic cod. Journal of the Fisheries Research Board of Canada **21**: 57-100

Law, R. 1979. Optimal life histories under age specific predation. The American Naturalist **114**: 399-417.

Law, R. and D. Grey. 1989. Evolution of yields from populations with age specific cropping. Evolutionary Ecology **3**: 343-359.

Macer, C. T. and M. Easey. 1988. The North Sea cod and the English Fishery. Lowestoft Laboratory Leaflet no. **61**. Directorate of Fisheries Research, Lowestoft.

McKenzie, W. D., D. Crews, K. Kallman, D. Policansky, and J. Sohn. 1983. Age, weight, and genetics of sexual maturity in the platyfish (*Xiphophorous maculatus*). Copeia **3**: 770-774.

Michod, R. E. 1979. Evolution of life histories in response to age specific mortality factors. American Naturalist **113**: 531-550 .

Naevdal, G., M. Holm, R. Leroy, and D. Møller. 1979. Individual growth rate and age at sexual maturity in rainbow trout. Fiskdir. Skr. Ser. Havunders **17**: 11-17.

Oosthuizen, E. and N. Daan. 1974. Egg fecundity and maturity of North Sea cod, (*Gadus morhua*. L), Netherlands Journal of sea Research **8**: 378-397.

Policansky, D. 1983. Size, age, and demography of metamorphosis and sexual maturation of fish. American Zoologist **23**: 57-63.

Poulsen, E.M. 1931. Biological investigations upon the cod in Danish waters. Meddr. Komm. Dann. Fisk-og. Havunders., Fisk. **9**: 148.

Reznick, D. A. and H. Bryga. 1987. Life history evolution in guppies (*Poecilia reticulata*): 1: Phenotypic change in an introduction experiment. Evolution **41**: 1370-1385.

Reznick, D. A., H. Bryga and J. Endler. 1990. Experimentally induced life history evolution in a natural population. Nature **346**: 357-359.

Ricker, W. E. 1981. Changes in the average size at age of Pacific salmon. Canadian Journal of Fisheries and Aquatic Sciences **38**: 1636-1656.

Rijnsdorp, A. D., 1990. The mechanism of energy allocation over reproduction and somatic growth in female North Sea plaice, *Pleuronectes platessa* L. Netherlands Journal of Sea Research **25**: 279-290.

Rijnsdorp A. D., N. Daan, F. A. von Beek and H. J. L. Heesen. 1991. Reproductive variability in North Sea plaice, sole, and cod. Journal du Conseil International pour l'Exploration de la Mer **39**: 271-287.

Roff, D. A. 1981. Reproductive uncertainty and the evolution of iteroparity: Why don't flatfish put all their eggs in one basket? Canadian Journal of Fisheries and Aquatic Sciences **38**: 968-977

Roff, D. A. 1982. Reproductive strategies in flatfish: A first synthesis. Canadian Journal of Fisheries and Aquatic Sciences **39**: 1686-1689

Roff, D. A. 1984. An allocation model of growth and reproduction in fish. Canadian Journal of Fisheries and Aquatic Sciences **41**: 1395-1404.

Schaffer, W. M. 1974. Selection for life histories: the effects of age structure. Ecology **55**: 291-303.

Turner, J. L. 1977. Changes in size structure of cichlid populations of Lake Malawi resulting from bottom trawling. Journal of the Fisheries Research Board of Canada **34**: 232-238.

Yorrell, W. 1859. A History of British Fishes. (3rd Edition) Sir John Richardson ed. John Van Voorst, London.

INFLUENCE OF HUMAN ACTIVITY ON PROPERTIES OF ATLANTIC SALMON POPULATIONS

A.F. Sharov[1] and A.V. Zubchenko[2]

[1] All-Union Research Institute of Marine Fisheries and Oceanography (VNIRO), V.Krasnoselskaya 17a Moscow USSR
[2] Polar Institute of Marine Fisheries and Oceanography (PINRO) Knipovicha 6 Murmansk USSR

ABSTRACT

Changes in traits of Atlantic salmon *Salmo salar* L. populations caused by hydroelectric power station activity and hatchery reproduction are described. Populations living in rivers with dams lost some biological groups and experienced a significant reduction in number of repeat spawners. The period of the spawning run became stretched out. As a result of hatchery reproduction the proportion of grilse in the River Kola significantly increased (up to 70 per cent). The proportion of females decreased by over twofold and the number of age groups decreased from 20 to 15. Mean weight of spawners decreased from 4.8 to 2.9 kg. The pattern of the seasonal run also changed significantly. Similar changes were observed in other rivers. A selection program for hatchery forms is suggested to restore the native population structure.

INTRODUCTION

In most of northern countries the abundance of Atlantic salmon *Salmo salar* L. has shown a strong tendency to decrease. The main cause of decline of salmon stocks is human activity, especially the effect agriculture, industry and overfishing. As in other countries many of the Atlantic salmon stocks of the USSR are depressed. In this paper these changes are documented for salmon stocks on the Kolsky Peninsula.

THE RIVER TULOMA

One of the serious factors negatively influencing salmon stocks is the activity associated with the generation of hydroelectric power. Hydroelectric power stations partitioned six rivers on the Kolsky Peninsula and in only two of them have Atlantic salmon populations survived. Populations of Atlantic salmon of the the River Tuloma were saved due to a well-constructed fishway in the lower Tuloma dam, small loss of riverbed in the reservoir, and the existence of spawning and feeding grounds in some tributaries. Nevertheless flooding of the riverbed spawning grounds of the Lotta and Notta tributaries, as a result of building the power station, decreased population abundance by 25,000 fish on average. This dam has seriously influenced the biological structure of the salmon stock. For example the autumn spawners, which were about 5 per cent of the stock before erecting the dam have almost entirely disappeared. Proportion of repeat spawners decreased from 7-8 per cent to 0.8 per cent. The time of kelt migration shifted to a later period. In native rivers the run of kelts stopped in July, but in the Tuloma it continued up to the middle of August. Spawners were delayed downstream from the dam of the power station. As a result, the run of summer salmon continued up to the time at which the fishway was closed in October, while in native rivers the run of summer fish finished usually at the beginning of August. Due to the long delay of spawners below the dam, the proportion of fish which had undergone the physiological changes for living in fresh water increased in some years up to 48 per cent. This led to a decrease in the market quality of the fish. The run of smolts from the reservoir to the sea finished 1.5 months later than in native rivers where it used to stop in July. In the River Tuloma smolts run up to the time at which it freezes. Some of the smolts do not reach the sea and stay in the reservoir until the next summer. This later run of smolts decreases the period of the marine stage of life and this is reflected in the mean weights of spawners, which for the River Tuloma is lower than in other rivers of Kolsky Peninsula (Vshivtsev 1990).

THE RIVER KOLA

The River Kola which enters the same bay in the neighbourhood of the Tuloma, is under another form of human influence. Here natural reproduction of Atlantic salmon was displaced by the hatchery at a farm situated 70 km from the river mouth. Before 1958 there was natural spawning of salmon along all the river (105 km) and its tributaries. Since 1959 the river has been partitioned off entirely by a net trap 25 km from the river mouth. All migrants were taken as catch, and some of them were used as spawners for the hatchery farm. Thus access to 75 per cent of the natural spawning grounds was prevented (Salmov 1981). Now, natural spawning takes place downstream from the trap net only. Offspring, grown at the farm, are released at age 2+ into the River Kola and its tributaries. All released smolts have been tagged by finclipping since 1971. This procedure allows the efficiency of the hatchery farm to be established by the ratio of natural to farm reproduction. It is estimated that about 1 per cent of released fish returned to the river, while 5.5 per cent of wild fish returned (Yakovenko 1976). It is notable that finclipped salmon returned at least two years later. This means that for the most part, fish released at age 2+ were delayed in the river by at least one year. Annual mortality of delayed fish in the river was estimated as 65 per cent (Petrenko 1976). It had been supposed that hatchery farming would increase the amount of spawning stock. But the results of this analysis showed that significant changes in salmon population structure came about, as follows.

1) the proportion of females decreased more than twofold, from 46.7 per cent to 20.8 per cent. Mean river age decreased from 3.52 to 3.43 years, marine period from 1.7 to 1.4, and total age from 5.2 to 4.8 (Salmov 1981).

2) the structure of the population changed markedly. The number of age groups decreased from 20 to 15. For the period 1945 to 1957 the proportion of fish at sea for 1, 2 and 3 years was respectively 29.8, 35.7 and 32 per cent (Azbelev 1960). During the period 1969 to 1973 it was 52.4, 26.2, 19.3 per cent (Bukshtansky et al. 1980), while during 1986 to 1988 the tendency of fish to return early became even stronger: 72.6, 15.1, 11.9 per cent (Figure 1). It is necessary to note that, in addition to the negative influence of the farm hatchery there was intensive marine fishing, which selectively removed the big fish.

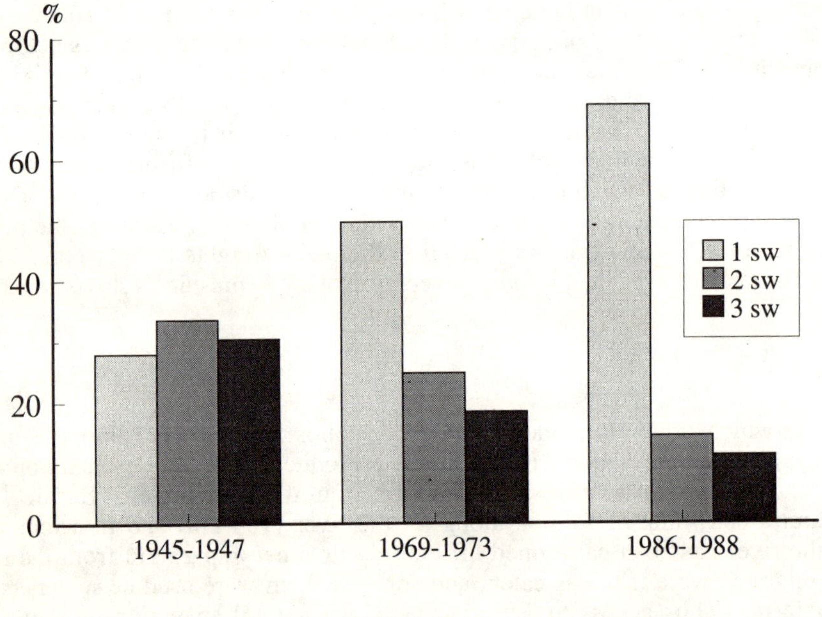

Figure 1. Sea age structure of Atlantic salmon in the River Kola (SW: seawinter)

3) the rates of linear and weight growth also decreased. Mean weight of hatchery spawners was 2.9 kg, and for wild fish 4.8 kg.

4) it seems especially important that changes came about in the time intervals of the spawning runs. It is very characteristic for the runs of wild salmon to have two peaks, at

the middle of June and July (Figure 2). Hatchery spawners have only one maximum, in July. Most of the finclipped spawners consisted of summer a biological group in which the one seawinter males were predominant.

Analysis of all these data showed that, due to hatchery farming which ignored the complicated genetic structure of salmon populations together with the marine fishery and illegal fishery in homewaters, serious destructive changes took place in the Kola salmon population. These changes were reflected not only on the structure at population level (ratio of age groups, changes in weight, growth of young fish proportion etc.), but also at the subpopulation level (changes in migration time and form).

One of the main reasons of such changes is that the complicated genetic structure of salmon populations was absolutely ignored in the hatchery process. Age, size, fecundity and other parameters of spawners were not taken into consideration in the breeding process.

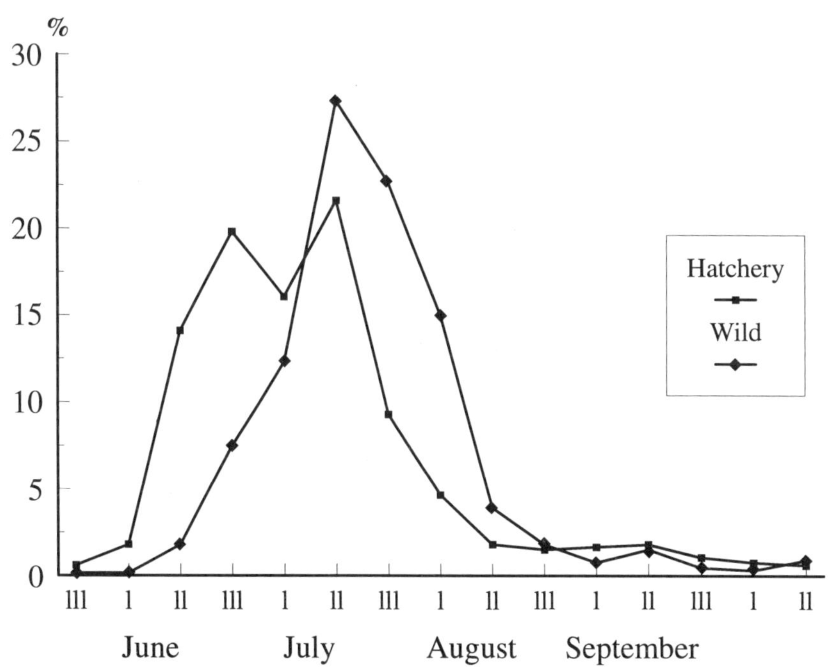

Figure 2. Dynamics of the seasonal run of Atlantic salmon in the River Kola in 1983 to 1987

OTHER RIVERS

Similar changes occurred for other stocks where stock enhancement programs also exist. Thus in the River Keret the proportion of grilse increased by up to 65 per cent in 1981 to 1985. This fact could be explained by the proportion of wild and reared fish, 93 and 7 per cent respectively in 1973 to 1975 and 48 and 52 per cent in 1983 to 1985 (Dirin

1990). The predominance of reared fish leads to the spawning migration of small fish at age 2.1+ and 3.1+ of which 98 per cent consist of grilse: 76 per cent of grilse in the Keret are males. Thus hatchery activity in the Keret has led to changes in the age and sex structure of the population.

Analogous changes have taken place at Neva hatchery farm. In 1929 to 1936 30 per cent of fish were age 3+ and 4+ and 70 per cent were 5+ and older, while in 1973 to 1975 the proportions were 66 per cent and 34 per cent respectively. The mean weight of fish also decreased (Dirin 1985).

The reasons for this phenomena are as follows: eggs for the hatchery were taken at one place, over a very short period (2 to 6 days), the intention being to get eggs that were as homogeneous as possible to provide simultaneous hatching, homogeneous juveniles, and to eliminate one and two year old fish with weight less than 1 and 5 g. The use of warm water to shorten the incubation period caused earlier maturation of fish, increased the proportion of precocious males and the proportion of grilse. The use of precocious males negatively influenced the results of reproduction, because their proportion at the second generation could increase ninefold (Thorpe et al. 1983).

An increase in the proportion of small young fish leads to a decrease in the total and relative fecundity, egg weight and juvenile survival. Thus, mean fecundity in females from the Neva with weight 7.17 kg in 1921 to 1923 was 8500 eggs, which is 22 per cent more than that in 1978 for females at a similar mean weight. Mean weight of eggs also decreased from 201 to 156 mg (Dirin 1982).

Therefore growing big smolts which, in the opinion of Kazakov (1982), should lead to a significant increase of coefficient of spawners returning, in fact caused earlier maturation, migration of small fish, which had a negative influence on the reproduction and yield from the fishery.

A significant decrease in fish sizes and proportion of older age groups, caused by intensive fishery pressure, also took place in most of the rivers with natural reproduction. In earlier years salmon at seawater age 3 years and older (5.5 to 30 kg) were very common in the north (Dirin 1990). Table 1 demonstrates the percentage of these age groups for different rivers.

Table 1. Proportion of fish with 3 or more sea years in the first half of the century (%)

Pechora	Mezen	Pinega	Vjig	Kem	Niva	Voronja	Kola	mean
22	28	45	64	73	46	42	35	44

Loss of 3 seawinter fish can cause a two or threefold decrease in the mean fecundity. It is very important that the mean weight of Atlantic salmon of Barents and White seas decreased from 5.2 to 3.9 kg already.

DISCUSSION

The complicated population structure of Atlantic salmon, defined by the size and age structure, sex ratio, fecundity and the existence of different seasonal biological groups, was formed during a long evolutionary period and defines the type of population dynamics. Intra and interpopulation variability of these parameters depends on the climatic and geographic conditions under which these populations evolved. But human influences have become more and more intensive. Their sharp increase has caused the shift in all parameters, led to the depression of populations and even their extinction. Attempts at hatchery reproduction, ignoring genetic structure of salmon populations, do not reconstruct the structure of stocks, but increase these negative phenomena.

The most important step in this situation is to reconstruct the native population age structure. In our opinion, regular selection work in this direction would help to solve the problem of the reduction in mean age and mean weight at age, the changing sex ratio and would facilitate the reconstruction of the time and structure of spawning run.

What are the grounds for such ideas? The rapid increase in Atlantic salmon cultivation has stimulated a number of investigations into genetic and phenotypic characters - phenotypic and genetic variances, heritabilities, correlations, etc. The breeding goals for farming purposes and wild population conservation are not the same, but some of the results could be useful for the latter. Thus it was shown that growth rates of salmon juveniles and age of smolting are heritable (Gunnes and Gjedrem 1978, Reftsie and Steine 1978, Thorpe and Morgan 1978, Thorpe et al. 1987). There is quite a good correlation between the duration of river and sea periods (Saunders 1967). The duration of the marine period of the life cycle is also known to be heritable, and a genetic correlation between egg size and fecundity has been shown (Naevdal 1983, Gjerde 1984). All these facts are promising for selection programs to reconstruct stocks.

As a first step to avoid destruction of the age structure, it is necessary to form the same age structure in spawners at the hatchery farms as in native populations, with emphasis in some cases on increasing the proportion of fish with sea age 2 and 3 sea winters.

The response to selection (R) depends on four parameters (Gjedrem 1976):

$$R = G = ih^2 \sigma / L.$$

where i is the selection differential, h^2 the heritability, σ the phenotypic standard deviation, and L the generation interval.

Because of the high productivity of salmon, very strong selection can be applied and a large selection differential is possible. Estimates of the heritability of age at sexual maturity in the sea are in the range 0.2 to 0.48 (Gjerde 1984, Gjerde and Gjedrem 1984). The results of data analysis showed a very large phenotypic variation in the multiseawinter fish portion (for example, the coefficient of maturation for the River Kola was estimated as 25 per cent). All these facts indicate that the opportunity to change this trait through selection is considerable. Constant monitoring of the population structure will indicate any changes necessary arising from genetic correlations in the breeding process.

Strict measures should be made to enable all biological groups of wild populations to pass to the spawning grounds so as to provide maximum genetic diversity.

REFERENCES

Azbelev V.V. 1960. Notes on the biology of Atlantic salmon of Kola Peninsula and its survival. Papers of PINRO **12:** 5-70. (In Russian).

Bakshtansky, A.L., Nesterov, V.D., and Y. A. Simukov. On the situation in salmon farming in Murmansk and Arkchangelsk regions. Pp. 192-206, in Salmonid fish. Leningrad. (In Russian).

Dirin, D.K. 1985. Inner factors of reproductive variability of salmon. Pp. 73-86, in Ecology and reproduction of anadromous salmonids in the basin of White and Barents seas. Murmansk. (In Russian).

Dirin, D.K. 1990. Depression of population and age polymorphism of salmon. In Symposium on Atlantic salmon. Sjikjivkar. (In Russian).

Gjedrem, T. (1976) Possibilities for genetic improvements in salmonids. Journal of the Fisheries Research Board of Canada **33**: 1094-1099.

Gjerde, B. 1984. Response to individual selection for age, sexual maturity in Atlantic salmon. Aquaculture **38**: 229-240.

Gjedre B. and T. Gjedrem. 1984. Estimates of phenotypic and genetic parameters for carcass traits in Atlantic salmon and rainbow trout. Aquaculture **36**: 97-110.

Gunnes, K. and T. Gjedrem. 1978. Selection experiments with salmon. 4. Growth of Atlantic salmon during two years in the sea. Aquaculture **15**: 19-33.

Kazakov, R.V. 1982. Biological basis of Atlantic salmon rearing. Moscow. 1982. (In Russian).

Loenko, A.A. 1986. Comparative analysis of salmon hatchery farms work of Murmansk region. State and perspectives of salmon farming of Europe North. (In Russian).

Naevdal, G. Genetic factors in connection with age at maturation. Aquaculture **33**: 97-106.

Petrenko, L.A. 1976. Efficiency of hatchery reproduction of Atlantic salmon. Papers of GOSNIORKCH. **163**: 24-33. (In Russian).

Reftstie, T., and T.A. Steine. 1978. Selection experiments with salmon. 3. genetic and environmental sources of variation in length and weight of Atlantic salmon in the freshwater phase. Aquaculture **14**: 221-234.

Saunders, R.I. 1989. Seasonal pattern of return of Atlantic salmon in the North West

Miramichi river, New Brunswick, Journal of the Fisheries Research Board of Canada. **24**: 2078-2083.

Salmov, V.Z. 1981. Role of natural and hatchery reproduction for Atlantic salmon of Kola River. Papers of PINRO **45**: 94-111. (In Russian).

Thorpe, J.E., J.G. Morgan. 1978. Parental influence on growth rate, smolting rate and survival in hatchery reared juvenile Atlantic salmon, *Salmo salar* L. Journal of Fish Biology **13**: 549-556.

Thorpe, J.E., Morgan J.G., Talbot C., M.S. Miles. 1983. Inheritance of developmental rates in Atlantic salmon, *Salmo salar* L. Aquaculture **33**:.110-128.

Vshivtsev, A.S. 1990. Efficiency of Taibol hatchery farm. In Biology of Atlantic salmon on European North of the USSR. Sjiktjivkar. (In Russian).

Vshivtsev, A.S. 1990. Influence of hydroelectric power stations on status and structure of stocks of Atlantic salmon (*Salmo salar* L.,). In Symposium on Atlantic salmon. Sjikjivkar. (In Russian).

PART II
REACTION NORMS

NORMS OF REACTION IN FISHES

David N. Reznick

Department of Biology, University of California, Riverside, California 92521, USA

ABSTRACT

The norm of reaction is defined as the 'phenotypic expression of a genotype in different environments' (Schmaulhausen 1949). This concept is of importance in fisheries management because it describes how morphology and life history can change in response to changes caused by fishing. Specifically, high fishing pressure often reduces population densities and increases resource availability to each individual that survives. I summarise experiments on guppies that show that increased resource availability causes an increase in the size at maturity, a decrease in the age at maturity, increased fecundity, increased frequency of reproduction and an increase in population growth rate. An independent experiment shows that increased mortality, a more direct consequence of fishing, selects for earlier maturity at a smaller size and higher fecundity early in life. Both of these changes can be expected in response to high fishing pressure, but the plastic response to food availability is much larger than the evolved response to high mortality rates. This environmental effect can thus overshadow the direction of life history evolution, even though the latter still occurs. If this were true of a natural fishery, then there would be active selection for smaller body size, an undesirable result, that would be masked by the environmental response to high resource availability. I also evaluate the generality of these results with respect to the published literature and discuss the possible value of investigations on such model systems for the management of economically more important, but experimentally less tractable species.

INTRODUCTION

Exploited populations of fish often show rapid changes in their life histories, as summarized by Policansky (this volume). As one example, Leaman (1991) reports that exploitation of rockfish (*Sebastes*) is associated with reduced abundance, increased growth rate, earlier maturity, and increased reproductive effort at early ages. The existence of such changes in natural populations of fish can be attributed to various causes. First, it is possible that the changes can be attributed to a phenotypic response to a reduction in density and hence an increase in food availability. Alternatively, it is possible that the changes are an evolutionary response to exploitation. Finally, there is always the possibility that the observed changes are influenced by factors independent of exploitation. It is generally not possible to distinguish among these alternatives when dealing with natural populations of large, long-lived species. Nevertheless, distinguishing among alternatives is important because of their implications for management. For example, phenotypic changes are readily reversible, while genetic changes are not. Furthermore, non-genetic responses to environmental effects potentially obscure evolutionary responses to fishing, yet these changes could still occur and still be deleterious to sustained yield.

It is possible to evaluate the potential magnitude of these effects with experiments on short-lived, non-commercial species. Such studies can in turn yield insights about the possible causes of changes seen in commercial fisheries. I will review such experiments on guppies (*Poecilia reticulata*). I will deal primarily with environmental effects on growth, age and size at maturation, and fecundity. I will then compare the magnitude of these effects with the results of long-term selection experiments which mimic the influence of a commercial fishery. This comparison illustrates how combined genetic and environmental influences might affect our perception of the impact of fisheries on the life history of exploited species.

ENVIRONMENTAL EFFECTS - THE NORM OF REACTION

Environmental influences on the phenotype can be expressed as a 'norm of reaction'. The formal definition of a norm of reaction is: the phenotypic expression of a genotype in different environments (Schmalhausen 1949). Plate 1 is a physical example of what is meant by this phrase. This photograph is the result of an experiment evaluating the effects of food availability on age and size at maturity in guppies. Each column is a group of brothers. Each row is a different level of food availability, with the highest food at the top and progressively lower levels of food as you move down each column. All males were preserved on the day they attained maturity. Food availability has a profound effect on the size at which a guppy attains sexual maturity, with the males on low food ration attaining maturity at much smaller sizes. I will show below that food availability also affects the age at maturity.

A norm of reaction is defined by the character measured and the environments in which it is evaluated. In the example of male guppies, the different environments were represented by different levels of food availability and the phenotype was represented by the size at maturity. Many other aspects of the environment could be singled out, such as water temperature or population density. One could also single out other aspects of the phenotype, such as fecundity or age at maturity. A given norm of reaction is therefore specific to the aspect of the environment that changes (the independent variable - e.g. food availability) and the aspect of the phenotype that is measured as a response to the change (the dependent variable - e.g. size at maturity).

In fisheries management, such norms are of interest because fishing changes the environment, and hence can alter the phenotype of the fish. These changes in phenotype can in turn influence the yield of the fishery, either through the effect on individual body size and growth rate, or on the capacity of the population to increase in number. Exploitation is particularly likely to result in a decrease in population density and hence an increase in food availability. For this reason, I will concentrate on the influence of food availability on the phenotype. First, I will describe some typical norms of reaction, as revealed in laboratory studies of fish. I will then consider the consequences of these norms for the short term response to fishing and compare their magnitude to the size of the change that can be realized under selection. Next I will consider whether these lab results are general and could apply to natural fish populations. Throughout these sections, I treat the 'norm of reaction' as the average response seen in a population of fish. Finally, I will consider genetic variation in and evolution of these norms and the long-term consequences of fishing pressure. This last part is particularly important because evolution will proceed in spite of the environmental response to fishing, even if the environmental response is much larger. The available data imply that this evolutionary response will have a deleterious effect on yield, even after fishing pressure is relaxed.

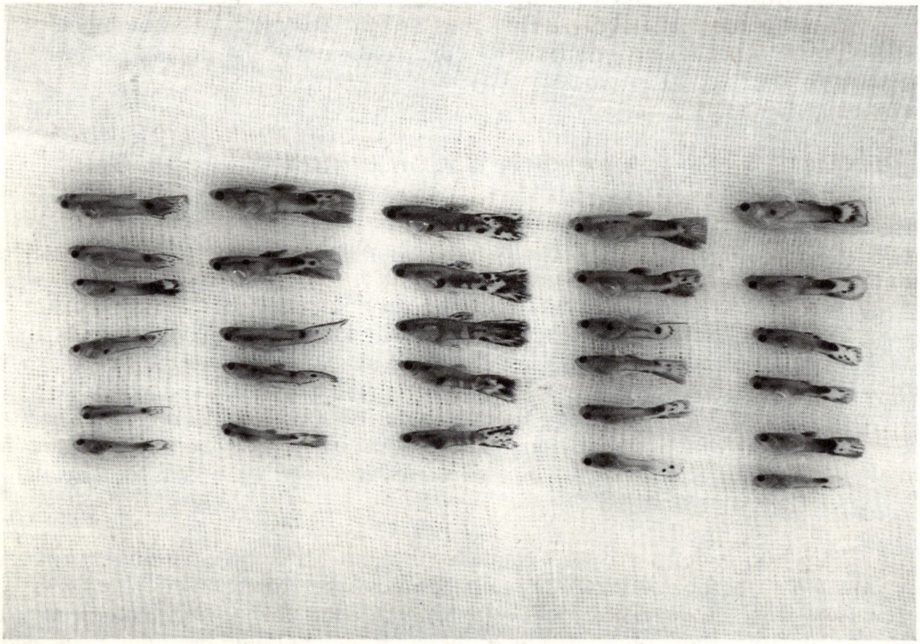

Plate 1. Male maturation experiment. These fish represent Experiment 1 in Reznick (1990). Each column is a different sibship, each row a different level of food availability, with 'ad lib' food availability for the top row and progressively lower levels of food availability for each row beneath it. The 'ad lib' treatment was not included in all litters, accounting for the unequal numbers in some rows.

METHODS

All of these experiments were done on guppies (*Poecilia reticulata*) that, with one exception, were derived from natural populations in Trinidad, then reared in the laboratory for two generations. The one exception is the study of male maturation; the experiment reported here contained a combination of domestic and wild-caught guppies. Guppies are small, stream dwelling fish. Males are generally 14 to 18 mm long, while mature females are generally 15 to 30 mm long. They attain maturity in as little as eight weeks, but will take longer under restricted food availability. Females bear live young, and will produce a litter every three to four weeks.

I will summarize results from four experiments, some being reanalyses of studies that have already been published, some being new experiments. The laboratory procedures were the same in all experiments and follow the methods detailed in earlier publications (Reznick 1982, Reznick 1983, Reznick and Bryga 1987), so I will only summarize them here.

Pregnant females were isolated in eight l aquaria until they produced a litter of young. Newborn young were reared in groups of five in eight l aquaria or groups of up to 20 in 38 l aquaria on ad lib food until they were old enough to sex. This period of 'ad lib' feeding varied among experiments from 15 to 35 days. Young were sexed, based on the presence (female) or absence (male) of melanophores at the base of the abdomen, then all individuals were weighed to the nearest milligram. Thereafter, individuals were reared in isolation on different levels of food availability.

For males, the dependent variables were age and size at maturity. Maturity is scored as the completion of the metamorphosis of the anal fin to form the gonapodium (Turner 1941), the intromittent organ. For females, the dependent variables were female age, female size, and number of offspring for the first three litters.

The male maturation experiment was a two-way, unreplicated design, with food availability and litters as the independent variables. Litters were kept together on the same shelf, with food availability assigned at random, so the litter effect confounds genetic and environmental effects. The appropriate F-ratio for food effects is the litter by food interaction. Age and size at maturity were evaluated with a bivariate analysis of variance, then a discriminant function analysis was used to evaluate the independent contribution of age and size to the initial significant result (see Reznick 1990 for further details). The female experiments had the same basic design, but each dependent variable was evaluated separately (see Reznick 1983 and Reznick and Bryga 1987 for details). As much as possible, I concentrate on the qualitative nature of the results in the discussion that follows. Unless stated otherwise, all analyses satisfied the assumptions of the analysis of variance (normally distributed residuals, homogenous variances) and all were significant with a probability of less than 0.05.

RESULTS

MALES

I found that, with increased food availability and growth rate, there was a substantial increase in the size at maturity and a substantial decrease in the age at maturity (Figure 1). In this figure, the steepest line, on the left, represents the mean growth trajectory for males on the highest food level. Progressively less steep lines represent progressively lower food levels. The points joined by the curve are the mean ages and weights at maturity for each food level; this curve therefore represents the norm of reaction for age and size at maturity where the environmental (independent) variable is food availability. The magnitude of the effect of food availability on age and size at maturity varies among experiments, but in general a fish on the highest food level can be half the age and three times the size of one on the lowest level.

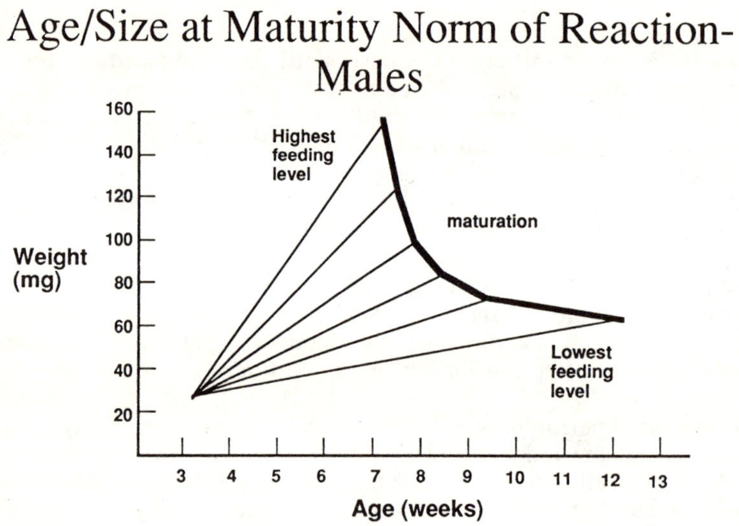

Figure 1. The male age/size at maturity norm of reaction for the fish illustrated in Plate 1 (figure modified from Reznick 1990).

FEMALES

Females lack an external cue that indicates sexual maturity, so I instead used the age and size at first parturition as dependent variables. As in males, better fed, more rapidly growing females matured at an earlier age and larger size than their more slowly growing siblings (Figure 2). Figure 2 includes the results of two different experiments, illustrating the generality of this pattern. More importantly, it illustrates the extreme lability of age and size at maturity. There is up to a 45% increase in age at first parturition and up to a 30% decrease in the size at first parturition for the low food treatment in comparison to the high food treatment. These differences could be greatly increased if the experiments were initiated with younger fish, rather than rearing them on ad lib food for 3-4 weeks, and if a wider range of food availability was employed.

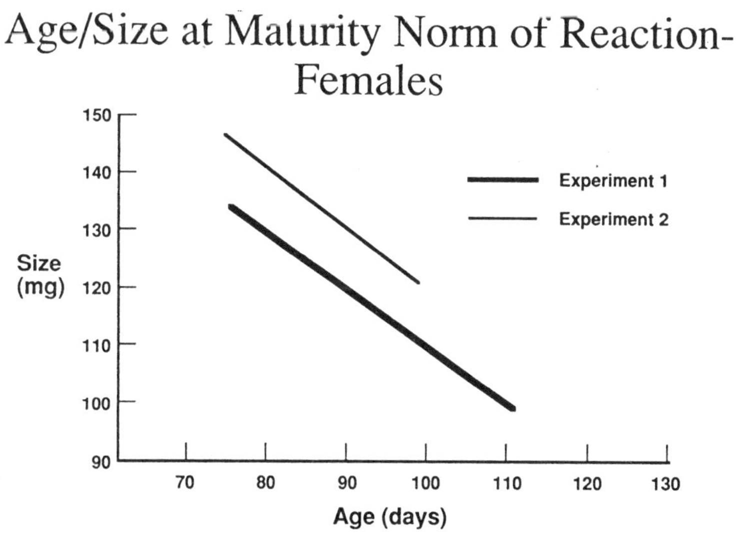

Figure 2. The female age/size at maturity norm of reaction. Experiment 1 summarizes some of the results reported by Reznick and Bryga (1987). Experiment 2 is an unpublished study of the same design.

The influence of food availability on fecundity can also be dramatic (Figure 3). High food females produce far more offspring by any given age because they begin to reproduce at an earlier age and because they are larger and hence produce larger litters. In these experiments, the litter size difference remains even when the body size effect was removed. High food females also have shorter intervals between broods, although this latter effect was not significant in all experiments.

One way of summarizing the combined influences of food availability on the female life history is to consider its influence on the rate of increase of population size. Doing so is important because each of the changes that you see in response to increased food availability will cause an increase in population growth rate. The population growth rate therefore represents the combined influence of these different environmental responses. To illustrate this point, I have constructed a comparison in population growth rates based on a six age-class life history which approximates the values for age at maturity and fecundity for the two studies that illustrated the female maturity norm of reaction (Figure 4). I simplified the life table by assuming that each age unit equals 25 days and that the high and low food lines differ in the age at first parturition by 25 days (close to the real values in the experiments summarized in Figure 2). I further simplified the illustration by assuming 100% survival until the third litter is born. This is clearly not realistic, but the intention of this example is simply to depict the cumulative effects of differences in the timing of reproduction plus differences in fecundity. Finally, I used the real data for the number of female offspring produced in the first three litters for one of the experiments reported above. In doing so, I have assumed that young are produced in a 1:1 sex ratio and have used one half of the average litter size for each of the three litters. Population growth rate therefore just considers the rate of production of female offspring. This simplified life history will slightly underestimate the real differences in population growth rate because it does not include the shorter interbrood intervals of the high food females.

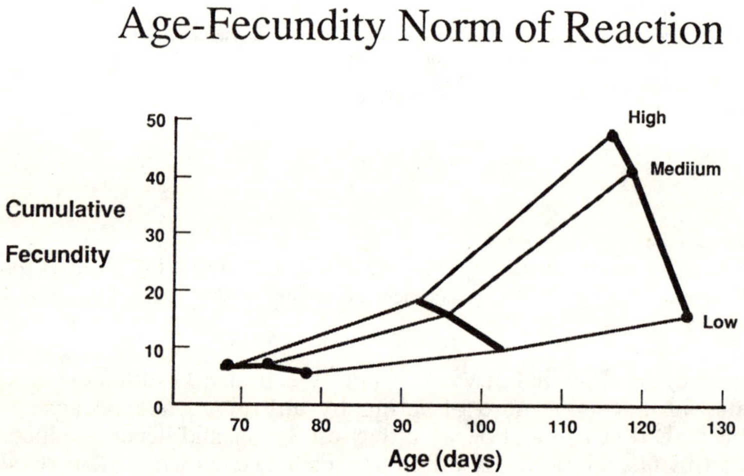

Figure 3. The female fecundity norm of reaction. These data are derived from Experiment 1 in Reznick (1983). This experiment was chosen, rather than those illustrated in Figure 2, because it employed a wider range of food availability and hence had a more dramatic result. The results for the experiments summarized in Figure 2 were qualitatively the same and were statistically significant. The one exception to this statement is interbrood interval; it is always the case that higher food treatments have shorter intervals, but this difference has not always been statistically significant.

	High Food		Low Food	
	l_x	m_x	l_x	m_x
0	1.0	0	1.0	0
1 (25days)	1.0	0	1.0	0
2	1.0	0	1.0	0
3	1.0	1.5	1.0	0
4	1.0	3.6	1.0	1.1
5	1.0	6.1	1.0	2.6
6			1.0	4.8
λ =		1.59		1.40

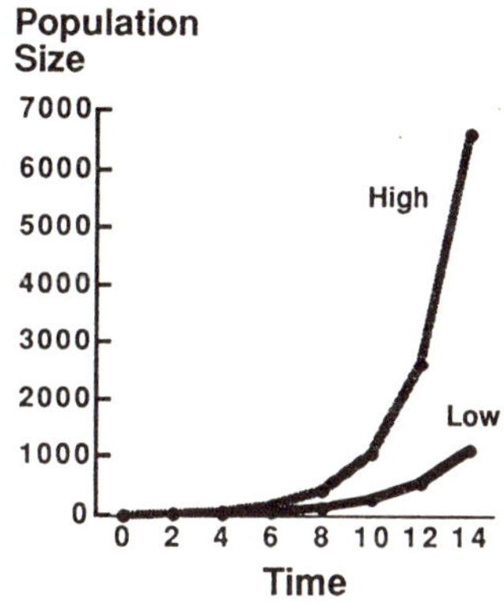

Figure 4. Model life tables for high and low food treatments, with the resulting population growth trajectory. The life tables are modelled after the differences observed in the experiments summarized in Figure 2. See text for further details.

The high food treatment has about a 50% greater potential rate of increase per 25 day period (Figure 4). This difference is greater than the difference among treatments for each of the components of the life history, thus illustrating their cumulative influence on population growth rate. This difference in population growth rate results in a five-fold difference in population size after only twelve time intervals, or three to four generations.

To summarize, higher food availability results in earlier maturity at a larger body size in both males and females. It also results in the production of more offspring in each litter and in shorter intervals between successive litters. The combined effects of all of these changes in the female life history is a large increase in the potential growth rate of the population. This increase in individual and population growth rate represents what is sometimes referred to as 'reproductive compensation' in exploited fish populations, or the tendency for the production rate to increase in response to exploitation.

How large are these environmental effects in comparison with the potential evolutionary change in life history under exploitation? The example of guppies allows me to make such a comparison because I have observed life history evolution of natural populations in a separate experiment (Reznick et al. 1990).

There are differences among natural populations in guppy life history patterns, depending upon the species of predators that guppies co-occur with. When guppies co-occur with predators that prey preferentially on large, sexually mature size classes of guppies, the guppies attain maturity at an earlier age and smaller size than when these predators are absent (Reznick and Endler 1982; Reznick 1982; Reznick 1989). The role of predators in selecting for these changes was evaluated by introducing guppies over a barrier waterfall that had served as the limit to the upstream dispersal of guppies and predators. This introduction therefore released the guppies from predation.

When released from predation, the guppies evolved changes in their life history and came to resemble the life history associated with guppies that normally do not co-occur with such predators (Table 1). This change was assessed after eleven years, or 30 to 60 guppy generations. The change amounted to a 10 to 20% increase in age and size at maturity and a 25% decrease in the number of young produced in the first litter. In different, as yet unpublished, experiments we found that there is a two-fold difference in mortality rate between sites with and without the key guppy predators and that this difference applied equally to all size classes of prey. While this is a very rapid evolutionary response, it is a great deal smaller in magnitude than the non-genetic, environmental response that can be achieved within a single generation. For example, the effects summarized in Figure 1 for male age and size at maturity are four to ten times as large as the evolved responses reported in Table 1. It therefore seems likely that a genetic change in an exploited population could be completely obscured by environmental effects, yet it can still be expected to occur (see below).

HOW GENERAL ARE THESE RESULTS?

The available data on the age-fecundity norm of reaction suggest that what I found in guppies will also be seen in other species (Table 2). All studies reported a trend towards higher fecundity and more frequent reproduction when fish are fed more. This is only the qualitative result. It is also of interest to consider the quantitative aspects of how fish partition extra resources between growth and reproduction. Ware (1980) noted that there could be large differences depending on the nature of the life history of the species. For example, species with a high extrinsic mortality rate might be predicted to devote a larger proportion of the extra resources to reproduction and less to growth. For those with a lower extrinsic mortality rate, we anticipate a larger proportion of the extra resources to go into somatic growth. The available data do not allow an assessment of such quantitative differences in the response to food availability.

Fewer experimental studies consider the age/size at maturity norm of reaction (Table 2). McKenzie et al (1983) report the same pattern for age and size at maturity in male *Xiphophorus maculatus* (*Poeciliidae*) as I observed in guppies; however, their study differed from the mine because they simultaneously varied other factors, such as temperature, density, and water quality. Wootton (1973) noted that fish on lower food took longer to begin reproducing. Sticklebacks rarely survive to reproduce in more than one season, so the experimental subjects in Wootton's (1973) experiments were almost certainly reproducing for the first time. The differences in time of first reproduction are

Table 1. The following is a subset of the data included in Reznick, Bryga and Endler (1990) and Reznick (1982). The earlier data set represents a comparison between guppies from two Rivulus and two Crenicichla localities. The data from Reznick, Bryga and Endler compare the descendants of guppies that were transplanted from a Crenicichla site (the control) to a Rivulus site (the introduction site) eleven years earlier. The Rivulus site did not have any resident guppies prior to this introduction. All of these comparions were made on the second generation, lab-reared descendants of wild-caught fish, which removes environmental influences on the life history; all remaining differences are assumed to have a genetic basis.

Aripo Tributary Introduction
Means (SE)

Life HistoryTrait	Reznick, Bryga and Endler (1990)		Reznick (1982)	
	Control(Cren.)	Intro.(Riv.)	Crenicichla	Rivulus
Male Age at Maturity (days)	48.5 (1.2)	58.2 (1.4)**	51.8 (1.1)	58.8 (1.0)**
Male Size at Maturity (mg-wet)	67.5 (1.2)	76.1 (1.9)**	87.7 (2.8)	99.7 (2.5)**
Female Age at First Parturition (days)	85.7 (2.2)	92.3 (2.6)*	71.5 (2.0)	81.9 (1.9)**
Female Size at First Parturition (mg-wet)	161.5 (6.4)	185.6 (7.5)**	218.0 (8.4)	270.0 (8.2)**
Brood Size - Litter 1	4.5 (0.4)	3.3 (0.4)*	5.2 (0.4)	3.2 (0.5)**
Offspring Size (mg-dry) - Litter 1	0.87(0.02)	0.95(0.02)†	0.84(0.02)	0.99(0.03)**
Reproductive Effort (%)	22.0 (1.8)	18.5 (2.1)ns	25.1 (1.6)	19.2 (1.5)*

ns - not significant
† - 0.05 < p < 0.10 *
* - p < 0.05 **
** - p < 0.01

similar to the results for guppies. Scott (1962) notes that significantly fewer females matured when they were maintained on a low ration, which is also consistent with delayed maturity. Alm's (1959) data were for fish reared in ponds without controlled levels of food availability. It was almost always true that, within a stock, larger, faster growing fish matured at an earlier age. In some of his studies and in his literature review, he found all possible relationships between growth rate and age at maturity, with slower growers sometimes maturing at a smaller size and earlier age, or larger size and later age. This diversity in the maturity norm of reaction reported by Alm partly motivated the theoretical studies of the evolution of norms of reaction by Stearns and Crandall (1984) and Stearns and Koella (1986).

Table 2. Published experiments which have considered the effects of food availability on the fecundity and maturity norms of reaction.

A. Fecundity norm of reaction. All of the following studies yielded results similar to those reported here for guppies.

a. Scott, 1962 - Rainbow trout (*Salmo gairdneri*)

b. Hester, 1964 - Guppies (*Poecilia reticulata*)

c. Wootton, 1973 - 3-Spined stickleback (*Gasterosteus aculeatus*)

d. Hislop, et al. 1978 - Haddock (*Melanogrammus aeglefinus*)

e. Hirshfield, 1980 - Medaka (*Oryzias latipes*)

B. Maturity norm of reaction. All of the following studies, except Alm, yielded results similar to those for guppies. See text for details.

a. Alm, 1959 - various species

b. Scott, 1962 - Rainbow trout (*Salmo gairdneri*)

c. Wootton, 1973 - 3-Spined sticklebacks (*Gasterosteus aculeatus*)

d. McKenzie, et al. 1983 - Platyfish (*Xiphophorus maculatus*)

I see two possible explanations for the diversity in the maturity norm of reaction reported by Alm. The first is the added influence of seasonal reproduction. Guppies have an unusual life history in comparison to most exploited species because there is little or no seasonality in their reproductive cycle. This means that age and size at maturity are expressed continuously, since individuals will begin to reproduce as soon as they are capable of doing so. Most fish of commercial importance have seasonal reproduction, so that age (and size) at first reproduction is expressed in discrete units of one year. This discontinuity in the reproductive cycle can alter the shape of the maturity norm, even if the underlying mechanism is the same. For example, imagine that a fish has to attain a critical size by a critical time of year in order to mature in that year. If the critical size is not attained by the critical time then, even if the individual is capable of maturing and reproducing, it will delay reproduction by one year and continue to grow during that period. By the time it does attain maturity, it may well have exceeded the average size of the more rapidly growing individuals that attained maturity a year earlier.

Data from Vinegar (1975) on the lizard *Sceloporus virgatus* illustrate this kind of phenomenon (Figure 5). This figure is based on a mark-recapture study of a single cohort of wild lizards. It divides the cohort into individuals that attained maturity when one year old versus those that took an additional year to mature. Individuals that mature when one year old are larger at the end of the first year than those that delay maturity. However, by the time this slower group attains maturity the following year, their size at maturity is greater than the size at maturity for the faster growers that matured a year earlier. The late maturing individuals have also almost surpassed the early maturing lizards in size at two years of age, possibly because the growth function changes at maturity and/or the energetic costs of reproduction reduce the growth rate. The norm of reaction for age/size at maturity would join the two points that represent age and size at first reproduction for the two subsets of the cohort. It would have a positive slope (later maturing lizards are larger), rather than the negative slope seen for guppies. This sort of phenomenon can explain Alm's observation that slower growing, later maturing individuals can be as large as or larger than those that grew fast and matured at an earlier age.

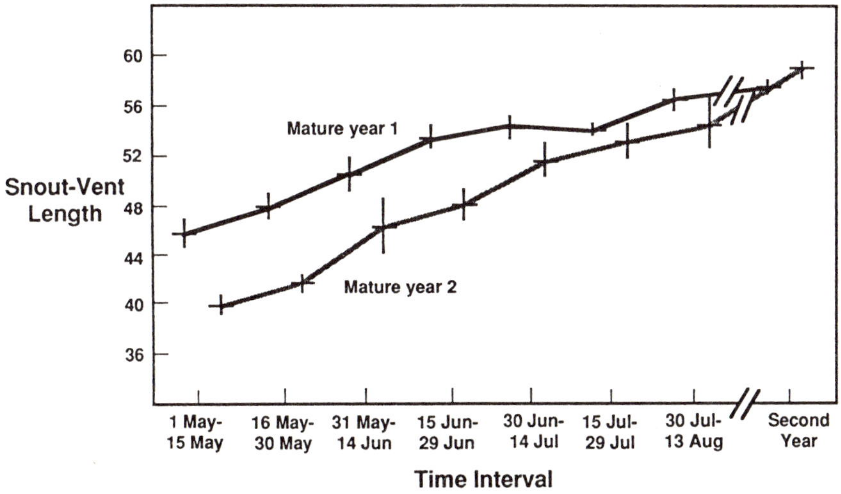

Figure 5. Data simplified from Figure 2 of Vinegar (1975). These results represent two groups of lizards from the same cohort. One group attained maturity and began to reproduce when one year old, while the other attained maturity when it was two years old. The later maturing group was significantly smaller when one year old, but these size differences disappeared by the second year. The vertical line equals plus and minus one standard error. The age/size at maturity norm of reaction would consist of two points, representing the average age and size at maturity for the two halves of the cohort. The norm of reaction would have a positive slope, rather than the negative slope seen for guppies.

A second possible explanation for the diversity of maturity norms of reaction reported by Alm is that many of the Alm's comparisons confound multiple factors that influence age and size at maturity. For example, he and other investigators often compared age and size at maturity among populations of fish. In one case, lake populations of trout grew faster, but attained maturity at a later age and larger size than river populations (Alm 1949). Alm therefore confounds differences among populations in the age and size at maturity with the environmental responses present within a population. Genetic differences among populations do not constitute a norm of reaction and potentially obscure the nature of environmental variation present within each population.

With the exception of the added complication of seasonality, I have not seen any results that would challenge the generality of my laboratory results for guppies for the influence of food availability and growth rate alone on the age and size at maturity. Stearns and Crandall (1984) and Stearns and Koella (1986) predict other forms of the age/size at maturation norm of reaction on theoretical grounds. I am not aware of an empirical study that documents the existence of such norms, independently of confounding influences or of seasonality. There are certainly other environmental factors which could influence the age and size at maturity, but which are not included here. For example, higher temperatures can cause a decrease in the age and size at maturity in female *Poecilia latipinna* (Trexler 1990). As a second example, when fish are reared in groups, the presence of mature males can inhibit the development of their tank mates, causing an increase in the age and size at maturity (*Xiphophorus variatus*: Borowsky 1973; *Gambusia manni*: Sohn 1977).

THE EVOLUTION OF NORMS OF REACTION

Thus far, I have treated norms of reaction in terms of the average properties of a population, without considering the possible role of genetic variation. There is a growing body of theoretical and empirical studies that document the existence of genetic variation in norms of reaction in natural populations and consider the evolutionary significance of this variation (e.g. Via 1984; Wade 1990).

Two aspects of genetic variation for the norm of reaction are illustrated in Figure 6. This figure depicts hypothetical family means for a given dependent variable (in this case age at maturity, the y-axis) expressed in two environments (food availability, the x-axis). The y-axis is expressed as a logarithmic transformation because I assume that there will be a proportional response to a change in food availability; i.e. that each family will change age or size at maturation by a given percent, rather than some absolute value. Each line represents the reaction norm for an individual family. In Figure 6a, families differ in their y-intercepts, or in their average value for the age at maturity. The rate of response to a change in food availability is the same in each family, so the norms of reaction are parallel. In Figure 6b, families differ both in their average value and in the rate of change in age at maturity with changes in food availability. This kind of rate difference results in non-parallel norms of reaction and is referred to as a genotype by environment interaction.

The presence of such differences among genetic lines in the response to environmental variation can result in evolution of the norm of reaction. Falconer and Latyszewski (1952) illustrated this when they selected for increased growth rate in mice. They selected on the body weight at six weeks of age under two different regimes. In one regime, food availability was unlimited. In the second, food supplies were restricted. Their goal was to see if the two environments would select for different properties and different levels of performance in a given test environment.

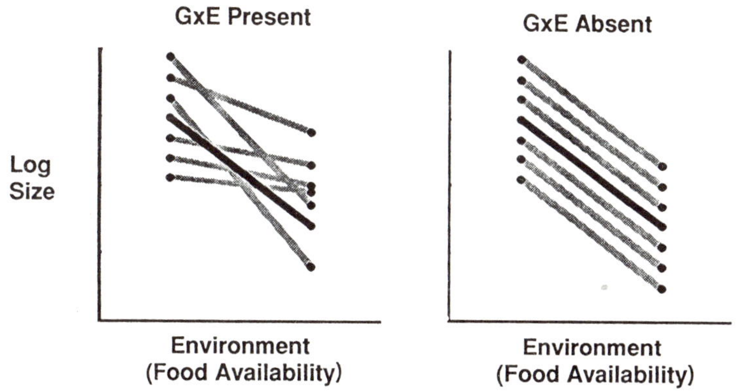

Figure 6. Genetic variation for the norm of reaction. Each line represents a hypothetical norm of reaction for one family. Differences among the lines represent genetic variation for the norm of reaction.

After seven generations of selection, they evaluated the two lines under either a full ration or a restricted ration (Figure 7). They found that the mice selected for rapid growth under a full ration did better when the two lines were compared in those circumstances. When the two lines were compared under a restricted ration, the line selected for rapid growth under those circumstances did better. This result illustrates that the phenotypic trait of rapid growth has a different underlying mechanism when it is selected for in different environments. More generally, it shows that the norm of reaction evolved as a consequence of the two lines being subjected to selection in different circumstances, or simply that norms of reaction are not fixed properties of the population. This result has applied value. It suggests that, if a domesticated line is being developed for some ideal trait, such as rapid growth rate or large size at maturity, selection should be carried out in the same circumstances that the strain will later be reared. Other studies, some in natural populations, also suggest the evolution of the norm of reaction (e.g. Barbato et al. 1983; Hrbackova and Hrbacek 1979; Wohlfarth et al 1983).

In my second example, I consider hypothetically what might happen to the norm of reaction if a population were subjected to strong directional selection, such as selection against large individuals, as seen in some commercial fisheries (Figure 8). Here I implicitly assume that each type of environment (high versus low food) is equally common and that selection acts on the average of the body sizes in the two environments. The profiles with dashed lines are those that are at a selective disadvantage. The profiles with solid lines are those families that are at a selective advantage. The figure at the right illustrates what I would expect after an interval of intense selection. For the sake of argument, I assume that the three lineages with a large average body size in both environments will be eliminated, while those with a low average body size will be retained. This will also result in a shift in the norm of reaction. In this particular example, the norm is shifted to a lower value, with little change in slope, although this would not always be the case.

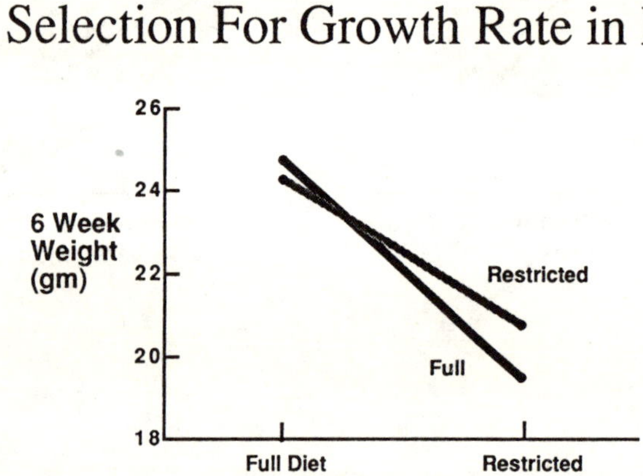

Figure 7. Data from Falconer and Latyszewski (1952) illustrating an evolutionary change in the norm of reaction for weight at six weeks of age in mice. This figure is derived from the data contained in their Figure 1.

The closest example of an experimental result that yields this kind of response was presented by Scheiner and Lyman (1991) on fruit flies. They found that selection on the population mean can change the norm of reaction and, conversely that selection on the norm of reaction can change the population mean. These features of the population are therefore linked to some extent. The important feature of their result is that it addresses a common assumption about organisms with a high degree of phenotypic plasticity. It is generally assumed that such plasticity represents all purpose genotypes which can perform well in a variety of environments. It is further assumed that this plasticity can serve as a substitute for genetic adaptation to different environments. Scheiner and Lyman's own experiments and their review of the literature does not support this assumption. Instead, their results suggest that selection will favour specialists in any given environment.

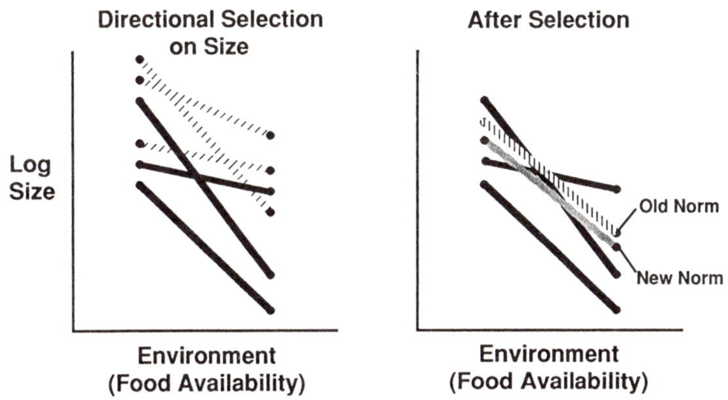

Figure 8. Hypothetical evolutionary response of the norm of reaction under directional selection. a) These are the same six families illustrated in Figure 6b. Families at a selective disadvantage under directional selection against large body size are depicted with dashed lines. Those at a selective advantage are depicted with solid lines. b) The appearance of the population after directional selection. For the sake of simplicity, families at a selective disadvantage have been deleted. the line labelled 'Old Norm' represents the former norm of reaction, while the line labelled 'New Norm' represents the new norm of reaction.

The main conclusion here is that, in spite of the short term compensation to exploitation implied by norms of reaction, we should still expect evolutionary changes in the mean values of life history traits in the longer term; the sorts of deleterious effects on yield predicted by Law and Grey (1989) are thus still likely to be seen. However, our ability to evaluate these effects may be obscured by the nature of the norm of reaction, as implied by the combined information on reaction norms and the evolution of life history patterns in guppies (Figures 1-3 versus Table 1). In that case, the magnitude of the genetic differences in life histories was small in comparison to the environmental effects. Nevertheless, exploited populations will evolve and the nature of the genetic changes, such as earlier maturity at a smaller size, will be deleterious to the yield of the fishery. It will also take a prolonged relaxation of fishing pressure for these changes to be reversed (Law and Grey 1989).

Finally, there is a growing body of empirical and theoretical literature which proposes that the norm of reaction may represent an evolved response to an unpredictable environment (this argument and some empirical studies are reviewed by Stearns (1989)). The general message of these papers is that the norm results in phenotypes that are best suited to each of the environments where the individual might commonly be found. Stearns and Koella's contribution serves as an example of a theoretical approach to the issue. Their model is structured around Euler's equation for describing population growth in an age-structured population and uses 'r', the Malthusian parameter, as an index of fitness. They propose that the main features of the environment that mould the norm of reaction are adult and juvenile mortality rates, both of which will be influenced by growth rate (the response to changes in the quality of the environment). The response is also moulded by the internal constraints of the organism, such as the relationships between size and fecundity or size and offspring quality. They predict a limited number of possible norms of reaction, given these external and internal constraints, and suggest that the shape of the norm of reaction can reflect the circumstances under which it evolved. Other theoretical approaches (e.g. Via and Lande (1985), De Jong (1990)) are based on quantitative genetic models. These authors use the genetic variation among individuals in the norm of reaction (the genotype by environment interaction) as the raw material for the evolution of the population mean response. Among other things, it is possible to consider the conditions necessary for a population to evolve a norm of reaction which results in the optimum phenotype in different environments.

SUMMARY AND CONCLUSIONS

A frequent consequence of the exploitation of natural populations will be a reduction in population density and hence an increase in resource availability per individual. Laboratory experiments with guppies demonstrate that an increase in food availability will result in higher growth rate, earlier maturity at a larger size, more frequent reproduction, and higher fecundity. All of these responses will result in a compensatory response to fishing pressure. The magnitude of these environmental effects is far greater than the evolved changes that have been observed in long-term selection experiments on the same species. Because such evolutionary changes can be assumed to continue, even in the face of such strong environmental effects, they may well go undetected in observations on natural populations of exploited species.

These results illustrate the potential of working with laboratory models to develop an understanding of why populations change under exploitation, but they also demonstrate the limitations. One limitation is that the species used in the laboratory reproduced continuously, while natural, exploited populations generally have seasonal reproduction. This difference can alter the shape of the maturity norm of reaction. In the example cited here, slower growth in a seasonal environment resulted in delayed maturity at a larger size, while guppies in the lab have delayed maturity at smaller sizes. When making comparisons among species, one might expect differences in the quantitative response to resource availability, such as in the degree to which extra resources are devoted to growth versus reproduction. There is also an inherent narrowness in the choice of experimental conditions. In the examples above, I just varied food availability.

Such experiments do not incorporate other types of interactions that could be affected by increased fishing pressure, such as behavioural interactions. Never-the-less, such systems are the only ones which will allow one to isolate and evaluate features of the environment, plus compare environmental and evolutionary effects. There is thus an inherent trade-off between experimental tractability and the immediate relevance to exploited populations.

REFERENCES

Alm, G. 1949. Influence of heredity and environment on various forms of trout. Rep. Inst. Freshwater Res., Drottingholm **29**: 29-34.

Alm, G. 1959. Connection between maturity, size, and age in fishes. Rep. Inst. Freshwater Res., Drottingholm **40**: 1-145.

Barbato, G. F., P. B. Siegel and J. A. Cherry. 1983. Selection for body weight at eight weeks of age. Restriction of feed and water. Poultry Science **62**: 1944-1948.

Borowsky, R. L. 1973. Social control of adult size in males of *Xiphophorus variatus*. Nature **245**: 332-335.

de Jong, G. 1990. Quantitative genetics of reaction norms. Journal of Evolutionary Biology **3**: 447-468.

Falconer, D. S. and M. Latyszewsky. 1952. The environment in relation to selection for size in mice. Journal of Genetics **51**: 67-80.

Hester, F. J. 1964. Effects of food supply on fecundity in the female guppy, *Lebistes reticulatus* (Peters). Journal of the Fisheries Research Board of Canada **21**: 757-764.

Hislop, J. R. G., A. P. Robb and J. A. Gauld. 1978. Observations on effects of feeding level on growth and reproduction in haddock, *Melanogrammus aeglefinus* (L.) in captivity. Linnean Journal of Fish Biology **13**: 85-98.

Hirshfield, M. F. 1980. An experimental analysis of reproductive effort and cost in the Japanese Medaka, *Oryzias latipes*. Ecology **61**: 282-292.

Hrbackova, M. and J. Hrbacek. 1979. Rate of the post-embryonic development in several populations of the group of the species *Daphnia hyalina* Leydig at various concentrations of food. Vestnik Ceskoslovenske Spolecnosti Zoologicke Svazek **XLIII**: 253-259.

Law, R. and D. R. Grey. 1989. Evolution of yields from populations with age- specific cropping. Evolutionary Ecology **3**: 343-359.

Leaman, B. M. 1991. Reproductive styles and life history variables relative to exploitation and management of *Sebastes* stocks. Environmental Biology of Fishes **30**: 253-271.

McKenzie, W., D. Crews, K. D. Kallman, D. Policansky and J. J. Sohn. 1983. Age, weight, and the genetics of sexual maturation in the platyfish, *Xiphophorus maculatus*. Copeia **1983**: 770-774.

Reznick, D. N. 1982. The impact of predation of life history evolution in guppies: the genetic components of observed life history differences. Evolution **36**: 1236-1250.

Reznick, D. N. 1983. The structure of guppy life histories: the trade-off between growth and reproduction. Ecology **64**: 862-873.

Reznick, D. N. 1990. Plasticity in age and size at maturity in male guppies (*Poecilia reticulata*): an experimental evaluation of alternative models of development. Journal of Evolutionary Biology **3**: 185-203.

Reznick, D. N. and H. Bryga. 1987. Life history evolution in guppies: 1. Phenotypic and genotypic changes in an introduction experiment. Evolution **41**: 1370-1385.

Reznick, D. N., H. Bryga and J. A. Endler. 1990. Experimentally induced life-history evolution in a natural population. Nature **346**: 357-359.

Schmalhausen, I. I. 1949. Factors of Evolution. Blakeston, Philadelphia, PA.

Scott, D. P. 1962. Effect of food quantity on fecundity of Rainbow Trout. Journal of the Fisheries Research Board of Canada **19**: 715-731.

Scheiner, S. M. and R. F. Lyman. 1991. The genetics of phenotypic plasticity. II. Response to selection. Journal of Evolutionary Biology **4**: 23-50.

Sohn, J. J. 1977. The consequences of predation and competition upon the demography of *Gambusia manni* (*Pisces: Poeciliidae*). Copeia **1977**: 224-227.

Stearns, S. C. 1989. The evolutionary significance of phenotypic plasticity. Bioscience **39**: 436-446.

Stearns, S. C. and R. E. Crandall. 1984. Plasticity for age and size at sexual maturity: A life-history response to unavoidable stress. Pp. 13-33 in Fish Reproduction, G. Potts and R. J. Wootton eds. Academic Press, London.

Stearns, S. C. and J. C. Koella. 1986. The evolution of phenotypic plasticity in life-history traits: predictions of reaction norms for age and size at maturity. Evolution **40**: 893-914.

Trexler, J. C., J. Travis and M. Trexler. 1990. Phenotypic plasticity in the sailfin molly, *Poecilia latipinna* (*Pisces: Poeciliidae*). II. Laboratory experiment. Evolution **44**: 157-167.

Turner, C. L. 1941. Morphogenesis of the gonapodium in *Gambusia affinis affinis*. Journal of Morphology **69**: 161-185.

INDIVIDUAL VARIATION IN ACQUISITION/ALLOCATION REACTION NORMS

Keith Nelson

University of California, Bodega Marine Laboratory, Bodega Bay, CA 94923, USA

ABSTRACT

Fisheries impose upon populations selective forces that are probably proportional to fishing mortality, even when that is not size-selective. Whether there is an evolutionarily response in a population depends upon the amount of genetic variation in traits associated in particular with allocations to growth and reproduction. Genetic change in such traits is obscured and confused by the purely phenotypic changes in acquisition/allocation schedules resulting from fishery-induced changes in density. Such linked reaction norms of allocation in iteroparous species may be described in terms of allometry. Individual variation in allocation reaction norm in the North Sea sole (*Solea solea*) was reported by de Veen (1976). In responding to changes in food availability the faster-growing females displayed a steeper reaction-norm allometry between ovarian and somatic weight. Reaction norms to temperature also show allometric differences. Relative gonad size in smaller northern anchovies (*Engraulis mordax*) is more sensitive to temperature, even to the point of negative allometry during an El Niño year. We may reasonably summarize such variation in a bipartite growth function expressing the relationship of surplus, growth, and reproductive energies as power functions of size (Ware 1980). Both the evolutionary response to selection and its detection depend upon the sources of individual variation in the joint reaction norms of these quantities to changes in food supply, density, and temperature. Only when these joint reaction norms have been identified can their evolutionary change be detected.

INTRODUCTION

In many longer-lived ectotherms growth rate, size and age at maturity, and fecundity are often functions of the history of variables such as temperature and resource availability. Such patterns of phenotypic plasticity have been called *norms of reaction* (cf. Caswell 1983). Indeed, viewed developmentally a genotype is an integrated system of such functions, a *reaction system* as understood by Goldschmidt (1940). Thus an evolving population with overlapping generations consists of a collection of reaction systems, each

on its own developmental trajectory. The evolution of such a population involves selection among the individual variant reaction systems.
The representation of an individual in ensuring generations is mediated by a summary set of life history variables describing the accumulation and expenditure of energy in growth and growth-dependent reproduction. This may be called the *acquisition/allocation* (A/A) reaction system. Ware (1980, 1982) modelled postmaturity A/A systems in terms of *surplus energy* available after metabolic needs were satisfied, as a forward difference equation for the energy of growth:

$$W_{x+1} - W_x = a_1 W_x^{b_2}, \tag{1}$$

where W_x is the weight at age x and the terms on the right summarize the weight-dependence of surplus and reproductive (e.g. ovary weight) energies respectively. Previously such bipartite growth functions had often been interpreted with growth as the difference between 'anabolic' and 'catabolic' energies (cf. Pauly 1979), but they also allow simple descriptions of individual and temporal variation in A/A and their consequences for population dynamics and potentials for different kinds of selection. For example, with Ware's (1982) model parameters and mortality a convex function of a_2, density-dependence in both a_1 and a_2 (instead of in a_1 alone) appeared advantageous only for long-lived fishes.

Reaction-norm changes often appear *cogradient* (Levins 1968) with the evolutionary response to a selective regime which is predicted by some model, and may be confused with it. Deciding which is the likelier explanation for a particular set of changes in a population's allocation pattern might be greatly facilitated by comparing not just the mean observed response, but also individual A/A variation and covariation, structured by both age and size. But when changes in allocation variables other than age and size at maturity are reported for iteroparous fishes, size and age are generally confounded, making interpretation difficult.

This chapter approaches the subject of evolution in exploited populations in the context of the first studies to be specifically concerned with individual variation in age-specific A/A reaction norms in commercially important fishes. Changes in female North Sea sole (*Solea solea*) growth and reproductive schedules in the years 1962-1973 were originally described by the late J.F. de Veen (1976) in an unjustly neglected paper. These changes were almost surely primarily due to changes in individual food supply, probably brought about by fishing mortality-related density shifts, and by increase in available prey density as well (Rijnsdorp and van Beek 1991). Year- and age-specific female growth rates adjusted for weight had high correlations with density, previous year's phosphate input, fishing mortality and bottom disturbance (beam trawl hours) respectively of r = -0.53, +0.51, +0.62, and +0.71 (d.f. 45, $p < 0.001$; unpublished calculations from data in de Veen 1976, Table 3 and Tables 7, 9, 10, and 11). In this study, temperature had little correlation with growth (de Veen 1976, Table 13).

Although in other cases the possibility of mistaking a phenotypic response to temperature change for an evolutionary response to exploitation may have often been present, a probable result of global warming is an increase in the likelihood of such errors. My second example concerns changes in individual variation in reproductive allocation as a function of size, in the response to the 1982-83 California El Niño (CEN) event by the California-Northern Baja population of Northern anchovy (*Engraulis mordax*; K. Nelson et al. in prep.)

VARIATION IN ALLOCATION REACTION NORM ALLOMETRY IN SOLE

De Veen's group recorded body and ovary weights from weekly market samples of otolith-aged sole landed in several Dutch ports. In the sole, ovary weight is more highly correlated with size than is fecundity (Deniel 1984). De Veen's (1976) figures and tables detail the changes in A/A in relation to size. Somatic and ovarian growth are both cyclic, with somatic growth leading in phase. From a minimum following spawning in May and early June, each mature female having spawned several times during the brief season (Houghton et al. 1985), weight increases in summer and fall to peak at the beginning of the next year (the amplitude of the oscillation amounts to a full 40% of the yearly average weight). Ovarian growth begins in fall and peaks typically the following April. Yearly growth in maximum weight can be computed in terms of increase (trough-to-peak) and decrease in average weight, $I(W_x)$ and $D(W_x)$ (de Veen 1976, Figure 19):

$$\begin{aligned} max(W_{x+1}) - max(W_x) &= I(W_x) + D(W_x) \\ &= 8.17\ max\ (W_x)^{0.525} - 0.00028\ max(W_x)^{1.734}. \end{aligned} \tag{2}$$

The terms on the right are related to *gross* surplus and reproductive energy (including winter and spawning metabolism, migration, ovarian replenishment), rather than to energy *substantiated* in somatic and ovarian growth as in Equation (1). However, also from de Veen's Figure 19, average peak ovary weight is proportional to $D(W_x)$ and we may use it as an estimator of total reproductive investment, as a function of the preceding maximum weight:

$$max(O_x) = 0.306\ D\ (W_x)^{1.03} = 0.00066\ max\ (W_x)^{1.805}. \tag{3}$$

The similar nonlinearities in the relations of $D(W_x)$ and $max(O_x)$ to $max(W_x)$ reflect *ontogenetic reproductive allometry* in female sole. The exponents in the relations to weight of total surplus energy (estimated as $I(W_x)$) and substantiated surplus energy are not as similar. The latter, estimated as average yearly growth increment $(min(W_x) - min(W_{x-1}))$ plus average ovarian decrement $D(O_x)$, is essentially independent of weight.

The salient relationships concerning variation in A/A reaction norm among individuals can be described in terms of the allometric relation of individual peak ovary weight to somatic weight (or to the cube of length). De Veen (1976, Figure 10) gave curves representing year- and age-specific (y, x) *individual correlated variation* in April length and ovary weight, which we may designate somewhat imprecisely as curves of *static reproductive allometry* (cf. Lande 1984). In Figure 1 several of these curves are redrawn on logarithmic coordinates, for ages 4-6 in the years 1964 (after several surplus-energy-poor years; dashed lines) and 1971 (after several very good years; solid lines). We may obtain from each curve a static allometry regression

$$\ln\ (O_{y,x,j}) = \ln(a_{y,x}) + b_{y,x} \ln\ (L_{y,x,j}) + r_{y,x,j},$$

where j is one of a number of points along the curve and $r_{y,x,j}$ its residual (in reality each (y, x)-sample should yield an elongate bivariate distribution of points but the original data are not available). The slope $b_{y,x}$ of the static allometric regression was correlated highly significantly with average growth rate of that (y, x)-sample (r = + 0.67), and also with density, phosphate, fishing mortality, and bottom disturbance (r = - 0.51, +0.56, +0.67, +0.56 respectively; d.f. 45, $p < 0.001$). $b_{y,x}$ was always greater than 3; thus the larger (more rapidly growing) members of a (y, x)-sample had relatively as well as absolutely larger ovaries. Average ontogenetic allometry, estimated from these curves as the regression of ovary size at mean length $\ln(O_{y,x,L})$ upon mean length $\ln(L_{y,x})$, had a slope of 6.0; that is, in the 'average female', peak ovary weight increases with age approximately as the square of somatic size.

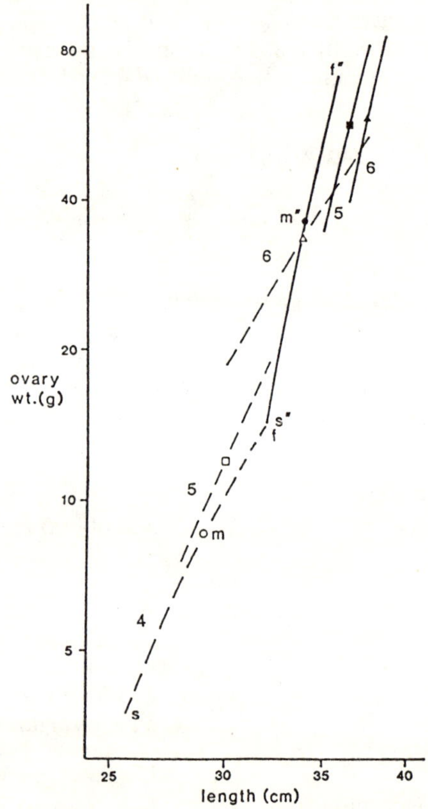

Figure 1. North Sea sole, year- and age-specific relationships between April length and ovary weight, from de Veen (1976), Figure 10, redrawn on log scales for ages 4-6 in years 1964 (dashed lines) and 1971, s, m, f and s", m", f": slow-, medium-, and fast-growing 4-year-olds in the two years respectively.

The residual of this last regression, the relative elevation of the curve of static allometry, was correlated neither with $b_{y,x}$ nor with growth rate, nor with the environmental variables associated with growth and static allometry slope. That is, *average reaction norm allometry* had approximately the same slope and intercept as average ontogenetic allometry. Individual variation in age-specific reaction-norm allometry is shown almost diagrammatically in Figure 1. The reaction norm of a slow-growing 4-year-old (s-s") is practically on the curve of static allometry for a time of low surplus energy, 1964 (dashed line); that for a fast-grower (f-f") is similar to the 4-year-old static curve for the good year (solid line). The slopes of the regressions of the age-specific minima and maxima for all years 1962-1973, estimating the reaction norms of slow- and fast-growers respectively, are significantly different (t = 2.70, d.f. 3, p 0.05, one-tailed). How many reaction norms are there in the sole population? One possibility is that there is just one convex reaction norm for each age, incorporating those of slow-, intermediate-, and fast-growing females, as in Figure 2A. A female's position on a reaction norm would depend solely upon her surplus-energy history. But then we expect the ontogenetic trajectory of average females in average times (half-filled symbols, Figure 2A) to be measurably lower in elevation than in poor and good years (empty and filled symbols respectively.) They were not. In fact ontogenetic and reaction-norm allometries for average females were remarkably similar, and similar as well to the ontogenetic allometries of both slow- and fast-growers in years of intermediate surplus energy, as in Figure 2B. Marked deviations in elevation from this trajectory occurred only in slow- and fast-growers in trophically deviant years. In summary, in female sole there seems to be a continuum of allocation allometry reaction norms to trophic conditions, coupled to the ability - or disposition - to accumulate surplus energy. This raises a question as to whether this correlated variation has a genetic basis, to which we return below.

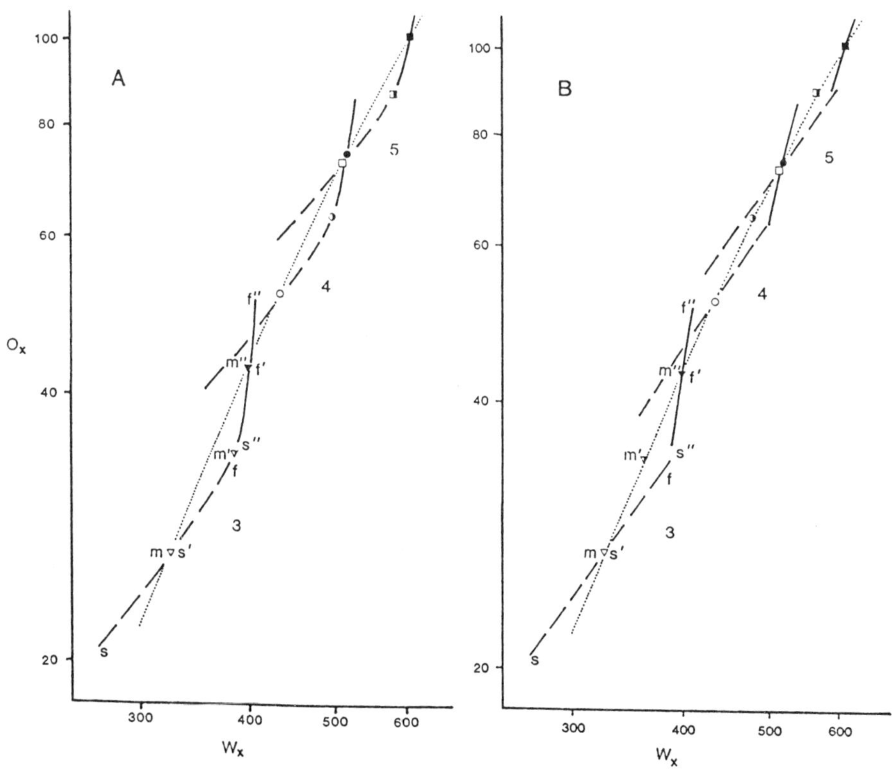

Figure 2A. The hypothesis that for each age, both individual and interannual variation in female sole reproductive allocation follow a single convex reaction norm. W_x, somatic weight; O_x, ovary weight. Dotted line, curve of average ontogenetic allometry in trophically poor and rich times. Slow-, medium-, and fast-growing 3-year-olds differ only in acquisition history. **2B.** The hypothesis that slow-, medium-, and fast-growing females have different reaction norms, here similar to the curves of static allometry for trophically poor, average, and good periods respectively. Dotted line, average ontogenetic allometry in all years and reaction-norm allometry of average female sole.

One further point. Had de Veen not carefully distinguished age and size, but instead confounded them in a single curve for each year, there would have been the appearance of a pronounced change in the *slope* of population reproductive allometry (compare the years 1964 and 1971 in Figure 1). We then might have concluded, incorrectly, that change in this slope was also a reaction norm of female sole.

GONADOSOMATIC ALLOMETRY VARIATION IN THE ANCHOVY

We have studied variation in allozymes, morphometrics, and reproductive parameters in anchovies collected within the Southern California Bight in Jan-Mar 1984-1985 on NOAA/CalCOFI spawning biomass surveys (cf. Hedgecock et al. 1989; D. Hedgecock et al. in prep.; K. Nelson et al. in prep.) An important question concerns whether the anchovies remain in the same localities during a northward CEN water mass excursion, or whether instead they move with the water mass (we would then be sampling from a somewhat more southerly group in 1984.) In either event, 1985 samples and those from more northwesterly coordinates in both 1984 and 1985 were maturing in cooler waters. Sole and anchovy share certain features, but differ in many ways. Anchovies are smaller and shorter-lived; they have a much longer spawning period and more spawning episodes per individual. Thus neither fecundity nor gonad size are necessarily representative of reproductive output. Estimates of batch fecundity, the number of oocytes in the ripest of several modes, are preferred for fishery work (Hunter et al. 1985). Nevertheless, individual variation in A/A system reaction norm to temperature can be shown in the relations of gonadosomatic index, GSI: (dried gonad weight)/(wet body weight minus gonad); dry gonad weight is used because of the doubling of ovary volume which occurs in females with hydrating oocytes. Surprisingly, only 0.7% of the individual variance in GSI is attributable to differences between the sexes, and they are here combined. Figure 3 displays regressions of ln GSI on

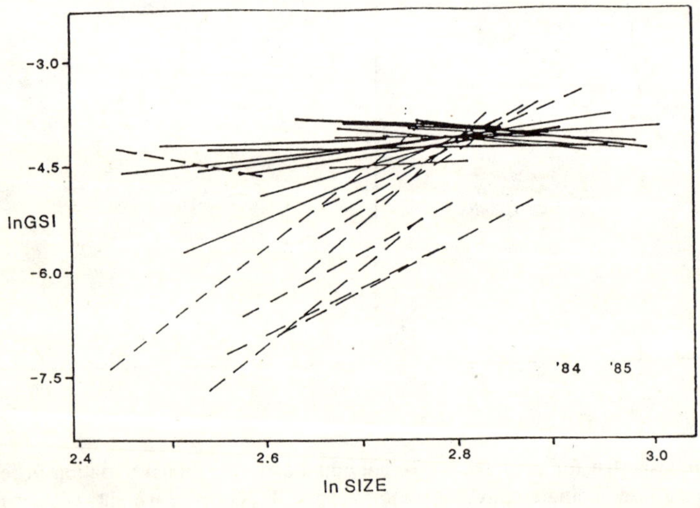

Figure 3. Northern anchovy locality, age-subsamples, 1984-85. Regression of logged individual gonadosomatic index ln GSI, ln (dried gonad wt/wet body wt. minus gonad) on ln SIZE, the mean of logs of 13 morphometric variables. Broken lines, collections containing no females with hydrating eggs.

ln SIZE (the mean of logs of 13 morphometric measures) for all 1984-1985 (locality, age)-subsamples with $n > 13$. Except for a few subsamples of immature fish, the regression lines seem to fan out from a region of ln SIZE near 2.8 (corresponding to 105-115mm standard length).

Subsamples containing females with hydrating eggs (from spawning aggregations) display a significant negative relationship between the slope of the ln GSI - ln SIZE regression (reflecting allometry; a slope of zero represents isometry) and mean GSI ($r = -0.74$; d.f. 18; $p < 0.001$). The correlation appears to be with temperature (slope vs station sea surface temperature, $r = -0.60$, d.f. 35; $p < 0.001$). Maturing in cooler times and areas, smaller anchovies in a subsample have lower GSI; under warmer conditions, with higher average GSI, the GSIs of the smallest fish are highest (mean SST for subsamples with females with hydrating oocytes and positive slopes, 13.8°C; for negative slopes, 15.3°C, $t = 31.6$, d.f. 18; $p < 0.001$). It is not unusual that smaller fish of an age are slower to mature and may fail to spawn at all. But one does not then expect the smallest fish to develop the highest GSI, simply because of an increase in temperature. Smaller fish are more sensitive to temperature in both maturation rate and in final GSI. For the spawning population as a whole, dry gonad weight, body weight, and the cube of linear dimensions are isometric. From regressions we can also determine that temperature-correlated deviations from reproductive and condition isometry are associated with deviations from mean GSI and condition, in a way reminiscent of trophic reaction-norm allometry in the sole.

MODELLING ACQUISITION/ALLOCATION ALLOMETRY

Allocation can be managed in different ways, depending upon which activity is given priority. (1) We can base growth upon a continuous function such as

$$\dot{W}^1 = a_1 W^{b_1} - a_2 W^{b_2}, \tag{4}$$

with terms on the right interpreted as instantaneous values of quantities related to surplus and reproductive energy. (2) Alternatively, we could interpret Equation (4) simply as a growth function, with reproductive allocation the difference between yearly growth and some arbitrary surplus energy accumulation. In Figure 4, $b_1 = 0.5$ and $b_2 = 1.0$. Surplus energy accrues independent of size much as in the sole, once the point of inflection W_i *(i.e. size at maturity) has been reached.* In this instance it accumulates at a rate equal to the maximum growth rate $\dot{W}^1{}_1 = a_1^2/4a_2$ characteristic on the inflection point. Surplus energy is thus a convex function of a_1 but an inverse function of a_2 (similar to the metabolic models; Pauly 1979), yet reproductive allotment remains an increasing function of a_2 (as in Ware's model, Equation 1), with allometry not too different from female sole.

Around each point on the ontogenetic trajectory are drawn age-specific (a_1, a_2) coordinate frames of ± 10%. Such coordinate frames may be called *auximetric grids*, after Pauly (1979). The auximetric grids allow us to interpret ontogenetic, static, and reaction-norm allometries in terms of the coefficients in Equation (4). Figure 4B superimposes auximetric grids on the 3-year-old curves from Figure 2B: with this model, the static allometry relative difference in parameters between slow- and fast-growers shifts from ($0.2a_1$, $0.05a_2$) in trophically poor times to (0.1, 0.1) in rich ones. Variation in reaction norm allometry must be similar, with a_1 and a_2 more sensitive to changing trophic conditions in slow- and fast-growing sole respectively, according to their respective predilections for change in allocation to growth and reproduction. In summary, individual differences among A/A reaction systems can be expressed by correlated differences in relative sensitivities of the parameters of a growth equation. Other bipartite growth models such as Model (1) give qualitatively similar results, but some product better fits to the changes in length at maturity without corresponding changes in maturation age which were reported by de Veen (1976). The age-specific auximetric grids for the anchovy ln GSI - ln SIZE regressions in Figure 3 would be overlapping, horizontally elongate trapezoids oriented with their a_1 axes parallel to the x-axis. Again, differences in size within a subsample would imply differences primarily in a_1. Increase in GSI would reflect increase in a_2. Fish of similar size but of different ages and perhaps origins cluster together in a trawl sample, reflecting size-selective schooling behaviour and perhaps also an CEN effect upon year-class growth rates (Fiedler et al. 1986). In such samples fish in an older subsample must therefore have grown more slowly on the average. Notwithstanding such subsample growth rate difference uncertainties, the differences in GSI and in its allometry remain and imply temperature-correlated differences at least in a_2 and in a higher sensitivity of a_2 to temperature in smaller, less rapidly growing anchovies within a subsample.

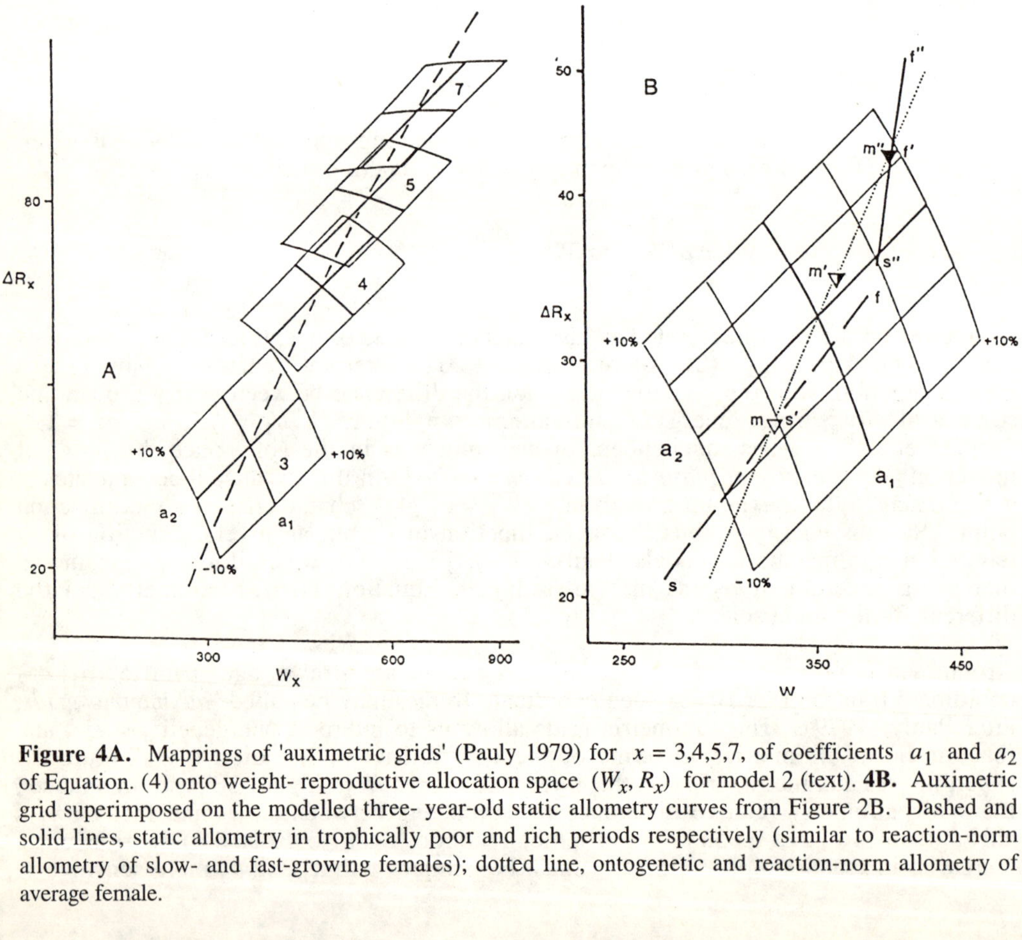

Figure 4A. Mappings of 'auximetric grids' (Pauly 1979) for x = 3,4,5,7, of coefficients a_1 and a_2 of Equation. (4) onto weight- reproductive allocation space (W_x, R_x) for model 2 (text). **4B.** Auximetric grid superimposed on the modelled three- year-old static allometry curves from Figure 2B. Dashed and solid lines, static allometry in trophically poor and rich periods respectively (similar to reaction-norm allometry of slow- and fast-growing females); dotted line, ontogenetic and reaction-norm allometry of average female.

COMPARISON WITH OTHER STUDIES

Dygert (1986) found individual variation in reproductive allocation to be positively correlated with growth and condition in 'English' sole (*Parophrys vetulus*), but not in herring (*Clupea harengus pallasi*). In North Sea sole, there was similar individual correlated variation in body and ovary size, but variation in *time* in average size-specific reproductive allocation was uncorrelated with other allocation or environmental variables (there was, however, an increase in average GSI during the years in question, due to larger size-at-age combined with the reproductive allometry). Rijnsdorp (1990) found a lack of correlation of individual relative fecundity or GSI with growth rate back-calculated from otolith annuli in commercially-caught plaice (*Pleuronectes platessa*). His findings in particular are in conflict with much experimental evidence of decreased reproductive allocation with reduced ration, which he reviews. He suggests that only when ration fails to build up or to sustain prespawning condition above a size-related threshold will an effect upon reproduction be seen.

A/A reaction norm variation has behavioural and social aspects. A fish confined to a tank is unable to compensate for low rations (and for any competitive disadvantage aggravated by crowding) by increasing its foraging efforts or changing its location (MacCall 1990). We must distinguish between 'density' and 'food limitation', especially when discussing individual variation. Thus competition might exacerbate differences among individuals particularly when density results in food limitation; alternatively a territorial species may show increased variance in growth in the artificial situation of high density and excessive food (Grosse 1982, discussed by Dygert 1986).

An explanation for species differences in response to trophic change may lie in differences in reproductive allometry. In species such as North Sea sole with strongly allometric reproductive ontogeny, increase in a_1 brings about increase in relative reproductive allocation as a matter of course, and a_2 change may be superfluous. Therefore one testable hypothesis is that density dependence in a_2 is more likely to be found in species with isometric or weakly allometric reproductive ontogeny. Thus, some of the field observations of increase in average reproductive allocation may reflect higher surplus energy, but some may be correlated with temperature change. This may reflect broader correlation with the *rate* of development of an allocation schedule. Pauly (1979) demonstrated among populations in many species of fishers positive correlations of temperature with parameter K of the Pütter-Bertalanffy growth function, a special case of Equation (4), here given in bipartite differential form:

$$\dot{W}^1 = HW^{2/3} - 3KW. \qquad (5)$$

The 'anabolic' and 'catabolic' parameters H and K are analogous to a_1 and a_2 in Equation 4. Pauly interpreted variation in K in terms of rate of oxidative metabolism. Strong correlation of K with age at maturity x_m among populations and species of fishes is common (Jensen 1985; Gunderson and Dygert 1988; other data from the literature reanalyzed, personal observation). We can note with Jensen and K and a_2 (in Figure 4) are each reciprocally related to the age of inflection x_i of their respective growth curves, thought by many to be associated with x_m (Jensen 1985; Sebens 1982; Sibly et al. 1985).

Latitudinal or interannual differences in age at maturity or fecundity are often in agreement with the expectation of temperature-induced variation in reproductive allocation representable by variation in a_2 (e.g. Leggett and Carscadden 1978;

Tanasichuk and Ware 1987). But the latter authors found that although fecundity variation, primarily interannual, was correlated with winter temperature in pacific herring (*Clupea harengus pallasi*), ovary size was not. They interpreted this correlation instead as a temperature-dependent tradeoff between egg size and egg number. Another complication arises from correlation between temperature and surplus energy accumulation, which may be positive e.g. due to increase in growing season duration (Le Cren 1958), or negative as in Pauly's interpretation or as suggested for production during CEN events (Fiedler et al. 1986). Such complications underline the importance of comparative study of individual reaction-norm variation in both acquisition and allocation in different species, densities, and temperature regimes. Until this is done, we may only conclude that reaction norms to population density and food supply can be described as changes at least in a_1,and sometimes in a_2 as well, whereas reaction norms to temperature involve at least a_2, and sometimes also a_1!

REACTION SYSTEM VARIATION AND EVOLUTION

We cannot foresee the long-term effects of exploitation upon individual reaction-norm variation without determining what part of the correlated variation is genetically-based, but the phenotypic variation we see sets probable upper limits. The trophic changes acting in the past upon the population of sole reaction systems probably resulted mainly from runs of year-class failures and successes, variation typically in juvenile mortality, although occasional catastrophes may also have reduced crowding for a time (Rijnsdorp and van Beek 1991). It has been suggested that such a history might lead to 'bet-hedging': late maturation, low reproductive effort, and long life (Murphy 1968). On the other hand, fishery-induced changes would be primarily in the adult mortality schedules, which like growth have a cumulative influence upon expected lifetime reproduction of a recruit, R.

An idea of the forces involved may be obtained by examining 'fitness profiles', expected viability and fertility selection gradients of R on a_2 (Figure 5A; density and hence a_1 and juvenile mortality held constant). In curve 1, for model (1) (Equation 4, with $b_1 = 0.5$, $b_2 = 1.0$, and ovary weight equal to $0.25a_2W_x$), we see the effects of a constant annual mortality of 0.3. A population with mean $a_2 \sim 0.5$ would see a gentle stabilizing fertility-selection gradient. Increasing annual mortality to 0.5 greatly reduces $max(R)$ and shifts it from $a_2 \sim 0.5$ to ~ 1, but with little change in curvature (curve 3). The population would now be positioned on a moderate gradient in R. Density-dependence in juvenile mortality and a_1 may allow population persistence; with sufficient additive genetic variation the mean a_2 would be expected eventually to increase. Thus even when not size-selective, a change in annual mortality can result in evolution. A change to model (2), with reproduction the difference between growth (Equation 4) and a constant surplus energy (= W_i; $b_1 = 0.5$; $b_2 = 1$), increases a_2 at $max(R)$ and increases the curvature (curve 2). Changing from model (1) to model (2) is equivalent to changing the exponent of reproductive allometry from 1 to $\sim$ 2, which effectively delays the reproductive schedule with respect to the growth schedule by about a year (cf. Laird et al. 1968). Increasing the age at maturity to > 1 year in model (1) has much the same effect.

Curve 4 traces the selection gradient for model (1) with age-dependent annual mortality A_x increasing from $A_1 = 0.3$ to $A_{21} = 1.0$. The curve is intermediate in position between curves 1 ($A = 0.3$) and 3 ($A = 0.5$), and there is little difference in curvature. This is still not parental viability selection, but such age-dependent mortality is in one sense size-selective (Figure 5B). The effect of true size-selection can be illustrated by applying the size-mortality schedule of such age-dependent mortality for $a_2 = 0.688$ (Figure 5B), near the maximum of curve 4, to all a_2 variants (curve 5, Figure 5A). Now the lift history with $a_2 = 0.688$ lies on a steep selective gradient.

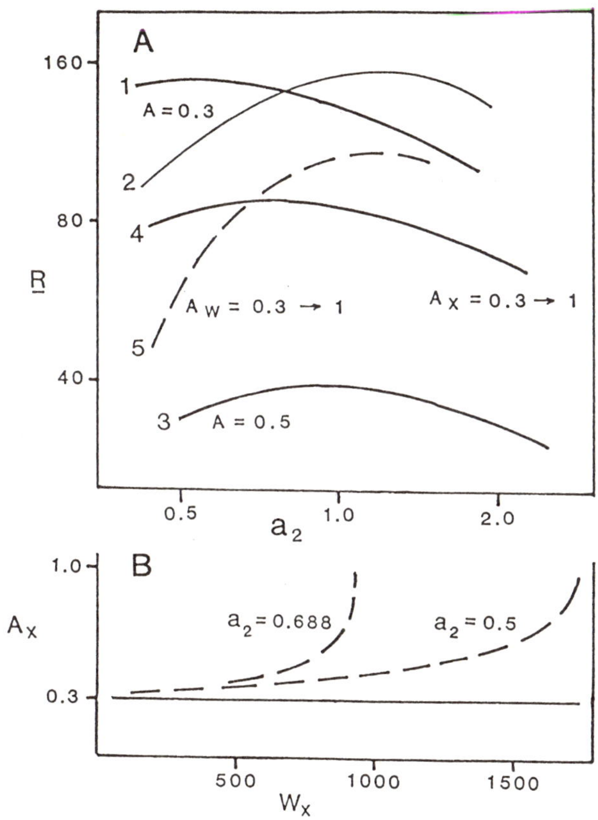

Figure 5A. Expected lifetime reproduction (R) of a recruit as a function of growth coefficient a_2 under different allocation and mortality schedules. $\dot{W}^1 = a_1 W^{0.5} - a_2 W$. Curves 1,3,4,5: Model 1 (text); annual reproductive allocation $R_x = 0.25 a_2 W$. Curve 2: Model 2. Curves 1 and 2, annual mortality A = 0.3. Changing from Model 1 to Model 2 effectively doubles the exponent of reproductive allometry. Curve 3, A = 0.5. Increasing A shifts max(R) from a_2 fl 0.5 to fl 1.0. Curve 4: A_x increases linearly from $A_1 = 0.3$ to $A_{21} = 1.0$. This produces the size-specific schedules shown in Figure 5B. Curve 5: the result of imposing the size-specific mortality schedule in Figure 5B for $a_2 = 0.688$ upon all growth schedules. **5B.** Size-specific mortality schedules result from imposing an increasing age-specific mortality schedule (Figure 5A, curve 4).

Models (1) and (2) were made equivalent in the proportion of asymptotic ovarian to somatic weight for a given value of a_2; with an a_2 of 2.0, almost half of an older individual's body would be gonad at spawning time. In all these examples, viability selection as an additional cost of reproduction (cf. Bell and Koufopanou 1986) might be expected to become important particularly with more extreme values of a_2, resulting in a lower a_2 at $max(R)$ and once again an increase in curvature. There is a chicken-egg aspect to this: increasing mortality < - > increasing a_2. An increase in mortality, even one that itself is not viability-selective, if severe enough might lead eventually to viability selection as the population evolves.

Whatever the path of causality, correlations among species of fishes between natural mortality and a_2 (in the form of K, Equation 5), age at maturity, and post-maturation growth and reproduction have often been noted. Gunderson and Dygert (1988; cf. also Charnov and Berrigan 1991) review this literature, and they suggest that the relatively low variance of estimation of gonadosomatic index, GSI, makes it more reliable than K or similar candidates as a predictor of natural mortality. In their method of estimating natural mortality, Myers and Doyle (1983) incorporate additional nonlinear mortality cost of reproduction discussed above or a 'concave egg conversion function', as in the inverse dependence of surplus energy upon a_2 in model 2.

In each example in Figure 5, mutation and stabilizing selection might allow a population to maintain genetic variation in a_2 of ± 5-10% about the peak value with little difficulty, as the selection gradients near the maxima are not steep. An increase in a_1 is essentially equivalent to a shift in the ordinate scale of Figure 5A, and other things being equal will be favoured. Models of a_1-selection of the type employed by Favro et al. (1979), but with fecundity dependent not upon age but instead linearly upon weight, result in rapid fixation of alleles of fastest growth even in the presence of severe size-selective mortality (Nelson and Soulé 1987). It is difficult therefore to imagine that individual variation in a_1 in sole and anchovies is largely a result of additive genetic variation. In sole, marked reproductive allometry and positive phenotypic covariation of a_1 and a_2 make this seem even less likely, because one is led to expect generally negative genetic covariance among such life-history characters (Lande 1982). While rapid growth and/or a large ovary might be associated with some extraordinarily disadvantageous property, such as a very high mortality cost to reproduction, in the sole there is no evidence for this. However, Houle's (1991) simulations suggest that if 'acquisition' loci are many times more frequent than 'allocation' loci, positive genetic covariance between such variables might possibly be maintained at mutation-selection equilibrium.

In the face of such uncertainties it would be foolhardy to predict the joint evolutionary response of reaction-norm variation in a_1 and a_2 to some mortality regime newly-imposed upon a population. It will involve interactions between selective surfaces and contoured response surfaces, both visualizable as sets of ridge-contoured auximetric grids. The outcome will depend upon the extent to which the individual variant A/A reaction norms to mortality-induced changes in density are *cogradient* or *contragradient* to the mortality-induced changes in the selective surfaces. Evolutionary forces are expected to be ameliorated or exaggerated to the extent that the former or latter relationships prevail (Levins 1968; Caswell 1983).

There are several hierarchical aspects to the responses of a population to environmental change. Bateson (1963; see Caswell 1983) notes that phenotypic plasticity might evolve in service to those functions whose stability is essential for the continuation of life; one may imagine that the allocation of a limited 'currency' of plastic response is governed by the priorities of homeostasis. We are then tempted to ask what A/A system properties are conserved in responding to the changes in density that follow upon increased mortality. One such character might be the slope and elevation of ontogenetic ovarian allometry in a mature female sole of average size or in average times; isometry may prevail in an average mature anchovy. Via and Lande (1985) observe that such invariance is merely 'a special case of [the] more general evolutionary problem' we confront when faced with phenotypic plasticity. Perhaps a very special case; the reaction norm of 'no change' suggests an optimal solution tot he 'life-history problem'. Bateson also suggests that the promptness, speed and reversibility of response should match that of the environment. Those rapid changes we call 'behaviour' are at one extreme of this temporal spectrum. A/A system compensatory response to density change may begin immediately, but further effects upon growth and reproduction may lag behind and there are effects on both that are cumulative over a lifetime. Even in the absence of other constraints, genetic response to chronic change in mortality schedule may be much delayed by, and obscured by, the many A/A systems' reactions to the new density regime, and may take many generations to be revealed.

For the sole there has been no suggestion that the changes seen in the 1960s were genetic, even though they have evidently persisted to the present (Rijnsdorp et al. 1991; Rijnsdorp and van Beek 1991). I think this is in part because there was little change in mean a_2: average age at maturity and reproductive allocation relative to size underwent little change whereas growth rate and size at maturity increased. This is contragradient to one's expectations for fisheries selection, and might exacerbate it (the size-selection embodied in Figure 5, curve 5 implies selection against both low a_2 and high a_1.) But suppose a reaction norm to density decrease were to include substantial increase in a_2, as might be the case in species with reproductive isometry. Then a correlated and cogradient decrease in age and perhaps even size at maturity might easily suggest a genetic response to exploitation, but might actually serve to ameliorate it. Inferences that such changes were genetic in nature have been made many times (references in Nelson and Souleæ 1987). In evaluating these proposals, the above-noted temporal hierarchy of response could be kept in mind and a kind of Occam's Razor employed: Do not attribute short- term changes in long-lived species to long-term evolutionary processes, when reaction-norm changes are more likely.

ACKNOWLEDGEMENTS

I thank D. Hedgecock for the opportunity to study anchovies with him, and E. Hutchinson, F. Sly, and G. Li for their partnership in the project. Support for research and for travel to the Jülich meeting was granted by the California Sea Grant College Program. I am grateful to the editors and to those at KFA who made the meeting possible.

REFERENCES

Bateson, G. 1963. The role of somatic change in evolution. Evolution **17**: 529-539.

Bell, G. and V. Koufopanou. 1986. The cost of reproduction. Oxford Surveys in Evolutionary Biology **3**: 83-131.

Caswell, H. 1983. Phenotypic plasticity in life-history traits: demographic effects and evolutionary consequences. American Zoology **23**: 35-46.

Deniel, C. 1984. La reproduction des poissons plats (Teæleæosteæens - Pleuronectiformes) en Baie de Douarnenez. II. Cycles sexuels et fécondité des Soleidae; *Solea vulgaris vulgaris, Solea lascaris, Buglossum luteum* et *Microchirus variegatus.* Cahiers de Biologie Marine **25**: 257-285.

de Veen, J. F. 1976. On changes in some biological parameters in the North Sea sole. Journal du Conseil International pour l'Exploration de la Mer **37**: 60-90.

Dygert, P. H. 1986. Management implications of variability in reproduction and growth of commercial marine fishes. Ph.D. dissertation, University of Washington, Seattle.

Favro, L. D., P. K. Kuo, and J. F. McDonald 1979. Population-genetic study of the effects of selective fishing on the growth rate of trout. Journal of the Fisheries Research Board of Canada **36**: 552-561.

Fiedler, P. C., R. D. Methot, and R.P. Hewitt 1986. Effects of the California El Niño 1982-1984 on the northern anchovy. Journal of Marine Research **44**: 317-338.

Goldschmidt, R. 1940. The material basis of evolution. Yale University Press, New Haven, Connecticut.

Grosse, D. J. 1982. An experiment in the artificial rearing of lingcod (*Ophiodon elongatus* Girard) for purposes of enhancement. Not found; cited by Dygert 1986 as Washington Department of Fisheries, Technical Report **10**. 92p.

Gunderson, D. R. and P. H. Dygert. 1988. Reproductive effort as a predictor of natural mortality rate. Journal du Conseil International pour l'Exploration de la Mer **44**: 200-209.

Hedgecock, D., E. S. Hutchinson, G.Li, F. L. Sly, and K. Nelson. 1989. Genetic and morphometric variation in the Pacific sardine, *Sardinops sagax caerulea*: comparisons and contrasts with historical data and with variability in the northern anchovy, *Engraulis mordax*. U.S. National Marine Fisheries Service Fishery Bulletin **87**: 653-671.

Houghton, R. G., J. M. Last, and P. J. Bromley. 1985. Fecundity and egg size of sole (*Solea solea* (L.)) spawning in captivity. Journal du Conseil International pour l'Exploration de la Mer **42**: 162-165.

Houle, D. 1991. Genetic covariance of fitness correlates; what genetic correlations are made of and why it matters. Evolution **45**: 630-648.

Hunter, J. R., N. C. H. Lo, and R. J. H. Leong. 1985. Batch fecundity in multiple spawning fishes. NOAA Technical Report, National Marine Fisheries Service **36**: 67-77.

Jensen, A. L. 1985. Relations among net reproductive rate and life history parameters for lake whitefish (*Coregonus clupeaformis*). Canadian Journal of Fisheries and Aquatic Sciences **42**: 164-168.

Laird, A. K., A. D. Barton, and S. A. Tyler. 1968. Growth and time: an interpretation of allometry. Growth **32**: 347-354.

Lande, R. 1982. A quantitative genetic theory of life history evolution. Ecology **63**: 607-615.

Lande, R. 1984. Genetic and evolutionary aspects of allometry. Pp. 21-32 in W. Jungers ed., Size and scaling in primate biology. Plenum Press, New York.

Le Cren, E. D. 1958. Observations on the growth of perch (*Perca fluviatilus* L.) over twenty-two years with special references to the effects of temperature and population density. Journal of Animal Ecology **27**:287-334.

Leggett, W. C. and J. E. Carscadden. 1978. Latitudinal variation in reproductive characteristics of American shad (*Alosa sapidissima*): Evidence for population specific life history strategies in fish. Journal of the Fisheries Research Board of Canada **35**: 1469-1478.

Levins, R. 1968. Evolution in changing environments. Princeton University Press, Princeton, New Jersey.

MacCall, A. D. 1990. Dynamic geography of marine fish populations. Washington Sea Grant, University of Washington Press, Seattle.

Mousseau, T. A. and D. A. Roff. 1987. Natural selection and the heritability of fitness components. Heredity **59**: 181-197.

Murphy, G. I. 1968. Patterns in life history and in the environment. American Naturalist **102**: 391-403.

Myers, R. A., and R. W. Doyle. 1983. Predicting natural mortality rates and reproduction-mortality trade-offs from fish life history data. Canadian Journal of Fisheries and Aquatic Sciences **40**: 612-620.

Nelson, K. and M. Soulé. 1987. Genetical conservation of exploited fishes. Pages 345-368 *in* N. Ryman and F. Utter, eds. Population genetics and fisheries management. Washington Sea Grant, University of Washington Press, Seattle.

Pauly, D. 1979. Gill size and temperature as governing factors in fish growth: a generalization of von Bertalanffy's growth formula. Bericht Inst. Meereskunde (Kiel) **63**.

Rijnsdorp, A. D. 1990. The mechanism of energy allocation over reproduction and somatic growth in female North Sea plaice, *Pleuronectes platessa* L.. Netherlands Journal of Sea Research **25**: 279-290.

Rijnsdorp, A. D., N. Daan, F. A. van Beek, and H. J. L. Heessen. 1991. Reproductive variability in North Sea plaice, sole, and cod. Journal du Conseil International pour l'Exploration de la Mer **47**: 352-375.

Rijnsdorp, A. D. and F. A. van Beek. 1991. Changes in growth of plaice *Pleuronectes platessa* L. and sole *Solea solea* (L.) in the North Sea. Netherlands Journal of Sea Research **27**: 441-457.

Sebens, K. P. 1982. The limits to indeterminate growth: an optimal size model applied to passive suspension feeders. Ecology **63**: 209-222.

Sibly, R., P. Calow, and N. Nichols. 1985. Are patterns of growth adaptive? Journal of Theoretical Biology **112**: 553-574.

Tanasichuk, R. W. and D. M. Ware. 1987. Influence of interannual variations in winter sea temperature on fecundity and egg size in Pacific herring. Canadian Journal of Fisheries and Aquatic Sciences **44**: 1485-1495.

Via, S. and R. Lande. 1985. Genotype-environment interaction and the evolution of phenotypic plasticity. Evolution **39**: 505-522.

Ware, D. M. 1980. Bioenergetics of stock and recruitment. Canadian Journal of Fisheries and Aquatic Sciences **37**: 1012-1024.

Ware, D. M. 1982. Power and evolutionary fitness of teleosts. Canadian Journal of Fisheries and Aquatic Sciences **39**: 3-13.

REACTION NORMS FOR REPRODUCTIVE TRAITS IN BROOK TROUT AND THEIR INFLUENCE ON LIFE HISTORY EVOLUTION AFFECTED BY SIZE-SELECTIVE HARVESTING

Jeffrey A. Hutchings

Institute of Cell, Animal and Population Biology, University of Edinburgh, West Mains Road, Edinburgh EH9 3JT, UK.

ABSTRACT

Using life history data on two unexploited Canadian populations (Freshwater and Cripple Cove rivers) of brook trout, *Salvelinus fontinalis*, I constructed reaction norms for female age, effort and body size at first reproduction, and then used these to predict how size-selective fishing mortality might effect evolutionary changes in the shapes of the reaction norms and ultimately influence harvestable biomass and mean reproductive age. The fitness, *r*, associated with five levels of relative effort (based on gonad weight/total body weight, or *GSI*) were calculated for females growing at 5 different rates and reproducing initially from 2 to 5 yr. The five effort levels reflected proportional (*GSI*-related) changes in fecundity. Maturation was assumed to reduce growth rate by an amount proportional to the female's expected size-specific GSI at reproduction. Age-specific fecundity estimates included the positive effects of maternal size on egg size and of egg size on early juvenile survival. Survival costs of reproduction (decrements in overwinter survival attributable to reproduction) ranged from nil for the '0' effort level (size-specific *GSI* minus 1 σ of the population mean) to death for the '1' effort level (size-specific *GSI* plus 1 σ) and were related to intermediate levels of effort (0.25, 0.50, 0.75) by a series of empirically-based sigmoidal functions whose inflexion points increased with body size.

Reaction norms reflected optimal trait values across a environmental gradient varying in prospects for individual growth. In both populations, increased growth favoured earlier reproduction and, for Freshwater females, increased reproductive size. Freshwater females were favoured to maximize effort across most environments whereas Cripple Cove females were favoured to increase effort from minimum to maximum levels as environmental quality increased. The main evolutionary effect of size-selective mortality in Cripple Cove was to change the shape of the reaction norm for effort to favour maximum effort across most environments. This adaptive response to fishing pressure was predicted to reduce harvestable biomass and mean reproductive age. If spatial environmental heterogeneity is sufficient to generate size-dependent differences in the optimal allocation of reproductive effort, then the fitness consequences associated with different age-effort strategies should be incorporated into models that predict the evolutionary consequences of harvesting on natural populations.

INTRODUCTION

The phenotypes of all organisms are modified, to some extent, by the environment. Within populations, such phenotypic modification can effect substantial variation in life history. The capacity of a genotype to alter its phenotype in response to environmental changes is termed phenotypic plasticity (Bradshaw 1965). Plasticity may reflect an environmentally-induced constraint on a genotype's physiology (e.g. reduced food supply decreases fecundity because of an organism's reduced nutritional state) or it can represent an adaptive response to an environmental factor influencing age-specific rates of fecundity or survival (e.g. reduced food supply results in increased age-specific fecundity because of the detrimental effects of reduced food on future age-specific survival). The systematic change with which a genotype alters its phenotypic expression across an environmental gradient is called a norm of reaction (Schmalhausen 1949).

The notion of adaptive phenotypic plasticity leads naturally to the concept of individual optimization of life history. For example, clutch size in several bird species depends largely on individual ability to acquire food, suggesting that clutch sizes may represent optimal responses by individuals to their local environment (references in Lessells 1991). Reaction norms can be used to describe patterns of individual optimization and to predict how changes in the environment can effect adaptive changes in life history. Stearns and Koella (1986) used this approach to construct general norms of reaction for age and size at maturity. They incorporated presumed effects of growth rate on juvenile and adult survival in their calculations of fitness, varied growth rate, and then calculated optimal ages at maturity as a function of body size.

Two conditions under which adaptive phenotypic plasticity can be expected to evolve occur when i) the spatial distribution of individuals across habitats is random with respect to genotype and when ii) habitat significantly influences fitness. These conditions probably exist for many stream-dwelling salmonid fish for which habitat variation is reflected by differences in water velocity, water depth, substrate composition, light intensity, and abundance and species composition of food. Following yolk sac resorption, juvenile salmonids such as brook trout, *Salvelinus fontinalis*, and Atlantic salmon, *Salmo salar*, emerge from the stream substrate in spring and are transported downstream to small, relatively discrete sections of slow-moving water where they spend most, occasionally all, of the summer months (Jones 1959, Hutchings 1990). The environment inhabited by these juveniles is probably random with respect to genotype because emerging salmonids are unable to swim against all but the weakest of water currents. Growth rate can vary significantly among juvenile brook trout within and among habitats (Hutchings unpublished data), effecting differences in fitness through size-dependent effects on life history traits (e.g. Roff 1984, Thorpe et al. 1984, Hutchings and Morris 1985, van den Berghe and Gross 1989).

In a recent study, Hutchings (submitted, in press) documented significant differences in a survival cost of reproduction (the decrement in overwinter survival of reproductive fish relative to non-reproductive fish of the same size and age) within populations of brook trout on Cape Race, southeastern Newfoundland, Canada. In addition to being positively associated with effort (approximated as the gonadosomatic index or GSI which equals gonad weight/total body weight), the cost was dependent upon body size. The smallest individuals experienced disproportionately high costs relative to their effort. This can be attributed to the faster rate at which small individuals utilize their proportionately lower lipid stores (Brett et al. 1969, Elliott 1976, Schmidt-Nielson 1984) relative to large individuals during the energetically demanding winter (Cunjak 1988). Similar size-related effects on reproductive costs have been documented for female elephant seals, *Mirounga angustirostris*, (Reiter and LeBoeuf 1991) and for male threespine sticklebacks, *Gasterosteus aculeatus*, (Dufresne et al. 1990).

If body size influences the relationship between cost and effort, individuals may allocate their own optimal effort subject to their ability to survive reproduction and subject to their prospects for increasing fecundity in subsequent spawnings. To explore this possibility, I used empirical data for two populations on Cape Race (Freshwater and Cripple Cove; detailed physical and life history data are in Hutchings 1990) to construct reaction norms for age, effort and body size at first reproduction for female brook trout. I calculated the fitness (r) associated with different body sizes, levels of reproductive effort, and ages at first reproduction (a) to evaluate how optimal values of these traits varied with growth rate (which reflected environmental variation). Given the primary importance to fitness of effort expended during an organism's initial reproductive bout (Gadgil and Bossert 1970, Charlesworth 1980), I assumed that effort varied only at α. To determine the optimal values for these traits (Stearns 1992 provides a good summary of such an analysis), one requires knowledge of the following: 1) the growth function relating size to age, 2) the relationship between fecundity and size (and to age via the growth function), 3) functions relating mortality to age/size for individuals reproducing initially at different ages and sizes, 4) the influence of reproduction on growth, and 5) the influence of maternal age/size on juvenile survival. Information on many of these relationships is available for the Cape Race populations (Hutchings 1990, in press). I reconstructed the reaction norms for three levels of size-selective mortality on the larger fish in Cripple Cove population to determine how evolutionary changes in reaction norms, effected by size-selective harvesting, might influence harvestable biomass and age at reproduction. I also evaluated the sensitivity of the harvesting simulations to the assumption that fish optimally allocate resources to reproduction on an individual, size-dependent basis.

METHODS

FITNESS CALCULATIONS

Fitness was defined as r, the intrinsic rate of natural increase, and was calculated from the discrete-time version of the Euler-Lotka equation

$$1 = \Sigma\, l_x\, m_x\, e^{-rx} \qquad (1)$$

where l_x represents female survival probability from the zygote to the beginning of the spawning season at age x, and m_x is the expected number of zygotes produced by a female of age x. To measure the fitness of a population of individuals having a specific maturation phenotype p, equation (1) becomes

$$1 = \Sigma\, l_x(p)\, m_x(p)\, e^{-rx} \tag{2}$$

Sibly and Calow (1983:545-546), following Charlesworth (1980:172-173, 200), have shown that fitness can be defined in this way for populations subjected to either density-dependent or density-independent population regulation. Calculating r in this manner is analogous to determining the rate of spread of an allele that has an effect on fitness (Charlesworth 1980), and Prout (1980) has shown that selection on genotypes can be density independent in a population regulated by density-dependence.

AGE-SPECIFIC BODY SIZES OF NON-REPRODUCTIVE AND REPRODUCTIVE FEMALES

The fitness associated with five relative levels of reproductive effort (*GSI*) were calculated for five sizes of female at each of several ages at first reproduction α (2-4 yr in Freshwater; 2-5 yr in Cripple Cove). To incorporate most of the natural variability in size at age in each population, I derived the five sizes of female at each a from the mean size, μL, at the end of the first year of life. (Length at age =1 was normally distributed in both populations. Ages were calculated from time of fertilization, i.e. October, so that integer values represent ages at the beginning of the spawning season.) The five size classes of females were

Size Class	Size
Small (*S*)	$\mu_L - 2\sigma$
Small-Medium (*S-M*)	$\mu_L - \sigma$
Medium (*MED*)	μ_L
Medium-Large (*M-L*)	$\mu_L + \sigma$
Large (*L*)	$\mu_L + 2\sigma$

where σ equals 1 standard deviation about each population mean. Beyond age = 1, females were assumed to grow at a rate that approximated the growth of non-maturing fish in each population. These growth rates were calculated from linear regressions which related length to age for non-reproductive fish aged 1.67 to 4.67 years collected in June, 1988 (details in Hutchings 1993).

During the year of initial maturation and for each year thereafter, growth rates were reduced to reflect the energetic demands that gonadal production places on somatic growth (cf. references in Bell 1980, Myers et al. 1986, Rowe and Thorpe 1990). I assumed that the proportion of surplus energy allocated to gonads was directly related to the proportional loss of surplus energy devoted to soma (Ware 1980). The annual growth rate of a female that would mature at age x was reduced by an amount $(1 - GSI_x)^\beta$ such that

$$L_{Mx} = L_{Ix}\,(1-GSI_x)^\beta \tag{3}$$

GSI increases with female body length (L, in mm) according to the regression equations (FW=Freshwater, CC=Cripple Cove)

$$\text{FW:} \quad GSI = 0.0006L + 0.1020 \ (\text{n} = 111, \ r = 0.33) \tag{4}$$

$$\text{CC:} \quad GSI = 0.0006L + 0.0248 \ (\text{n}= 48, \ r = 0.41) \tag{5}$$

To determine the age-specific lengths of reproductive females from the lengths they would have been had they not matured (i.e., determining L_{M_x} from L_{I_x}) and to incorporate the size-dependent effects on GSI, I used the general form of the linear regressions between GSI and body size, $GSI = \theta L_M + \varepsilon$, where θ and ε are the slope and intercept, respectively.

As stated in Eqn. (3), $L_{M_x}=L_{I_x}(1-GSI_x)^{\beta}$. Therefore,

$$L_M=L_I \ (1-(\theta L_M + \varepsilon))^{\beta} \tag{6}$$

The sums of squared deviations between observed and expected female lengths at reproduction were minimized in each population when $\beta= 1$. Thus,

$$L_M=L_I \ (1-\varepsilon)/(1 + \theta L_I)^{\beta} \tag{7}$$

Predicted age-specific lengths of reproductive and non-reproductive females are presented in Table 1.

AGE-SPECIFIC FECUNDITY

Age-specific fecundity was calculated from linear regressions between fecundity (m) and body length (L, in mm),

$$\text{FW:} \quad m = 1.18L - 74.26 \quad (\text{n} = 127, r = 0.81) \tag{8}$$

$$\text{CC:} \quad m = 2.48L - 251.73 \quad (\text{n} = 65, r = 0.84) \tag{9}$$

Five levels of GSI were considered for each of the five sizes of female at each α. For each size L, GSI was calculated from equations (4) and (5). The five levels of effort were $GSI_L - 1\sigma$, $GSI_L - .5\sigma$, GSI_L, $GSI_L + .5\sigma$, and $GSI_L + 1\sigma$, where σ is the standard deviation of the population mean GSI (0.03 and 0.02 for Freshwater and Cripple Cove, respectively). These GSI values corresponded to relative effort levels of 0, 0.25, 0.50, 0.75, and 1, respectively. Fecundity was assumed to be directly related to GSI. At an effort of GSI_L (i.e. a relative effort of 0.50), fecundity at size L, m_L, was calculated from equations (8) and (9). At other effort levels, m_L was increased or decreased by a proportion equal to the ratio of the corresponding GSI to GSI_L .

MATERNAL INFLUENCES ON EGG SIZE AND EARLY JUVENILE SURVIVAL

The implicit assumption of the Euler-Lotka equation that zygote survival is not influenced maternally is invalid for brook trout. Maternal fitness at age x depends on the size of egg produced because survival in early life is positively correlated with egg size and egg size (E, diameter in mm) increases with maternal body size (L, in mm) (Freshwater: $E = 0.012L + 3.203$, n=114, r=0.57; Cripple Cove: $E = 0.007L + 2.962$, n= 48, r=0.39) (Hutchings 1990, 1991). In the laboratory, Hutchings (1991) found post-emergence survival of brook trout (S) to be positively correlated with egg volume ($V = \pi E^3/6$, in mm^3) according to the equation $S = 0.013V + 0.026$ (non-*ad libitum* food conditions, n=11 broods of 20 juveniles each). Thus, for the fitness calculations, age-specific fecundities were multiplied by the expected post-hatching survival probabilities of each age-specific egg size.

Table 1. Calculated age-specific lengths (mm) of non-reproductive (*NR*) and reproductive (*R*) female brook trout from two populations on Cape Race, Newfoundland. Non-reproductive females were assumed to be virgins at each age. Lengths of reproductive females at each age are for individuals at their age of first reproduction, α.

			Age (yr)				
Population	*NR/R*	Size	1	2	3	4	5
Cripple Cove	*NR*	*S*	57.7	106.7	138.8	170.7	
		S-M	66.0	122.1	158.7	195.2	
		MED	74.3	137.5	178.7	219.8	
		M-L	82.6	152.8	198.7	244.3	
		L	90.9	168.2	218.6	268.9	
	R	*S*		97.8	124.9	151.0	176.6
		S-M		110.9	141.3	170.4	198.9
		MED		123.8	157.4	189.4	220.5
		M-L		136.5	173.1	207.8	241.4
		L		149.0	188.5	225.8	261.8
Freshwater	*NR*	*S*	50.0	73.0	88.3		
		S-M	56.3	82.2	99.5		
		MED	62.6	91.4	110.6		
		M-L	68.9	100.6	121.7		
		L	75.2	109.8	132.8		
	R	*S*		62.8	75.3	87.3	
		S-M		70.3	84.3	97.7	
		MED		77.8	93.1	107.8	
		M-L		85.2	101.8	117.8	
		L		92.5	110.5	127.6	

AGE- AND SIZE-SPECIFIC SURVIVAL

Age-specific survival of brook trout on Cape Race depends primarily on body size at the onset of winter (70-80% of the annual mortality occurs from October through April) and on reproductive effort (Hutchings a). In the absence of size-specific survival data for the summer (May-September), summer survival probabilities were set to 0.85 and 0.81, the estimated summer survival rates for age = 1.67 fish in Freshwater and Cripple Cove, respectively (Hutchings 1990). Prior to spawning for the first time, overwinter survival was equal to the mean overwinter survival rates for immature fish (0.36 in Freshwater and 0.58 in Cripple Cove). Survival from age = 0 to 1.67 equalled 0.19 and 0.05 in Freshwater and Cripple Cove, respectively (Hutchings 1993). Except for the winter following a female's initial spawning, size-specific overwinter survival was calculated directly from mark-recapture experiments conducted from October, 1988, to April, 1989 (Hutchings submitted). These data, and the curves used to predict size-specific, post-reproductive overwinter survival, are presented in Figure 1.

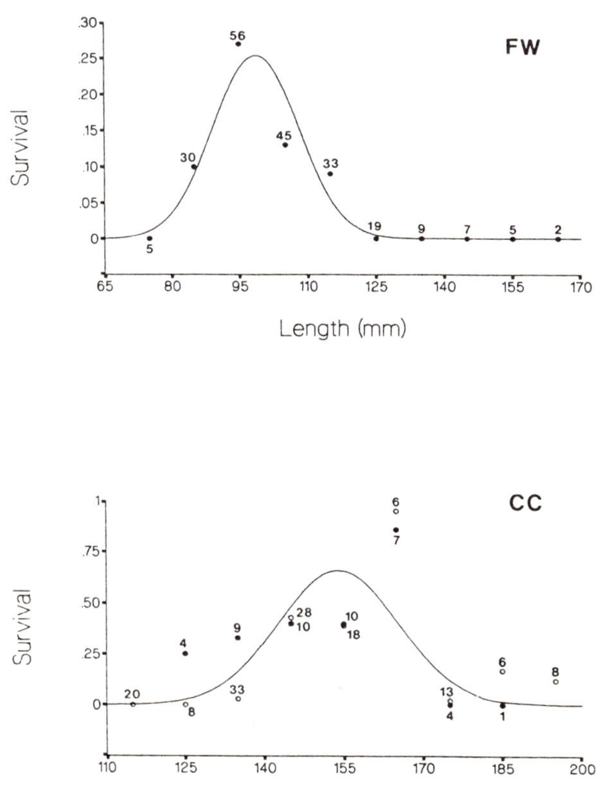

Figure 1. Relationship between post-reproductive overwinter survival and body size for brook trout in Freshwater River (FW: females only) and Cripple Cove River (CC: females - closed circles; males - open circles). Numbers of fish originally marked in each size class shown adjacent to each datum. Curves represent predicted relationships.

POST-REPRODUCTIVE SURVIVAL AS A FUNCTION OF EFFORT

At α, a female's probability of surviving winter was assumed to be a multiple of the survival probability she would have had if she had not matured (ϕ), i.e. if there was no survival cost to reproduction. Size-specific ϕ at age α was equal to the expected post-reproductive, overwinter survival divided by (1 - the average survival cost of reproduction for each population, i.e. 0.58 and 0.31 for Freshwater and Cripple Cove, respectively). I assumed that ϕ declined with increasing effort at rate $(1 - \Omega)$, where Ω equalled the decrement in overwinter survival attributable to reproduction. Values of Ω ranged from 0 (no survival cost) for the '0' effort level to 1 (i.e. death) for the '1' effort level. At the intermediate levels of effort (0.25, 0.50, 0.75), Ω was assumed to be related to effort by a series of sigmoidal functions whose inflexion points increased with body size (Table 2). Such a three-dimensional relationship incorporates the observations that survival costs increase with increasing effort and that cost per unit effort decreases with increasing body size (Hutchings submitted a).

Table 2. Hypothetical values of Ω (survival cost of reproduction, i.e. proportional reduction in overwinter survival attributable to reproduction) as a function of relative reproductive effort and female body size.

	Body Size				
Relative Effort	*S*	*S-M*	*MED*	*M-L*	*L*
0	0	0	0	0	0
0.25	0.77	0.29	0.08	0.08	0.08
0.50	0.86	0.86	0.50	0.15	0.15
0.75	0.98	0.98	0.98	0.98	0.23
1	1	1	1	1	1

REACTION NORMS FOR REPRODUCTIVE AGE, EFFORT AND BODY SIZE

Reaction norms were constructed by determining how optimal age, effort and size at first reproduction (i.e. the trait values that maximized fitness) varied with growth rate. Reaction norms describe how trait optima vary across an environmental gradient. I considered environmental variation to be reflected by differences in individual growth rate such that the slowest and fastest growers in an age class reflected the worst and best environments, respectively, in the environmental continuum. Thus, reaction norms were constructed by plotting the values of each trait that maximized fitness in each environment (i.e. individual growth rate). A single norm of reaction for reproductive age, effort and body size was constructed by plotting the optima for these traits in three dimensions for each population.

THE USE OF REACTION NORMS TO PREDICT THE EVOLUTIONARY CONSEQUENCES OF SIZE-SELECTIVE FISHING MORTALITY

Reaction norms for reproductive traits can be used to predict the evolutionary consequences of size-specific fishing mortality. This was examined by imposing different levels of fishing mortality on females in Cripple Cove population and determining how optimum reproductive age and effort varied with this external mortality. I simulated different levels of size-selective fishing pressure by subjecting all fish above a specific size to one of three levels of increased mortality. Low, medium, and high levels of fishing pressure corresponded to increases in mortality (relative to the observed data for the unexploited population) of 25 per cent, 50 per cent, and 75 per cent for fish larger than the length corresponding to the first quartile of reproductive female lengths (147 mm).

As indices of change in previously unexploited populations subjected to different levels of fishing intensity, I calculated the mean age of reproduction and the harvestable biomass (i.e. biomass of fish larger than the first quartile length, as determined from weight-length regressions in Hutchings 1990) under zero, low, medium, and high levels of fishing pressure. As a first approximation, these indices were calculated for a 'typical' cohort (individuals born during the same spawning season). I assumed that the five size classes of the cohort at age = 1 were normally distributed in both populations (supported by field data in Hutchings 1990). Size-specific survival rates were those described previously. Age and biomass were calculated in two ways. One set of estimates incorporated the assumption that individuals optimally allocate effort according to their size at a while the second set of estimates excluded this assumption. Inclusion or exclusion of this individual optimization assumption affected post-reproductive survival probabilities (and, as a consequence, the number of individuals in post-α age classes). Post-reproductive survival varied with effort when the reaction norm data were included. When these data were excluded, effort at body size and post-reproductive survival corresponded to those at the 0.50 level (i.e. at the *GSI* calculated for the size of interest).

RESULTS

RELATIONSHIPS AMONG FITNESS, EFFORT AND BODY SIZE

The fitness associated with different levels of effort varied with body size and with age at first reproduction, α. This was evident from the fitness functions at the optimal α for each population (3 and 4 yr for Freshwater and Cripple Cove, respectively; Figure 2). In Freshwater, for example, maximum fitness for females in the smallest two size classes was achieved when effort was maximized. Optimal relative effort for females in the upper three size classes ranged from 0.25 (*MED*) to 0.75 (*L*) and increased with body size. By contrast, optimal effort increased across all size classes of Cripple Cove females reproducing at α = 4 - the smallest females were favoured to minimize effort whereas females in the largest three size classes were favoured to maximize effort.

The relationships among optimal reproductive effort, reproductive size and age at first reproduction were qualitatively similar between populations (Figure 3). In general, females reproducing initially at the youngest or oldest ages were favoured to maximize effort whereas intermediate levels of effort were favoured for most size classes of females reproducing initially at intermediate ages.

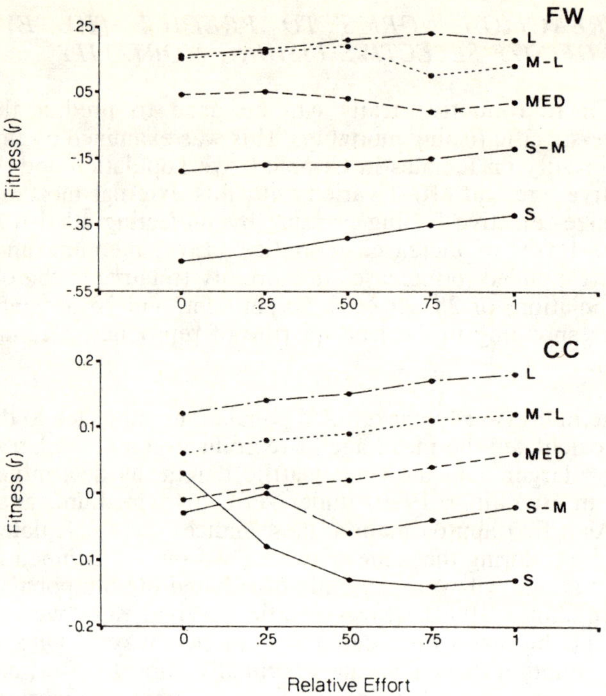

Figure 2. Fitness associated with five levels of relative effort at optimal age of first reproduction (α) for female brook trout of different sizes in Freshwater (*FW*) and Cripple Cove (*CC*) populations. Effort levels 0 and 1 result in no survival cost of reproduction and immediate post-reproductive death, respectively. Size classes (*S*-small; *M*, *MED*-medium; *L*-large) are defined in the text. α = 3 and 4 in Freshwater and Cripple Cove, respectively.

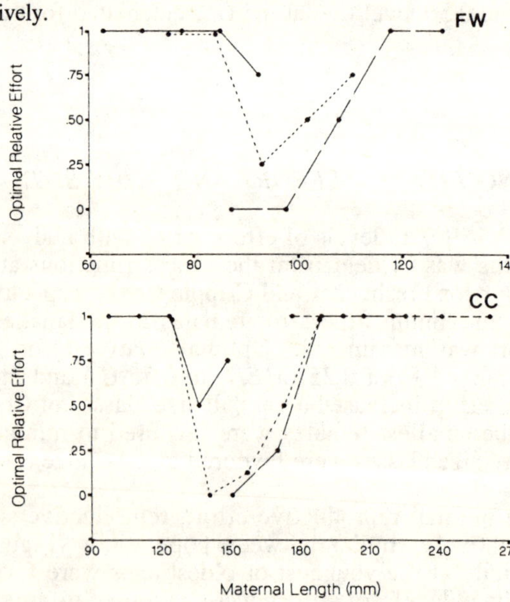

Figure 3. Optimal levels of relative effort as functions of body size and age at first reproduction (a) for female brook trout in Freshwater (FW) and Cripple Cove (CC) populations (——— α = 2, ------ α = 3, — — α = 4, - - - α = 5).

REACTION NORMS FOR REPRODUCTIVE AGE, EFFORT AND BODY SIZE

Environmental differences in the prospects for individual growth influenced optimal age at first reproduction, reproductive effort and body size at reproduction. For females in both populations, optimal α declined as environmental quality increased (Figure 4A). That is, increased growth rate favoured earlier reproduction. Maximal effort was favoured for Freshwater females across most environments whereas effort was favoured to be minimal for Cripple Cove females in poor environments and to increase to a maximum as environmental quality increased (Figure 4B). Optimal size at first reproduction increased with environmental quality in Freshwater but varied little across environments in Cripple Cove (Figure 4C).

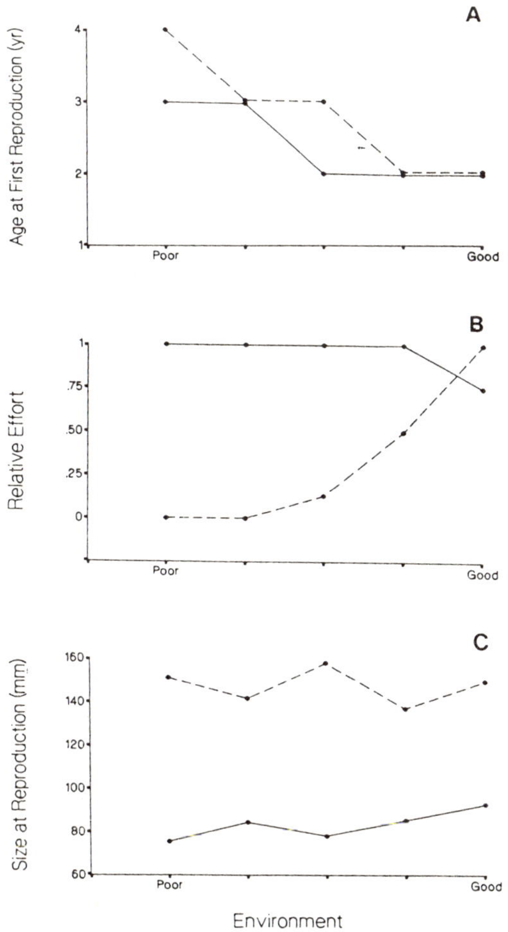

Figure 4. Norms of reaction for age (A), relative reproductive effort (B), and body size (C) at first reproduction for female brook trout in Freshwater (solid lines) and Cripple Cove (dashed lines) populations (environment ranges from poor to good prospects for individual growth).

Table 3. Optimal size-specific values of age at first reproduction and relative level of effort calculated for two brook trout populations subjected to four levels of size-selective fishing mortality (values of relative effort in parentheses).

Population	Fishing Mortality	Body Size *S*	*S-M*	*MED*	*M-L*	*L*
Cripple Cove	None	4 (0)	3 (0)	3 (0.12)	2 (0.5)	2 (1)
	Low	4 (0)	3 (0)	3 (0.25)	2 (0.5)	2 (1)
	Medium	4 (0)	3 (0)	3 (0.25)	2 (0.5)	2 (1)
	High	4 (0)	3 (1)	3 (1)	2 (1)	2 (1)
Freshwater	None	3 (1)	3 (1)	2 (1)	2 (1)	2 (0.75)

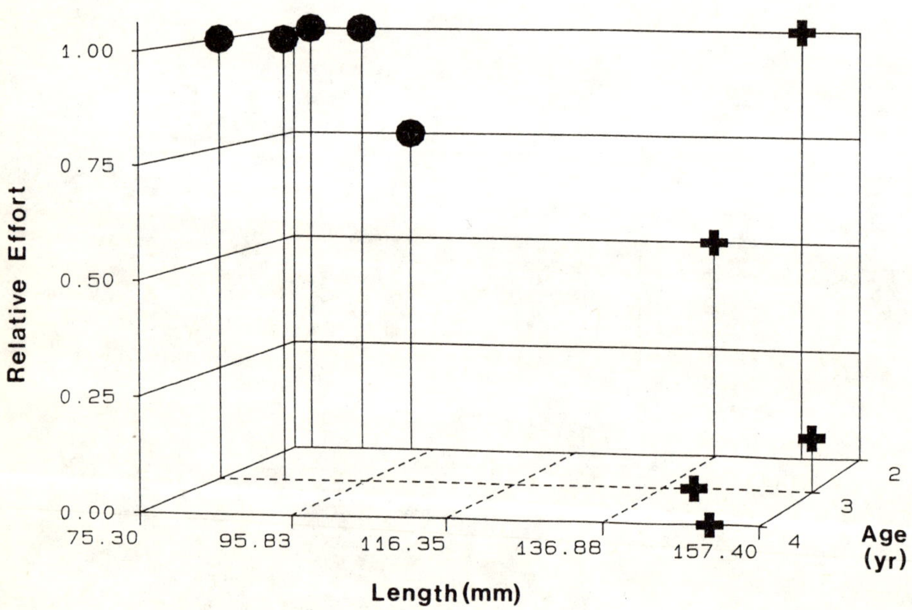

Figure 5. Three-dimensional norms of reaction for age, effort and size at first reproduction for female brook trout in Freshwater (circles) and Cripple Cove (crosses) populations.

Reaction norms among reproductive age, effort and size differed between Freshwater and Cripple Cove populations (Table 3; Figure 5). The dominant feature of the reaction norm for Freshwater was that, regardless of age and size at first reproduction, females were generally favoured to maximize effort. Alternatively, among Cripple Cove females, increased age at first reproduction was strongly associated with reduced effort although neither was associated with reproductive size.

EFFECTS OF SIZE-SELECTIVE MORTALITY ON REPRODUCTIVE AGE AND HARVESTABLE BIOMASS

The simulated influence of size-selective mortality on Cripple Cove population had its largest effect on reproductive effort. Compared with the unexploited population, in which only the fastest growing individuals were favoured to maximize effort, all but the slowest growing individuals were favoured to maximize effort in the population experiencing high fishing mortality (Table 3). Alternatively, mean age at first reproduction was relatively unaffected by increased fishing pressure regardless of whether reaction norm data (*RND*) were included (Figure 6A). Relative to the unexploited situation, mean reproductive age under high fishing pressure declined by 7 per cent when *RND* were included and by 3 per cent when *RND* were excluded. Harvestable biomass declined with increasing fishing mortality although the rate of decline depended largely on whether *RND* were included (Figure 6B). Relative to the unexploited population, harvestable biomass in a heavily fished population was predicted to decline 37 per cent when *RND* were included but by only 21 per cent when *RND* were excluded. The increased rate of decline was most evident between medium and high levels of fishing mortality as the decline in harvestable biomass when *RND* were included (22 per cent) was almost three times the decline in biomass when *RND* were excluded (8 per cent).

DISCUSSION

The reaction norms described here indicate that as the environmental prospects for individual growth improve, female brook trout should respond by reproducing earlier in life and by either maintaining (Freshwater) or increasing (Cripple Cove) reproductive effort to near-maximum levels. Increased growth rate is associated with increased size at reproduction in Freshwater but not in Cripple Cove. The reaction norm for age and size at reproduction for Freshwater females (i.e. rapid growth leading to reduced age and increased size) is a relatively common one (Stearns and Koella 1986).

It is not surprising that the three-dimensional reaction norms for reproductive age, effort and size differ between populations because they are based upon population-specific, demographic parameters. In Freshwater, where survival for reproductive and non-reproductive individuals is low, effort is predicted to be maximized regardless of age and size. Alternatively, in Cripple Cove, where adult survival is relatively high, increased age at first reproduction is associated with reduced effort. The greater variation in optimal effort across environments in Cripple Cove suggests that the scope for genetic change in the shapes of the reaction norms may be greater in this population than in Freshwater. Such a genetic change can be expected to occur when the population is exposed to increasing levels of size-selective fishing mortality. The main effect of these evolutionary responses is to change the shape of the reaction norm for effort such that maximum effort becomes favoured across most ages at first reproduction. The simulations suggest that this evolutionary shift in the shape of the reaction norm results in a reduction in harvestable biomass with increasing harvesting pressure that is greater than would occur if reaction norms for reproductive age, effort and size did not exist.

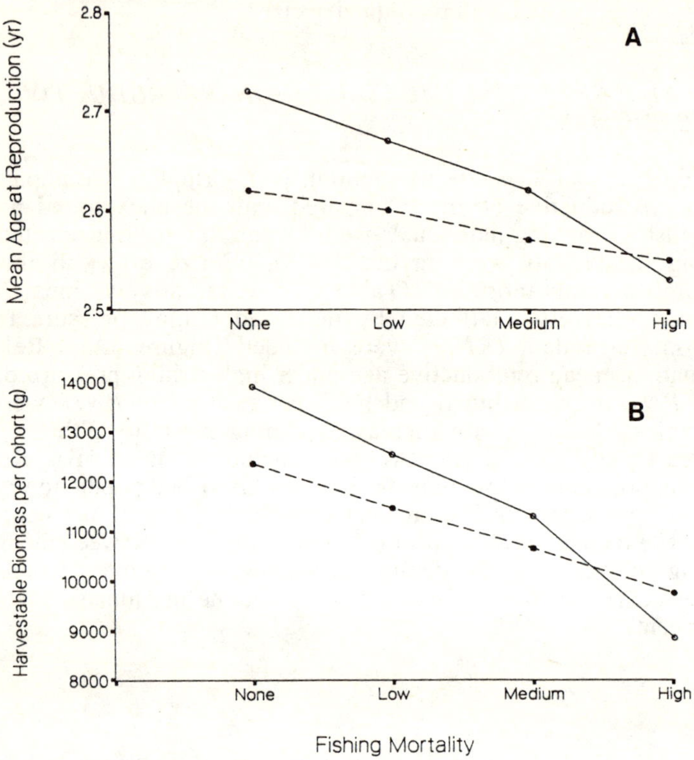

Figure 6. Hypothetical declines in mean reproductive age (A) and harvestable biomass (B) of female brook trout with increasing fishing mortality in Cripple Cove River (solid line - reaction norm data included; dashed line - reaction norm data excluded).

The existence of reaction norms for life history traits implicitly assumes that individuals can optimally time reproduction and allocate effort on an individual, growth-dependent basis. The environmental conditions necessary for the evolution of reaction norms certainly exist for brook trout, and probably for many other salmonids. At the beginning of exogenous feeding, trout emerging from the nest sites within the stream substrate are distributed largely at random with respect to genotype within and among spatially distinct juvenile habitats (small, shallow pools, separate from the main stream). For three Cape Race populations, this spatial environmental heterogeneity is sufficient to effect within-population differences in size among 2-month-old juveniles of 71 and 37 per cent within and among habitats, respectively (unpublished data). These habitat-induced differences in growth rate early in life lead to significant within-population differences in length by the end of the first year of life (e.g. 56 and 103 per cent for Freshwater and Cripple Cove, respectively). Environmental variation in juvenile habitat indirectly influences variation in fitness through associations between body size and fecundity, egg size, *GSI*, and breeding success (*inter alia* Thorpe et al. 1984, Hutchings and Morris 1985, van den Berghe and Gross 1989, Hutchings 1990). Body size also influences age-specific survival through differential competitive success (Hutchings 1991), ability to survive winter (Hutchings 1993), and susceptibility to predation (Reznick et al. 1990). Environmental uncertainty in juvenile habitat among generations coupled with its concomitant size-dependent effects on fitness should favour the evolution of reaction norms for reproductive age, effort and body size in stream-dwelling salmonids.

The evolution of norms of reaction for life history traits depends on the magnitude of the trait heritabilities and of the heritability of the phenotypic response of each trait, alone and in concert with other traits, to changes in the environment (Via and Lande 1985, Gomulkiewicz and Kirkpatrick 1992). Additive genetic variation exists for age and size at reproduction in salmonids and it seems improbable that such variation does not exist for reproductive effort, given that fecundity and egg size can also be heritable (cf. Gjedrem 1983, Thorpe et al. 1983, Gjerde 1984, 1986, Robison and Luempert 1984). In salmonids, the actual number of fully developed ova spawned by a female in autumn is low relative to the number of primary oocytes originally produced because most of the primary oocytes, those in which yolk is not produced, are resorbed by the female during the early stages of maturation (Vladykov 1956). It would seem reasonable to assume that reproductive effort could be under genetic control if there was a genetic 'switch' that controlled the production of a hormone such as vitellogenin (the hormone that initiates yolk production [Mommsen and Walsh 1988]). Such a switch could provide a wide range of effort levels. Vladykov (1956) calculated the *GSI* that would be expected of two female brook trout collected from two lakes in Quebec, Canada, assuming that every primary oocyte became a fully developed ovum at a size equal to the mean observed in these populations. His *GSI* estimates of 0.45 and 0.30 for these females, which measured 185 mm and 270 mm, respectively, were 50 to 225 per cent greater than the maximum *GSI* observed in each population. The assumption that there is additive genetic variation for the interaction between effort and body size is not unreasonable given that phenotypic plasticity in other organisms is genetically variable and will respond to selection (Gebhardt and Stearns 1988, Scheiner and Lyman 1989, Stearns 1992).

The reaction norms for reproductive effort described here are qualitatively valid for Freshwater and Cripple Cove populations. They suggest that the optimal strategy for Freshwater females is semelparity while that of Cripple Cove females is iteroparity. These predictions conform to empirical field data. Mean post-reproductive survival of Freshwater females (13 per cent) is 68 per cent less than that of Cripple Cove females (40 per cent). Furthermore, the low post-reproductive survival observed for small and large individuals in each population (cf. Figure 1) conforms with the predictions, based upon the shapes of the reaction norms, of high effort for young (small) and old (large) individuals.

The simulations of size-selective harvesting suggest that increasing fishing mortality among the larger size classes within a population will favour increased reproductive effort at a reduced age at first reproduction. These responses to fishing mortality support the predictions of life history theory that increased adult mortality (relative to juvenile mortality) favours reduced age at maturity and increased effort (Gadgil and Bossert 1970, Schaffer 1974, Law 1979). An abundance of correlational data suggest that long-term declines in age and size at maturity in commercially harvested salmonid populations can be attributed to size-selective harvesting practises (e.g. Ricker 1981, Bielak and Power 1986, Riddell 1986, Thorpe in press). Harvesting-induced increases in reproductive effort have been documented by Edley and Law (1988) who found that continuous removal of large individuals from clonal mixtures of *Daphnia magna* produced genetically-based increases in reproductive value at earlier ages at reproduction. Perhaps the best evidence that such responses can occur in natural populations was provided by Reznick et al. (1990). They used introduction experiments in the field to show that predator-induced mortality on large individuals selected for increased reproductive effort (higher *GSI* through increased fecundity) at reduced age at maturity within a population of guppies, *Poecilia reticulata*, in Trinidad.

It could be argued that reaction norms are unnecessary for modelling the effects of size-selective harvesting because models which exclude reaction norm data predict the same life history responses of harvested populations, i.e. increased effort at an earlier reproductive age. But while reaction norms may not alter the predicted effects of size-selective harvesting, they may affect the rate at which such changes are realized (and conversely the rate at which previously harvested populations return to their pre-harvested state).

The main effects of size-selective harvesting on a salmonid population are two-fold. The first effect will be a reduction in juvenile density resulting from the presence of fewer spawning adults (Myers et al. 1986, Thorpe in press). One immediate effect of reduced density is increased juvenile growth across all size classes. If reaction norms for reproduction exist, this increased growth rate might result in earlier age and increased effort at reproduction. The response should be an immediate one because the genetic framework for such a response - the norm of reaction - already exists. Secondly, reduced juvenile density may also increase juvenile survival. This increase in juvenile survival, coupled with harvesting-induced reductions in adult survival, will select for increased effort across all ages and sizes at reproduction (cf. Charlesworth 1980), resulting in an evolutionary change in the shape of the reaction norm. Such an evolutionary change may retard the rate at which a previously harvested population can return to its pre-harvested state. Thus, if reaction norms for reproduction exist, responses to changes in growth rate may be relatively rapid as they would be facilitated by a pre-existing framework for optimal changes in life history. Optimal evolutionary responses of an unexploited population to increased adult mortality and/or increased juvenile survival may be relatively slow because changes in age-specific survival would necessitate changes in the shapes of the pre-exploitation reaction norms.

In summary, the random distribution of genotypes across heterogeneous habitats can effect adaptive phenotypic plasticity in life history through selection for norms of reaction. Optimal, individual responses to changes in growth rate and mortality in natural populations can be predicted from the shapes of reaction norms for age, effort and size at first reproduction. Predicted evolutionary changes in reproductive age and effort in exploited fish populations are sensitive to the assumption that fish optimally allocate resources to reproduction on an individual, size-dependent basis. Empirical support for individual optimization in fish, its implications for demographic models of fish populations, and its importance to models of life history evolution warrant further study.

ACKNOWLEDGEMENTS

Financial support was provided by the Natural Sciences and Engineering Research Council of Canada (NSERC) through a Postdoctoral Fellowship and a Postgraduate Scholarship, and through an Operating Grant to Douglas W. Morris. An anonymous reviewer provided helpful comments on an earlier version of the manuscript.

REFERENCES

Bell, G. 1980. The costs of reproduction and their consequences. American Naturalist **116**: 45-76.

Bielak, A.T. and G. Power. 1986. Changes in mean weight, sea-age composition, and catch-per-unit-effort of Atlantic salmon (*Salmo salar*) angled in the Godbout River, Quebec, 1859-1983. Canadian Journal of Fisheries and Aquatic Sciences **43**: 281-287.

Bradshaw, A.D. 1965. Evolutionary significance of phenotypic plasticity in plants. Advances in Genetics **13**: 115-155.

Brett, J.R., J.E. Shelbourn and C.T. Shoop. 1969. Growth rate and body composition of fingerling sockeye salmon, *Oncorhynchus nerka*, in relation to temperature and ration size. Journal of the Fisheries Research Board of Canada **26**: 2363-2394.

Charlesworth, B. 1980. Evolution in age-structured populations. Cambridge University Press, Cambridge.

Cunjak, R.A. 1988. Physiological consequences of overwintering in streams: the cost of acclimatization? Canadian Journal of Fisheries and Aquatic Sciences **45**: 443-452.

Dufresne, F., G.J. Fitzgerald and S. Lachance. 1990. Age and size-related differences in reproductive success and reproductive costs in threespine stickleback (*Gasterosteus aculeatus*). Behavioural Ecology **1**: 140-147.

Edley, M.T. and R. Law. 1988. Evolution of life histories and yields in experimental populations of *Daphnia magna*. Biological Journal of the Linnean Society **34**: 309-326.

Elliott, J.M. 1976. Body composition of brown trout (*Salmo trutta* L.) in relation to temperature and ration size. Journal of Animal Ecology **45**: 273-289.

Gadgil, M. and W.H. Bossert. 1970. Life historical consequences of natural selection. American Naturalist **104**: 1-24.

Gebhardt, M.D. and S.C. Stearns. 1988. Reaction norms for developmental time and weight at eclosion in *Drosophila mercatorum*. Journal of Evolutionary Biology **1**: 335-354

Gjedrem, T. 1983. Genetic variation in quantitative traits and selective breeding in fish and shellfish. Aquaculture **33**: 51-72.

Gjerde, B. 1984. Response to individual selection for age at maturity in Atlantic salmon. Aquaculture **38**: 229-240.

Gjerde, B. 1986. Growth and reproduction in fish and shellfish. Aquaculture **57**: 37-55.

Gomulkiewicz, R. and M. Kirkpatrick. 1992. Quantitative genetics and the evolution of reaction norms. Evolution **46**: 390-411.

Hutchings, J.A. Submitted. Costs of reproduction within and among populations of brook trout, *Salvelinus fontinalis*.

Hutchings, J.A. 1993. Adaptive life histories effected by age-specific survival and growth rate. Ecology **74**: 673 - 684.

Hutchings, J.A. In press. Behavioural implications of intraspecific life history variation. In F.A. Huntingford and P. Toricelli eds. The Behavioural Ecology of Fishes. Harwood, London.

Hutchings, J.A. 1990. The evolutionary significance of life history divergence among brook trout, *Salvelinus fontinalis*, populations. PhD Thesis, Memorial University of Newfoundland, St. John's, Newfoundland.

Hutchings, J.A. 1991. Fitness consequences of variation in egg size and food abundance in brook trout, *Salvelinus fontinalis*. Evolution **45**: 1162-1168.

Hutchings, J.A. and D.W. Morris. 1985. The influence of phylogeny, size and behaviour on patterns of covariation in salmonid life histories. Oikos **45**: 118-124.

Jones, J.W. 1959. The salmon. Collins, London.

Law, R. 1979. The cost of reproduction in annual meadow grass. American Naturalist **113**: 3-16.

Lessells, C.M. 1991. The evolution of life histories. Pp. 32-68, in Behavioural Ecology (3rd. ed.) J.R. Krebs and N.B.Davies eds., Blackwell, Oxford.

Mommsen, T.P. and P.J. Walsh. 1988. Vitellogenesis and oocyte assembly. Pp. 347-406, in W.S. Hoar and D.J. Randall eds. Fish Physiology. Vol. XI. The Physiology of Developing Fish. Part A. Academic Press, New York.

Myers, R.A., J.A. Hutchings and R.J. Gibson. 1986. Variation in male parr maturation within and among populations of Atlantic salmon, *Salmo salar*. Canadian Journal of Fisheries and Aquatic Sciences **43**: 1242-1248.

Prout, T. 1980. Some relationships between density-independent selection and density dependent population growth. Evolutionary Biology **13**: 1-68.

Reiter, J. and B.J. Le Boeuf. 1991. Life history consequences of variation in age at primiparity in northern elephant seals. Behavioural Ecology and Sociobiology **28**: 153-160.

Reznick, D.A., H. Bryga and J.A. Endler 1990. Experimentally-induced life history evolution in a natural population. Nature **346**: 357-359.

Ricker, W.E. 1981. Changes in the average size and average age of Pacific salmon. Canadian Journal of Fisheries and Aquatic Sciences **38**: 1636-1656.

Riddell, B.E. 1986. Assessment of selective fishing on the age at maturity in Atlantic salmon (*Salmo salar*): a genetic perspective. Pp. 102-109, in D.J. Meerburg ed. Salmonid Age at Maturity. Canadian Special Publication of Fisheries and Aquatic Sciences **89**.

Robison, O.W. and L.G. Luempert. 1984. Genetic variation in weight and survival of brook trout (*Salvelinus fontinalis*). Aquaculture **38**: 155-170.

Roff, D.A. 1984. The evolution of life history parameters in teleosts. Canadian Journal of Fisheries and Aquatic Sciences **41**: 989-1000.

Rowe, D.K. and J.E. Thorpe. 1990. Differences in seasonal growth between maturing and non-maturing male Atlantic salmon (*Salmo salar* L.) parr. Journal of Fish Biology **36**: 643-658.

Schaffer, W.M. 1974. Selection for optimal life histories: the effects of age structure. Ecology **55**: 291-303.

Scheiner, S.M. and R.F. Lyman. 1989. The genetics of phenotypic plasticity. I. Heritability. Journal of Evolutionary Biology **2**: 95-107.

Schmalhausen, I.I. 1949. Factors of evolution. Blakiston, Philadelphia.

Schmidt-Nielson, K. 1984. Scaling: why is animal size so important? Cambridge University Press, Cambridge.

Sibly R. and P. Calow. 1983. An integrated approach to life-cycle evolution using selective landscapes. Journal of Theoretical Biology **102**: 527-547.

Stearns, S.C. 1992. The evolution of life histories. Oxford University Press, Oxford.

Stearns, S.C. and J.C. Koella. 1986. The evolution of phenotypic plasticity in life-history traits: predictions of reaction norms for age and size at maturity. Evolution **40**: 893-913.

Thorpe, J.E. In press. Impacts of fishing on genetic structure of salmonid populations. In Genetic Conservation of Salmonid Fishes, NATO.

Thorpe, J.E., R.I.G. Morgan, C. Talbot and M.S. Miles. 1983. Inheritance of developmental rates in Atlantic salmon, *Salmo salar* L. Aquaculture **33**: 119-128.

Thorpe, J.E., M.S. Miles and D.S. Keay. 1984. Developmental rate, fecundity and egg size in Atlantic salmon, *Salmo salar* L. Aquaculture **43**: 289-305.

van den Berghe, E.P. and M.R. Gross. 1989. Natural selection resulting from female breeding competition in a Pacific salmon (coho: *Oncorhynchus kisutch*). Evolution **43**: 125-140.

Via, S. and R. Lande. 1985. Genotype-environment interaction and the evolution of phenotypic plasticity. Evolution **39**: 505-522.

Vladykov, V.D. 1956. Fecundity of wild speckled trout (*Salvelinus fontinalis*) in Quebec lakes. Journal of the Fisheries Research Board of Canada **13**: 799-841.

Ware, D. M. 1980. Bioenergetics of stock and recruitment. Canadian Journal of Fisheries and Aquatic Sciences **37**: 1012-1024.

PART III
SELECTION RESPONSES

TRADE-OFFS AND GENETIC CORRELATIONS AMONG LIFE-HISTORY TRAITS: THEORY AND SIMULATION

R H Smith[1], *L A Linton*[2] & *R M Sibly*[3]

[1] School of Biological Sciences, University of Leicester, Leicester LE1 7RH, UK

[2] Department of Biological Sciences, The University of Calgary, 2500 University Drive NW, Calgary, Alberta, Canada T2N 1N4

[3] Department of Pure & Applied Zoology, University of Reading, Reading RG6 2AJ, UK

ABSTRACT

In this paper, we attempt to illuminate the relationship between life-history trade-offs and genetic constraints. We first review the main models that provide equilibrium solutions and then illustrate the dynamics of the evolution of genetic correlations using computer simulation.

INTRODUCTION

Genetic correlations have been well-known for many years to constrain the progress of artificial selection (Falconer, 1989). However, their role in constraining natural selection has only become clear more recently. In an authoritative series of papers considering the evolutionary implications of pleiotropy, Lande (1979, 1980) assumes weak stabilizing selection such that fitness falls off as the square of the distance of a phenotype from the optimum. In Lande's conception a population moves uphill on an adaptive landscape obtained by plotting fitness as a function of phenotypic variables. In a later paper Lande (1982) considers the implications of a constraint of negative correlation between two (or more) characters. Lande points out that (provided the population is not too small) directional selection will rapidly fix pleiotropic combinations of genes with positive effects on the components of fitness and leave segregating those genes with opposing effects on the components of fitness. However, Lande (1980, 1982) warns that additive genetic variance and covariance may change rapidly following a sudden alteration of the pattern of selection. Rapidly here means within a few thousand generations but one would expect the changes to be much faster if population size is small. Lande's work has led some (Reznick 1985, Reznick et al. 1986) to contend that genetic correlation is the only valid evidence for a cost of reproduction. Reznick et al. (1986) argue that in some circumstances genetic correlations represent fixed traits of a population.

However, Rose (1984) has shown that inbreeding may produce artifactual positive genetic correlations between fitness components and Service and Rose (1985) and Bell and Koufopanou (1986) argue that novel environments may produce the same result. In support of their claim Service and Rose (1985) demonstrate a reduction in negative correlation after fruit flies, *Drosophila melanogaster*, were introduced to a new environment differing in temperature, food and photoperiod from a 'standard' environment in which the flies had previously been kept for 80 generations. Effects such as these may explain the positive genetical correlations sometimes reported (eg Bell 1984a, 1984b, Bell and Koupofanou 1986).

Much of modern evolutionary ecology, and in particular life-history theory, is underpinned by the notion of optimisation within developmental and physiological constraints. Trade-offs are the constraints that limit life-history evolution and optimization. They mostly stem from the laws of biochemistry and physics but include also genomic, developmental, physiological and behavioural constraints on what is feasible for an organism in practice. However, although the trade-off concept is fundamental to most theories of life-history evolution, specifying exactly what the concept means, let alone demonstrating a trade-off empirically, has proved difficult (Reznick, 1985). Smith, Sibly and Møller (1987) distinguished between a tactical (physiological) and a genetic trade-off and suggested how a genetic trade-off might be investigated using the techniques of quantitative genetics (Falconer 1989; Reznick 1985; Rose and Charlesworth 1981a,b). Møller, Smith and Sibly (1989 a,b; 1990) provided evidence for both genetic and physiological trade-offs in the cowpea weevil *Callosobruchus maculatus* (F), using the techniques of quantitative genetics, environmental manipulation and laboratory natural selection.

It is unfortunate that there seems to have developed a gap in communication between evolutionary ecologists and population geneticists, apparently because of differences in the jargon used to describe what are essentially the same concepts. Smith, Sibly and Møller (1987) attempted to reconcile some of these differences by showing how the trade-off concept is related to patterns of genetic correlations and pleiotropy. In particular, it was suggested that the term *genetic trade-off* might usefully link the 'negative pleiotropy' of population geneticists (Rose and Charlesworth 1981) with the 'trade-off' of the evolutionary ecologists (Sibly and Calow 1986). More recently, Charnov (1989) and Charlesworth (1990) have demonstrated explicitly the links between mathematically defined trade-offs and the genetic variances and covariances at equilibrium between the fitness components involved in the trade-offs. Their results are discussed by Smith (1991) and will be summarised below by reference to the generalised life-cycle model of Sibly and Calow (1983, 1986). Both the optimisation approach and the genetic formulation make predictions about equilibrium conditions (evolutionary outcome). The approach (or otherwise) to equilibrium has not been described theoretically, except to the extent that very short-term predictions can be made from the selection gradient and the genetic variance-covariance matrix (see below) (Lande 1979, 1982; Lande and Arnold 1983; Caswell 1989a,b). As emphasised by Charnov (1989), the genetic covariance structure is not fixed and is likely to evolve as selection proceeds; this evolutionary process is studied numerically in the simulation study described after the next section.

MODELS OF LIFE-HISTORY EVOLUTION AT SELECTIVE EQUILIBRIUM

Both optimization and quantitative genetic models use the concept of Darwinian fitness: this is defined in the next section, and then the two types of model are outlined and contrasted.

FITNESS

The primary components of fitness are the probability l_x that an individual survives from age 0 to age x, and the fecundity m_x at age x ($0 \leq l_x \leq 1$ and $m_x \geq 0$). Feasible combinations of l_x and m_x define the options set (sometimes called the fitness set (Levins 1968) or trade-off (Charnov 1982)).

The combination of l_x and m_x that is expected to emerge under natural selection is that which maximises fitness. The general consensus is that the Malthusian parameter or intrinsic rate of increase r is the most appropriate measure of fitness (Fisher 1930; Charlesworth 1980, 1984; Lande 1982; Sibly and Calow 1983; Sibly 1989; Parker and Maynard Smith 1990), r being defined as the root of the Euler-Lotka equation:

$$1=\sum_{x} e^{-rx} l_x m_x \qquad (1)$$

Population geneticists conventionally refer to fitnesses of genotypes (eg Charlesworth, 1984) though Sibly and Calow (1986) specifically defined r as the rate of increase of a dominant gene. Fisher (1930) derived his Fundamental Theorem of Natural Selection using the average fitnesses of alleles (eg p37) and Crow (1986, p84) is one of the few to state explicitly that the genetic variance referred to by Fisher is the variance of the average fitnesses of alleles rather than of genotypes. It can be shown that equivalent equilibrium results are obtained whether models are formulated using average fitnesses or genes or fitnesses of genotypes (Sibly and Curnow, in press). Here, we will refer to fitness in relation to genes rather than genotypes.

OPTIMAL LIFE HISTORY

Consider a simplified life cycle where m female offspring are produced at a breeding attempt, survivorship of offspring to the first breeding attempt is s_1 and survivorship between breeding attempts is s_2. If the time between breeding attempts is the same as the time to first breeding, fitness r is related to the three life-cycle variables as follows (Charlesworth 1980, p226; Sibly and Calow 1983, 1986):

$$e^r = s_1 m + s_2 \qquad (2)$$

For simplicity we assume that males and females have the same life cycle and that juvenile survivorship s_1 is externally (environmentally) defined. From Equation (2), a given fitness r in a given environment characterised by juvenile survivorship s_1 is a linear combination of s_2 and m. Therefore lines of equal fitness with slope $-s_1$ can be plotted onto the (s_2,m) plane and the phenotype z^* with greatest fitness is the tangent of the fitness contour to the options set, whose boundary is defined by a trade-off curve or functional constraint $f_1(s_2,m)$ as illustrated in Figure 1a,b. Figure 1 shows that high juvenile survivorship should lead to evolution of semelparity, and low juvenile survivorship to iteroparity (Cole 1954; Bell 1980; Sibly and Calow 1983).

Formally, the optimal solution depends on both the trade-off constraint f_1 and the selection gradient ∇W (containing the partial derivatives $\partial r / \partial z_i$ of fitness r with respect to fitness components z_i). The selection gradient vector $\hat{u}W$ specifies the direction of steepest ascent up the surface of increasing fitness $r(z_i)$ and therefore relates to the progress as well as the outcome of natural selection. Charlesworth (1990) formally defined the optimal solution outcome (selective equilibrium at z^*) using the selection gradient ∇W and the constraints $f_i(z_i)$ on the life-history characters z_i.

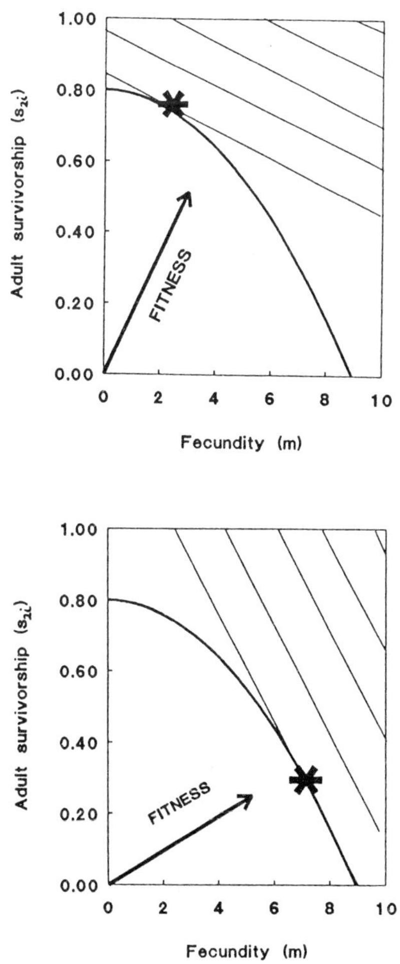

Figure 1. The options set is bounded by the trade-off curve. Optimal life history (starred) is the tangent of the fitness contour (straight lines with slope - s1) to the trade-off curve: (a) low juvenile survivorship

(s_1) favours low fecundity (m) and high adult survivorship (s_{2i}); (b) high juvenile survivorship (s_1) favours high fecundity (m) and low adult survivorship (s_{2i}).

GENETIC FORMULATION

A life-history character z_i that is a component of fitness r is assumed to be polygenically determined and to have an additive genetic variance V_{ii} and a covariance C_{ij} with any other characters z_j. The covariance of z_i with fitness is C_{ir}. Caswell (1989a) suggested that Robertson's (1968) Secondary Theorem of Natural Selection is a more useful generalisation than Fisher's Fundamental Theorem. The Secondary Theorem relates rate of change in mean phenotype to genetic covariance with fitness and shows why the selection gradient ∇W is important:

$$\Delta z_i = C_{ir} \qquad (3)$$
$$\cong V_{ii}(\partial r / \partial z_i$$

In the general multivariate case where **z** is a vector of mean values of a set of z_i and **G** is the genetic variance-covariance matrix of the z_i, Equation (3) becomes:

$$\Delta \mathbf{z} = \mathbf{G}\ \nabla \mathrm{W} \qquad (4)$$

At selective equilibrium, $\Delta \mathbf{z} = 0$ for all i which implies either that ∇W is a null vector (no selection) or that det $|\mathbf{G}| = 0$ (the genetic variance-covariance matrix is singular; Lande 1979, 1982; Lande and Arnold 1983). Charlesworth (1990) has highlighted some of the consequences of Equation (4).

OPTIMISATION VERSUS GENETICS

In the two variable case where there is a single functional constraint $z_2 = f(z_i)$, at selective equilibrium the population mean values are close to the optimisation predictions (Figure 1) and they should have a negative correlation of -1 as selection pushes up towards the constraint boundary. Charnov (1989) demonstrated this result using a linearised form of Fisher's Fundamental Theorem and obtained the same dynamical equations for phenotypic evolution as did Lande (1982) for the case of weak selection. The genetic variances and covariances must satisfy the form:

$$\mathrm{V}_{22} = \mathrm{V}_{11}\ (\partial f / \partial z_i)^{*2}$$

$$\mathrm{C}_{12} = -\mathrm{V}_{11}\ (\partial f / \partial z_i)^{*} \qquad (5)$$

where * denotes that a quantity is evaluated at selective equilibrium. Smith (1991) showed that for the simplified life-cycle above where fitness is defined by Equation (2), the theoretical equilibrium form of the genetic variances and covariances is:

$$\mathrm{Var},(s_2),= s_1^2\ \mathrm{Var}\ (m)$$

$$\mathrm{Cov},(s_2,m),= -s_1\ \mathrm{Var}\ (m)) \qquad (6)$$

This lays bare the link between the graphical optimality analysis of Sibly and Calow (1986) (see Figure 1) and the equilibrium genetic analysis. The slope of the genetic regression of s_2 on m near to equilibrium is $-s_1$, the same as the slope of the fitness contour that determines the optimal solution (Figure 1). Note that Equations (5) are equivalent to singularity of the genetic variance-covariance matrix for the two variable case (see Discussion). Putting together Equations (2) and (6), Smith (1991) showed that there is zero variance in e^r (and therefore in fitness r) at equilibrium even though there is genetic variance in the two components of fitness.

COMPUTER SIMULATION

In this section we use computer simulation to depict the progress of evolution pictorially (Figures 2-5) and to examine the changes that occur in genetic parameters in relation to the theoretical predictions described above. Most attention will be paid to the statistic used to describe the genetic trade-off empirically, namely the genetic correlation.

We also hope to illuminate previous discussion about genetic variance in fitness and its components (Smith, Sibly and Møller 1987). Gilbert (1984) argued that his demonstration of genetic variance in a component of fitness in the small white butterfly *Pieris rapae* contradicted Fisher's (1958) Fundamental Theorem of Natural Selection. However, Smith, Sibly and Møller (1987) noted that Fisher's arguments concerned genetic variance in fitness; because of negative genetic correlation (a genetic trade-off), there may be little or no genetic variance in fitness yet substantial genetic variance in component characters. This argument is presented in a simple pictorial way by Smith and Holloway (1988) and will be validated here using simple simulation.

The final aim also relates to variance in fitness. Since, for a given combination of life-history characters, fitness depends on environment, a change of environment can immediately and substantially increase the variance in fitness even though the components of fitness are not immediately affected. This will be demonstrated by two simulation experiments, one representing transplanting a population to a new environment and the other supplementing an environment with additional resources.

The basic structure of the deterministic simulation models is described by Linton, Sibly and Calow (1991). Adult survivorship between breeding attempts, s_2, is divided into two elements: an extrinsic element s_{2e}, which is the chance of adults surviving a constant, external cause of mortality; and an intrinsic element s_{2i}, which depends on genotype and is related to breeding effort (proportional to number of offspring per breeding attempt, m) by a trade-off function. Thus $s_2 = f(m)$.

OPTIONS SET

Unlike Linton et al. (1991), we consider genotypes within the options set in addition to genotypes along the boundary of the options set (the trade-off curve). We assumed that there were 100 codominant alleles at a single locus. This assumption was made for computational convenience and the results would have been qualitatively similar with, for example, 10 alleles at each of 10 loci provided the loci were unlinked and gene action was additive between alleles as well as between loci (i.e. no epistasis and no dominance). The diploid model was based on a curvilinear options set between adult survivorship s_2 and fecundity m, which for simplicity was taken as a quarter circle (Figure 2a). We defined three different environments, as follows:

Environment 1: $s_1 = 0.005$ - low juvenile survivorship;
Environment 2: $s_1 = 0.027$ - intermediate juvenile survivorship;
Environment 3: $s_1 = 0.047$ - high juvenile survivorship.

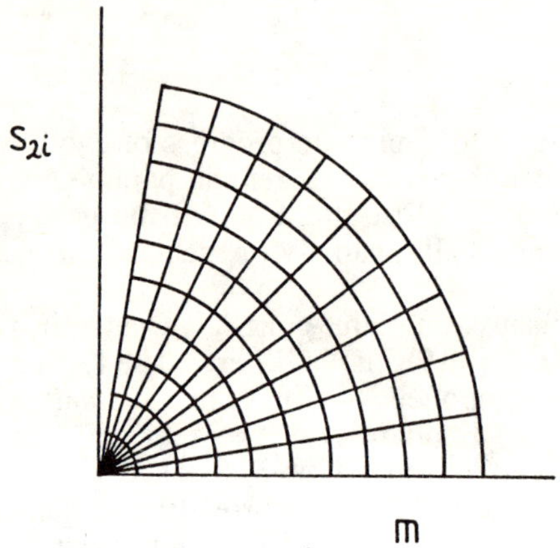

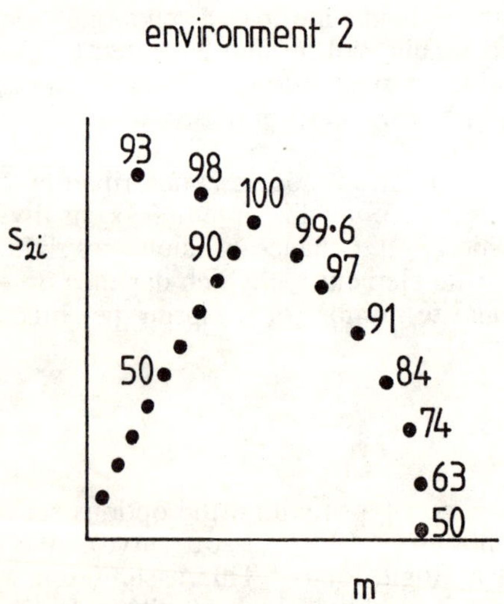

Figure 2. The options set used in genetic simulations: (a) shows 100 allelic values at the intersections of the chords and the arcs; (b) shows the relative fitnesses of homozygotes for 12 of the alleles, expressed as % of the fitness of the fittest homozygote in environment 2.

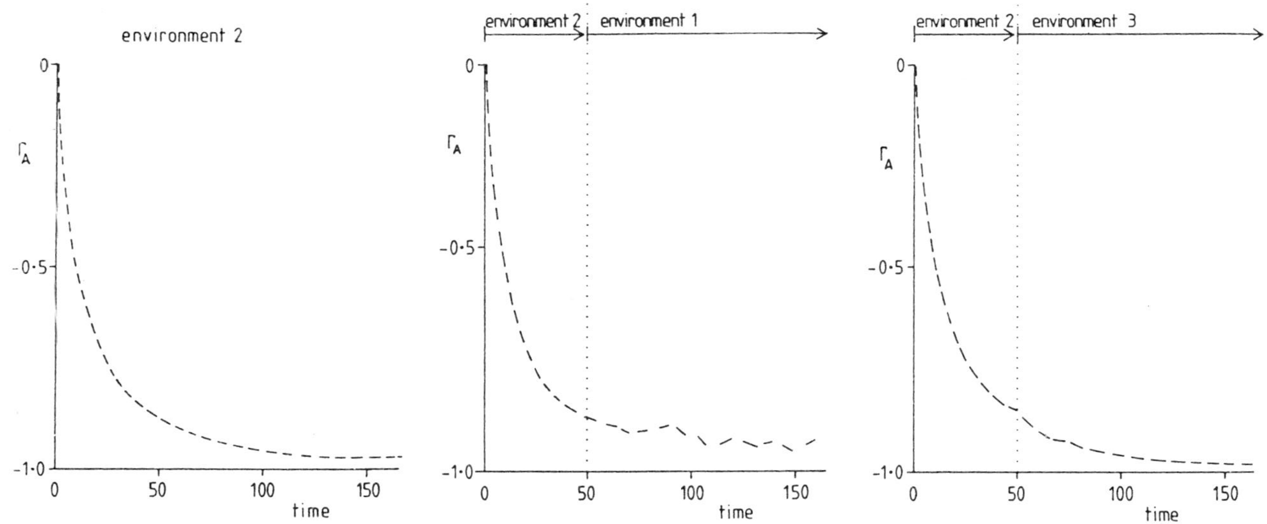

Figure 3. Evolution of the genetic correlation r_A through time: (a) environment 2; (b) a change in environment from 2 to 1 after 50 seasons; (c) a change in environment from 2 to 3 after 50 seasons.

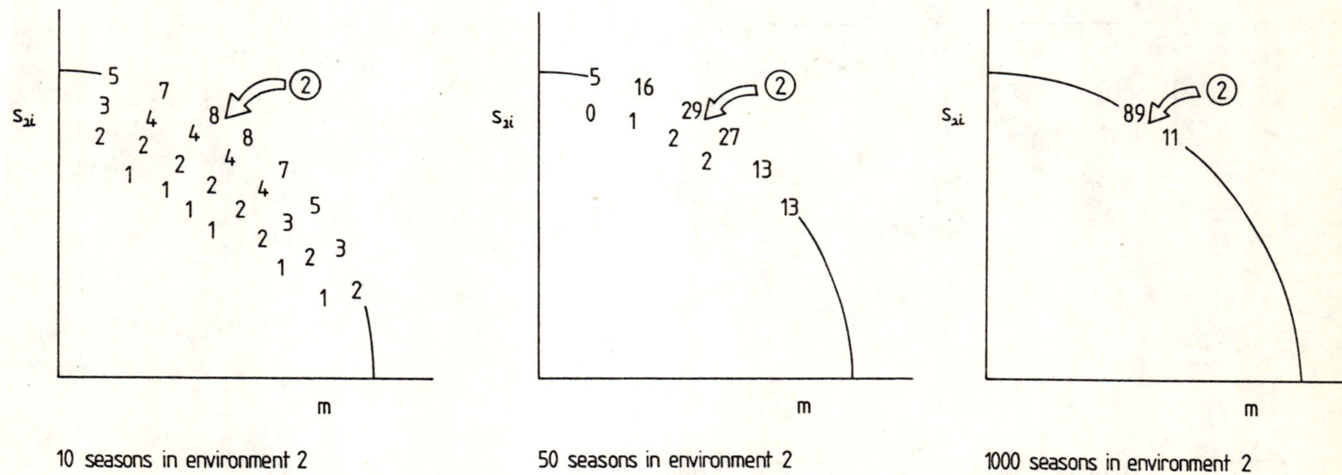

Figure 4. Changes in allele frequencies through time in environment 2. The numbers are frequencies (%) of all alleles present at 1per cent or higher after: (a) 10 seasons; (b) 50 seasons; (c) 1000 seasons. the optimal phenotype is shown by the arrow.

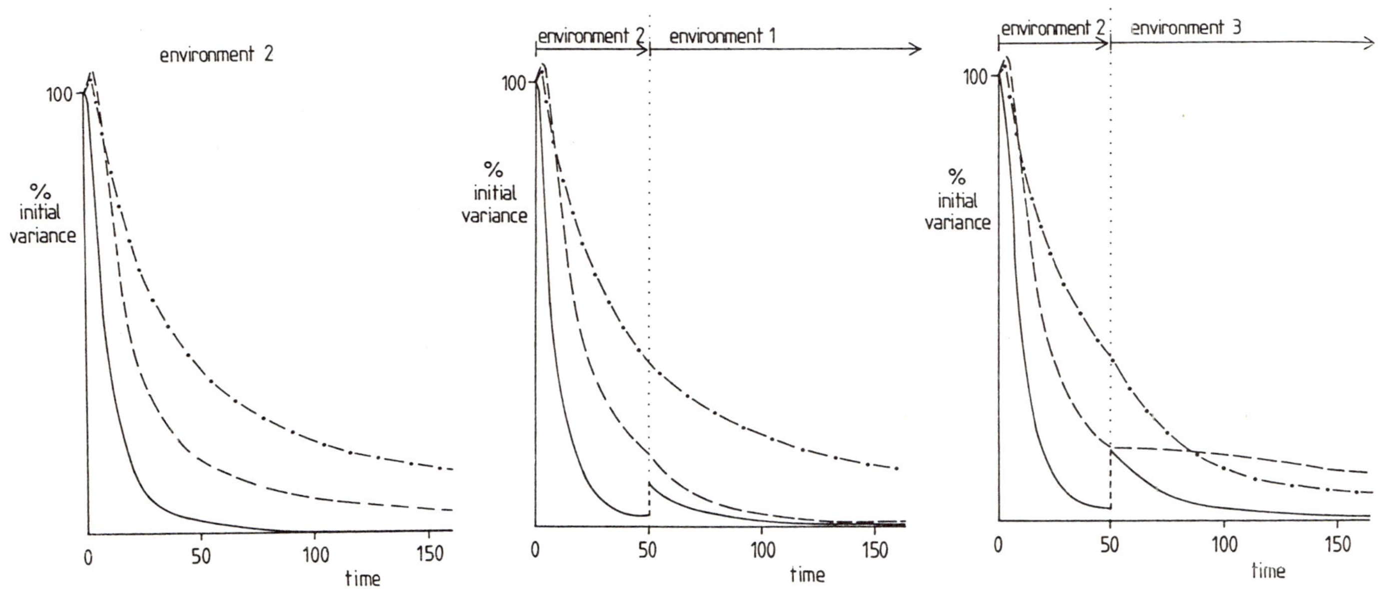

Figure 5. Changes in genetic variance through time (——— fitness (r); —·—· fecundity (m); — — adult survivorship (s_{2i})): (a) constant environment 2; (b) and (c) show the effect on genetic variances of changing environment to 1 and 3 respectively after 50 generations.

ALLELIC EFFECTS

The interpretation of the allelic effects is that an individual homozygous for a particular allele would have a phenotype (s_2, m) defined by the allelic position in Figure 2a. Alleles are codominant, and a heterozygote would have a phenotype exactly midway between its allelic positions. The 10 alleles giving $s_2 = 0$ correspond to equal intervals of m between 0 and the maximum value on the trade-off curve. The other 90 alleles correspond to equal interval positions on nine equiangular chords from the origin to the trade-off curve as shown in Figure 2a. The exact details are unimportant to the results; we simply chose to give even coverage of the options set along any line from the origin to the trade-off curve, and around any arc within the options set. Initially, all 100 alleles were present at a frequency of 0.01.

REPRODUCTIVE EFFORT AND SURVIVORSHIP

Reproductive effort was determined by the projection of the genotypic position onto the fecundity axis. No distinction was made between reproductive effort in male and female parents. Survivorship between time intervals (breeding attempts) was similarly determined. The relative contribution of an allele to the gene pool was determined by the fecundity m, the survivorship s_2 and the frequency of genotypes carrying that allele; thus there were no environmental effects but only additive genetic variation.

SIMULATION

The model was programmed in the same way as a 10 allele model described by Linton et al (1991). The computer program was essentially a massive book-keeping exercise and looped through a number of discrete time intervals (seasons) set by the user. Mating occurred at random with reproductive effort m of each of the 100^2 allele combinations averaged with reproductive effort of each possible partner to produce different offspring according to Mendelian principles. Zygotes were summed into a 100^2 array according to genotypes. For each genotype, the frequency in the next season was calculated from the number of zygotes expected to survive (according to the overall s_1) plus the number of adults expected to survive (according to the genotype-specific s_2). New allele frequencies were calculated from the frequency of genotypes surviving to the next time interval, standardised to sum to unity. The means, variances and covariances of s_2, m and fitness were calculated from allele frequencies, allelic values and Equation (1). Simulations were run for a set number of seasons (1000) rather than a set number of generations. Mean generation time varied with s_1,s_2 and population composition.

ENVIRONMENTAL CHANGE

Environmental change was simulated in two ways by a change after the first 50 seasons. In one, s_1 was altered as a representation of moving from one environment to another characterised by different juvenile survivorships. In another, m was doubled for each allele, representing a supplementation of an environment with additional resources that could be used in reproduction, i.e. a doubling of reproductive output for a given reproductive effort.

SIMULATION RESULTS

The results for the three constant environments were qualitatively the same and therefore only the results for environment 2 (intermediate juvenile survivorship) are presented. Figure 2b shows the relative fitnesses of the 10 different homozygotes (allelic values) along the trade-off curve and two along the chord between the origin and the homozygote with maximum fitness. Six of the 10 homozygotes along the trade-off curve have higher fitness than the homozygote nearest to the maximum inside the options set. It is therefore no surprise that a negative genetic correlation evolves rapidly as the alleles along the trade-off curve increase in frequency at the expense of those inside the options set.

The evolution of the negative genetic correlation through time is shown in Figure 3a. In all three environments the correlation was negative after one season and the value after 50 seasons in environments 1, 2 and 3 was - 0.66, -0.87 and -0.87 respectively, reaching -0.90 within 125 seasons even in environment 1 (where the negative correlation increased most slowly).

The way in which the negative correlations evolve is shown in Figure 4 for environment 2. Initially (time 0), all 100 alleles were present at equal frequency (1 per cent). After 10 seasons, 6 alleles on either side of the analytical optimum at the boundary of the options set are present in proportions exceeding 5 per cent, while another 21 alleles are present in proportions exceeding 1 per cent. After 50 seasons, only 4 alleles on either side of the optimum are present at frequencies exceeding 5 per cent and only 6 others at frequencies exceeding 1 per cent. As predicted, the probability of sampling a particular allele is becoming concentrated in a cloud centred on the theoretical optimum and aligned parallel to the trade-off curve at the optimum. After 1000 seasons, the allele closest to the optimum was present at a frequency of 88.5 per cent with the allele to the other side (lower survival, higher fecundity) at 11.5 per cent. The results for environments 1 and 3 showed the same patterns. In environment 3, where a single allele (the 'optimal' allele) coincided with the optimal combination of survival and reproductive effort, the frequency of the optimal allele reached 99.0 per cent after 1000 seasons. However, even after 500 seasons of selection, the frequency of the optimal allele had not reached 90 per cent and alleles on either side of the optimum were present at frequencies of 9.0 per cent and 1.8 per cent respectively.

The changes in genetic variances in environment 2 are shown in Figure 5a. The variances of the simulation data are plotted as percentages of their initial values. After initial transients (presumably while age structure is equilibrating), the variance in adult survivorship declines more rapidly than variance in fecundity, though both decline much less rapidly than variance in fitness. This shows how the apparent paradox of retention of variance in components of fitness (Gilbert 1984) may be resolved (Smith et al. 1987); the negative genetic correlation that evolves between components of fitness allows variance in those components to be retained while variance in fitness disappears under natural selection.

The effect of a changed environment was essentially the same whether the change was altered juvenile survivorship or supplementation of resources for reproduction; genetic variance in fitness increased immediately and substantially while the variance in components of fitness was of course not immediately changed. Fitness as defined in Equation (2) can be thought of as an index of selection (c.f. artificial selection; Falconer 1989). A change in one of the coefficients in that index (s_1) alters the relative strength of selection acting on the two variable components (s_2 and m). In consequence, the rate of decline of variance in s_2 and m changes (Figure 5b,c). For example, when juvenile survivorship increases after 50 seasons (environment 2 to environment 3; Figure 5c), natural selection has a relatively stronger effect on fecundity and the genetic variance in fecundity declines more rapidly than that of adult survivorship. The converse is true when juvenile survivorship decreases (Figure 5b). In all cases, however, variance in fitness declines the most rapidly; a demonstration of the evolutionary process described in Fisher's Fundamental Theorem.

The simulated dynamics of evolutionary change in an altered environment do have some unexpected features. The evolution of genetic correlations shows oscillations in time (Figure 3b,c) that are largest following a reduction in juvenile survivorship (environment 2 to environment 1). The oscillating dynamics may represent changes in equilibrium age structure. Environment 1 favours iteroparity, extending generation time and may be effectively delaying approach to a new equilibrium in a new environment. In these simulations the dynamics of evolution through time involve two dimensions in fitness space (Figure 1) and complex spatiotemporal dynamics are perhaps not too surprising.

Finally, it is of interest to compare the outcome of the simulations with the theoretical equilibrium solution (Equations (6); Charnov 1989). Table 1 shows that the linearised approximation is a good predictor of the simulation results except when the equilibrium is at the edge of the trade-off curve (Environment 1).

Table 1. Comparison of the observed and theoretical ratios of variances. The theoretical ratio is $\mathrm{Var}(s_2)/\mathrm{Var}\ (m) = s_1^2$ at equilibrium (Charnov 1989; Smith 1991). The observed variances in the simulations combine the intrinsic (genetically determined) value s_{2i} with the extrinsic element s_{2e} of adult survivorship. The ratios at time 0 differ in the different environments because of the inclusion of s_{2e} in adult survivorship.

Environment	Juvenile Survivorship (s_1)	Theoretical Ratio (s_1^2)	Time (Seasons)	Observed Ratio
1	0.005	0.025×10^{-3}	0	2.475×10^{-3}
			10	2.336×10^{-3}
			100	0.258×10^{-3}
			300	0.140×10^{-3}
			500	0.680×10^{-3}
2	0.027	0.729×10^{-3}	0	2.188×10^{-3}
			10	2.126×10^{-3}
			100	0.797×10^{-3}
			300	0.782×10^{-3}
3	0.047	2.209×10^{-3}	0	0.518×10^{-3}
			10	0.850×10^{-3}
			100	2.135×10^{-3}
			500	2.529×10^{-3}

DISCUSSION

We have chosen here to concentrate on constrained optimisation (the trade-off concept) as a cause of evolution of genetic correlations. The consequences of genetic correlations for evolution are considered here in relation to change in genetic variance and elsewhere in this volume in relation to fisheries and more general forms of population exploitation. Although a negative genetic correlation between components of fitness provides evidence of a trade-off constraint (Reznick 1985; Smith et al. 1987), recent theoretical developments show that a positive genetic correlation may arise when there are more than two components of fitness involved in trade-offs (Charlesworth 1990; Smith, 1991). In particular, models of hierarchical energy partitioning (de Laguerie et al. 1991; de Jong and van Noorduijk 1992; Stearns, de Jong and Newman 1991) show very elegantly how a sequential ordering of resource-allocation constraints in time can lead to positive as well as negative correlations between traits at equilibrium. Simple two-dimensional trade-off models miss these complexities completely (e.g. Sibly and Calow 1983) and it may be misleading to base general predictions about the signs of genetic correlations on two-trait models, as here. Our excuse here is that the computation involved in our two-trait simulations was difficult enough and that the objectives of our simulations were limited and well-defined. However, although a positive genetic correlation doe not disprove a trade-off hypothesis, a negative genetic correlation is good evidence of a trade-off constraint.

Selection always acts to reduce dimensionality (by reducing genetic variance in fitness) and indeed one test of selective equilibrium is that the genetic variance-covariance matrix of constrained components of fitness should be singular (Charlesworth, 1990; Smith, 1991). Smith (1991) calculated the genetic variance-covariance matrix from data presented by Møller, Smith & Sibly (1989a) for the cowpea weevil *Callosobruchus maculatus*. It is of interest to see whether the structure of the matrix indicates selective equilibrium and Richard Law and David Grey (pers comm) and R H Smith (unpublished) have independently used simple ordination to estimate the dimensionality of the cowpea weevil's genetic space (essentially the multivariate scatter of breeding values; Falconer 1989). Using correlations rather than covariances to overcome the problem of units and scaling, it turns out that the 4x4 genetic correlation matrix (Table 4(b) of Smith, 1991) only has one significant dimension, which accounts for 92 per cent of the total genetic variance! This reduction from four measured variables to one effective dimension is strong evidence for the operation of trade-off constraints and the progress of direct selection on any one of the four variables would be severely constrained by genetic correlations with the others. Holloway, de Jong and Ottenheim (in press) also found reduced dimensionality in the breeding value space of male ladybirds but not in females, suggesting either that the constraints affecting males did not all affect females (Maynard Smith et al. 1985) or that there were other components of fitness involved in trade-offs in addition to those measured. Thus genetic correlations both reflect previous evolution and constrain future evolution, though patterns of genetic correlations will be difficult to interpret if they are based on incomplete sets of fitness components.

ACKNOWLEDGEMENTS

The computer models were developed at the Universities of Sheffield and Reading while Larry Linton was working with Professor Peter Calow. We are grateful to Richard Law and Kevin Stokes for comments on our manuscript.

REFERENCES

Bell G 1980. The costs of reproduction and their consequences. American Naturalist **11:** 45-76.

Bell G 1984a. Measuring the cost of reproduction. I. The correlation structure of the life table of a plankton rotifer. Evolution **38**: 300-313.

Bell G 1984b. Measuring the cost of reproduction. II. The correlation structure of the life tables of five freshwater invertebrates. Evolution **38**: 314-326.

Bell G & Koufopanou V 1986. The cost of reproduction. Oxford Surveys of Evolutionary Biology **3**: 83-131.

Caswell, H. 1989a. Life-history strategies. Pp 285-307, in J. M. Cherrett ed. Ecological Concepts. Blackwell, Oxford.

Caswell, H. 1989b. Matrix Population Models. Sinauer Associates, Sunderland, Ma.

Charlesworth, B. 1980. Evolution in Age-structured Populations. Cambridge University Press, Cambridge.

Charlesworth, B. 1984. The evolutionary genetics of life-histories. Pp. 117-133, in B. Shorrocks ed. Evolutionary Ecology. Blackwell, Oxford.

Charlesworth, B. 1990. Optimization models, quantitative genetics and mutation. Evolution **44**: 520-538.

Charnov, E. L. 1982. The Theory of Sex Allocation. Princeton University Press, Princeton.

Charnov, E. L. 1989. Phenotypic evolution under Fisher's Fundamental Theorem of Natural Selection. Heredity **62**: 113-116.

Cole, L. C. 1954. The population consequences of life history phenomena. Quarterly Review of Biology **19**: 103-137.

Crow, J. F. 1986. Basic Concepts in Population, Quantitative and Evolutionary Genetics. Freeman, New York.

Falconer, D. S. 1989. Introduction to Quantitative Genetics, 3rd edition, Longman, Harlow.

Fisher, R. A. 1930. The Genetical Theory of Natural Selection (Revised edition 1958, Dover Publications, New York). Oxford University Press, Oxford.

Gilbert, N. E. 1984. Control of fecundity in *Pieris rapae*. I. The problem. Journal of Animal Ecology **53**: 581-588.

Holloway, G. J., P. W. de Jong and M. Ottenheim. 1993. The genetics and cost of chemical defence in the 2-spot ladybird (*Adalia punctaka* L.). Evolution (in press).

de Jong, G. and A.J. van Noorduijk. 1993. Acquisition and allocation of resources: genetic covariances, selection and life histories. American Naturalist (in press).

de Laguerie, P., I. Olivieri, A. Atlan and P.-H. Gouyon. 1991. Analytic and simulation models predicting positive genetic correlations between traits linked by trade-offs. Evolutionary Ecology **5**: 361-369.

Lande, R. 1979. Quantitative genetic analysis of multivariate evolution, applied to brain:body size allometry. Evolution **33**: 402-416.

Lande, R. 1980. The genetic covariance between characters maintained by pleiotropic mutations. Genctics **94**: 203-215.

Lande, R. 1982. A quantitative genetic theory of life history evolution. Ecology **63**: 607-615.

Lande, R. and S. J. Arnold. 1983. The measurement of selection on correlated characters. Evolution **37**: 1210-1226.

Levins, R. 1968. Evolution in Changing Environments. Princeton University Press, Princeton.

Linton, L., R. M. Sibly and P. Calow. 1991. Testing life-cycle theory by computer simulation. I. Introduction of genetical structure. Computing in Biology and Medicine **21**: 345-355.

Maynard Smith, J., R. Burian, S. Kauffman, P. Alberch, J. Campbell, B. Goodwin, R. Lande, D. Raup and L. Wolpert. 1985. Developmental constraints and evolution. Quarterly Review of Biology **60**: 265-287.

Møller, H., R. H. Smith and R. M. Sibly. 1989a. Evolutionary demography of a bruchid beetle. I. Quantitative genetical analysis of the female life history. Functional Ecology **3**: 673-681.

Møller, H., R. H. Smith and R. M. Sibly. 1989b. Evolutionary demography of a bruchid beetle. II. Physiological manipulations. Functional Ecology **3**: 683-691.

Møller, H., R. H. Smith and R. M. Sibly. 1990. Evolutionary demography of a bruchid beetle. III. Correlated responses to selection and phenotypic plasticity. Functional Ecology **4**: 489-483.

Parker, G. A. and J. Maynard Smith.1990. Optimality theory in evolutionary biology. Nature **348**: 27-33.

Reznick, D. N. 1985. Costs of reproduction: an evaluation of the empirical evidence. Oikos **44**: 257-267.

Reznick, D. N., E. Perry and J. Travis. 1986. Measuring the cost of reproduction: a comment on papers by Bell. Evolution **40**: 1338-1344.

Robertson, A. 1968. The spectrum of genetic variation. Pp. 5-16, in R C Lewontin ed. Population Biology and Evolution. Syracuse University Press, New York.

Rose, M. R. 1984. Genetic covariation in *Drosophila* life history: untangling the data. American Naturalist **123**: 565-569.

Rose, M. R. and B. Charlesworth. 1981a. Genetics of life history in *Drosophila melanogaster*. I. Sib analysis of adult females. Genetics **97**: 173-186.

Rose, M. R. and B. Charlesworth. 1981b. Genetics of life history in *Drosophila melanogaster*. II. Exploratory selection experiments. Genetics **97**: 187-196.

Service, P. M. and M. R. Rose. 1985. Genetic covariation among life-history components: the effect of novel environments. Evolution **39**: 943-945.

Sibly, R. M. 1989. What evolution maximizes. I **3**: 129-135.

Sibly, R. M. and P. Calow. 1983. An integrated approach to life cycle evolution using selective landscapes. Journal of Theoretical Biology **102**: 327-347.

Sibly, R. M. and P. Calow. 1986. Physiological Ecology of Animals: an Evolutionary Approach. Blackwell, Oxford.

Sibly, R. M. and R. Curnow. 1993. An allelocentric view of life-history evolution. Journal of Theoretical Biology (in press).

Smith, R. H. 1991. Genetic and phenotypic aspects of life-history evolution in animals. Advances in Ecological Research **2:** 63-120.

Smith, R. H. and G. J. Holloway. 1989. Population genetics relevant to insect introductions. The Entomologist **108**: 14-27.

Smith. R H, R. M. Sibly and H. Møller. 1987. Control of size and fecundity in *Pieris rapae*: towards a theory of butterfly life cycles. Journal of Animal Ecology **56**: 341-350.

Stearns, S., G. de Jong and R. Newman. 1991. The effects of phenotypic plasticity on genetic correlations. Trends in Ecology and Evolution **6**: 122-126.

THE EVOLUTION OF SIZE AND GROWTH IN HARVESTED NATURAL POPULATIONS

Mark Kirkpatrick

Department of Zoology, University of Texas, Austin, TX 78712 USA

ABSTRACT

Fishing and other forms of harvesting have genetic as well as demographic impact on natural populations. Quantitative genetics can provide predictions of the evolutionary changes in growth caused by harvesting. This paper briefly reviews recently developed methods useful for making these predictions. The approach is illustrated with data from rainbow trout. Results show that even moderate fishing pressure is expected to produce marked evolutionary changes in growth and fecundity within only a few generations. The changes can substantially decrease the economic value of the population. The quantitative genetic approach can be integrated with other current efforts to incorporate an evolutionary dimension into existing demographic models for the management of harvested natural populations.

INTRODUCTION

Civilization is entirely dependent on the products of artificial selection. Most of the world's food is derived from domesticated animals and plants which have been genetically modified through artificial selection to the point where they only vaguely resemble their wild progenitors. This process is ongoing: artificial selection continues to increase the means of traits in these populations at rates up to several per cent per year.

The remainder of our food is harvested from natural populations, primarily marine fisheries. Management of these natural populations presents a very different set of problems than does the harvesting of domesticated populations. In particular, we have little if any ability to manipulate the genetic composition of natural populations for our economic benefit. Major efforts have been made to understand the demographic consequences of harvesting, for example attempts to find the maximum sustainable yield that a population can withstand. The genetic dimension of human impact on these populations has received much less attention. There is, however, an immediate need for understanding here. In contrast to the situation in domesticated populations, where artificial selection produces large and continuing gains in the economic value of animals and plants, the genetic impact of harvesting in natural populations is often to reduce the value of the population. Fisheries, for example, selectively cull the largest individuals. This results in artificial selection for smaller size, which can cause evolutionary decreases in adult size. The selection produced by harvesting can also genetically alter life histories and other traits, leading to further decreases in the population's economic value.

To understand the genetic consequences of harvesting natural populations, we need a predictive theory of how growth trajectories and related traits evolve in response to harvesting. This paper outlines the methods that my colleagues and I have recently developed for studying the evolution of growth trajectories. The approach uses quantitative genetics to deduce the growth trajectories that will result from a given form of selection. Here I briefly review the model and the methods for obtaining the data needed to make predictions with it; details of the techniques can be found in the references. A future goal is to integrate this approach with recently developed models for management of natural populations that account for both the genetic and demographic effects of harvesting.

A MODEL FOR THE EVOLUTION OF GROWTH TRAJECTORIES

To predict the evolution of the growth trajectory in a population, our model requires knowing three things: the population's current mean growth trajectory, the genetic variation for the growth trajectory, and the form of selection acting on the growth trajectory. I will begin by describing the model assuming that this information is known, then discuss how those quantities can be estimated, and finally illustrate use of the model with data from rainbow trout.

The population's *mean growth trajectory*, denoted $\bar{z}(a)$, is simply the mean size of individuals at age a. The aim here is to predict how $\bar{z}(a)$ evolves. Following the standard approach of quantitative genetics (see Falconer 1981, Bulmer 1985), an individual's growth trajectory can be viewed as the sum of an additive genetic component and a second component consisting of environmental and nonadditive genetic effects. Both components are assumed to be normally distributed on some scale of measurement, perhaps after log-transforming the data (Kirkpatrick and Heckman 1989).

Genetic variation for the growth trajectory is be summarized by the population's *additive genetic covariance function*, $G(a_1, a_2)$. The value of (a_1, a_2) is equal to the additive genetic covariance in size between ages a_1 and a_2. A positive value of $G(a_1,a_2)$ reflects a positive genetic correlation for size at the two ages, while a negative value indicates the converse. The value of $G(a_1, a_1)$ equals the additive genetic variance for size at age a_1.

The pattern of directional selection on a growth trajectory is indicated by the *selection gradient function*, $\beta(a)$. The magnitude of $\beta(a)$ measures the strength of directional selection acting directly on size at age a; a positive value indicates selection for larger sizes while a smaller value shows selection favours smaller sizes (Lande and Arnold 1983).

Given these three quantities, the mean growth trajectory in the next generation is

$$\bar{z}'(a) = \bar{z}(a) + \int_0^{a_{\max}} G(a,x)\beta(x)dx \tag{1}$$

(Kirkpatrick and Heckman 1989). The integral on the right hand side represents the evolutionary change in the mean growth trajectory caused by one generation of selection. The amount of change is thus determined jointly by the amount and pattern of heritable variation for size (G) and by the strength and direction of selection acting on the growth trajectory (β).

An aside: While the discussion here focuses on growth trajectories, this model can also be used to study the evolution of reaction norms. Reaction norms share with growth trajectories the property that individuals can be represented by continuous functions (see Gomulkiewicz and Kirkpatrick 1992).

One of the strongest appeals of this model is that all the components of Equation (1) can be statistically estimated (Kirkpatrick et al. 1990, Kirkpatrick and Lofsvold 1992). Of the three quantities, the simplest is the current mean growth trajectory, $\bar{z}(a)$. The average size of a sample of individuals at each of several ages is calculated. A smooth interpolation through these means (using a method such as splines or polynomials) then produces a continuous estimate of $\bar{z}$. A small complication is that $\bar{z}$ represents the mean among unselected (unculled) individuals. To estimate $\bar{z}$ for a population that is being harvested, one can either use data on an unharvested subset of the population or statistically correct for the effect that harvesting has on the size distribution.

Estimation of the genetic covariance function G begins with the standard experimental and statistical methods that have been developed by quantitative geneticists. Data on growth trajectories of known relatives, for example parents and offspring, are used to estimate the additive genetic variances for size at each of several ages, and the additive genetic covariances between these ages (see Becker 1984). These give an estimate of the additive genetic covariance matrix. To adopt this matrix for use in Equation (1), a smooth function is fit to the matrix to produce an estimate of G, the additive genetic covariance function. For certain technical reasons there are strong advantages to using orthogonal functions such as Legendre polynomials for this purpose (see Kirkpatrick and Heckman 1989, Kirkpatrick et al. 1990, Kirkpatrick and Lofsvold 1992).

Obtaining the data on relatives needed for these genetic estimates is logistically impractical for many natural populations. In some populations, however, these data can be obtained from captive stocks or from recaptures of hatchery-bred individuals. Salmonids, for example, have great potential for this kind of study. It may even be possible to use the genetic data from well-studied species such as these in models for other species whose genetics cannot be studied directly. This approach would be justified if the patterns of genetic variation reflected in G remain stable for evolutionarily significant periods of time.

The final element needed in Equation (1) is the selection gradient function, $\beta(a)$. Two routes are available to estimate β. It can be directly measured from data on how size influences survival and reproductive success (Kirkpatrick et al. 1990). The returns from tagging programs can be used to determine the effect of size on survival, while size-fecundity relations can be used to estimate the impact of size on reproductive success.

Alternatively, the selection gradient can be calculated from a model of how size-selective harvesting culls individuals of different sizes. Fisheries typically catch only individuals above a size threshold. This results in a form of truncation selection whose selection gradient is:

$$\beta(a) \approx -F\,p(T,a) \Big/ \int_0^T p(x,a)\,\mathrm{d}x\,, \tag{2}$$

where F is the fishing mortality rate of individuals larger than T, $p(x, a)$ is the frequency of individuals of size x alive at age a, and T is the harvesting size threshold (see Falconer 1981 pp 174-176). Equation (2) assumes $F \ll 1$.

One of the practical advantages of our model is that the ages at which $\bar{z}$, G, and β are estimated do not need to be the same. There are many logistical reasons why it may be impossible to get estimates of genetic variation and of selection (for example) at the same ages. This is no problem: we interpolate between the ages at which the data were obtained to get continuous estimates of $\bar{z}$, G, and β, then use Equation (1).

The detailed methods for estimating $\bar{z}$, G, and β are discussed in detail in the references cited above. We have developed computer programs that, together with any standard statistics package, do these analyses. The software is distributed free on request.

EVOLUTION OF GROWTH IN RESPONSE TO FISHING

An example would be helpful at this point to illustrate how this model works. I have developed a simulation showing how fishing pressure might cause the growth trajectories of a fish population to evolve. Unfortunately, I have not found a data set on a harvested natural population that has all the appropriate information, so I have used a combination of real genetic data from rainbow trout and a hypothetical fishery. My hope is that this example will encourage fisheries biologists to use the approach with other populations.

Growth in rainbow trout has been studied by McKay, Ihssen, and Friers (1986*a*, *b*). They used a breeding design involving more than 1700 individuals in 100 families to estimate the additive genetic variances and covariances for weight at 7 ages from egg through 4 years. Dr. L.R. McKay very kindly provided data to supplement the published results that allowed me to analyze this data set. The mean growth trajectory, $\bar{z}$, for the trout is shown in Figure 1. McKay et al. estimated the 7 x 7 additive genetic covariance matrix for weight at these ages. By interpolating the values of this matrix, I estimated G, the additive genetic covariance function (see Kirkpatrick et al. 1990), shown in Figure 2 (top). As McKay et al. point out, the relatively small number of families used in this breeding design leads to fairly large standard errors around the estimates. I will therefore not use the methods we have developed to construct confidence limits for estimates (Kirkpatrick et al. 1990). Facilities that could generate sufficiently large data sets to estimate the genetic quantities with great precision are, however, available to hatcheries and commercial aquaculture operations.

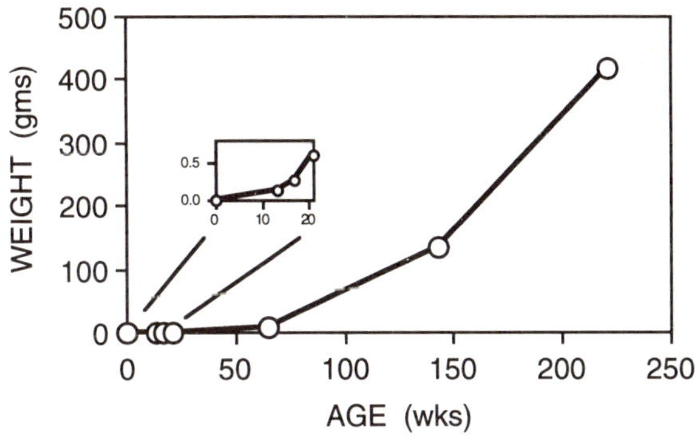

Figure 1. The mean growth trajectory of rainbow trout from egg to 4 years of age based on the data of McKay et al. (1986*a*, *b*).

We can use the estimate of *G* to determine the family of growth trajectories that selection could produce based on the population's current patterns of additive genetic variation (see Kirkpatrick et al. 1990). Analysis of McKay et al.'s data suggests that the vast majority of genetic variation in their population (more than 90%) corresponds to simple changes in the mean growth trajectory that shift it upwards or downwards at all ages, as shown in Figure 2 (bottom).

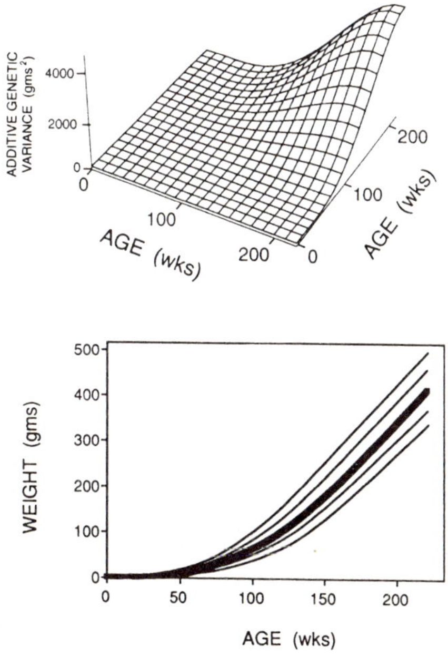

Figure 2. *Top*: The additive genetic covariance function *G* for trout. The methods of Kirkpatrick et al. (1990) were used to fit fourth degree Legendre polynomials to the 7 x 7 additive genetic covariance matrix estimated by McKay et al. (1986*a*, *b*). *Bottom*: Examples from the family of growth trajectories that selection could produce using current genetic variation based the eigenfunctions of *G* (see Kirkpatrick et al. 1990). The population's current mean growth trajectory is the heavy curve.

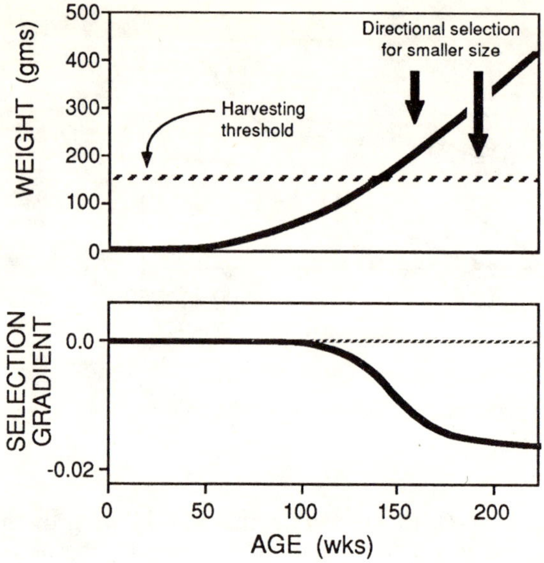

Figure 3. The selection gradient, $\beta(a)$. *Top*: The current growth trajectory (bold curve) and the size threshold for harvesting (dashed line); only fish larger than the threshold are at risk of being caught. *Bottom*: The resulting selection gradient.

Given these estimates for $\bar{z}$ and G, the final quantity needed is the selection gradient. I have used Equation (2) to calculate the selection gradient that would result from a hypothetical fishery harvesting these trout. In this fishery, only individuals greater than 150 gms are at risk of being caught. The fishing mortality rate is 0.2, a value that is probably substantially below real values for many major fisheries. The selection gradient, shown in Figure 3, is negative because larger individuals have higher mortality from the fishery.

To predict the evolution of the growth trajectory, four simplifying assumptions have been used here. At present we lack all of the data needed to make more realistic assumptions, but the model can be easily extended given the appropriate data. First, the population is assumed to be semelparous: all reproduction takes place at the end of the fourth year. Second, I will assume that the only selection acting on the growth rate is caused by the fishery. In any real population, other selection pressures (from life history effects, predation, etc.) will always be present. By ignoring these additional factors, we will see the maximum possible genetic impact of the fishery. More refined predictions can be obtained by adding these additional forms of selection to the model. Third, the genetic covariance function G will be assumed to remain constant. This assumption is probably reasonable for at least a moderate numbers of generations (Turelli and Barton 1990). Fourth, I will assume that the environmental component of growth remains constant in time. Experience from some species has shown that growth rates can accelerate for environmental reasons as population sizes decline under strong fishing pressure, since survivors have more food per capita (e.g. North Sea plaice: Rijnsdorp this volume). This kind of nongenetic, density-dependent effect is neglected in this simulation, but could be added to it.

Using these data and assumptions, the evolution of the population's mean growth trajectory is easily calculated on a computer using Equation 1. The results are shown in Figure 4. The population rapidly responds to the harvesting regime, evolving towards a slower growth rate and smaller final size. Marked changes in the growth trajectory are seen after only a few generations. With greater fishing mortality, the evolutionary changes in the population would be proportionally faster.

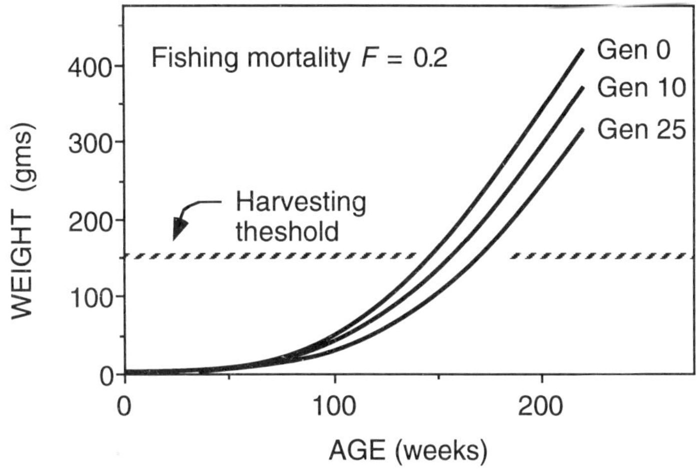

Figure 4. Simulation showing the evolutionary decrease in trout growth trajectories resulting from fishery; details are given in the text.

These genetic changes will of course result in demographic changes. In addition to reductions in the population produced directly by fishing, the smaller size of adults will generally reduce the population's reproductive output. How large of an impact this will have on the standing crop of mature fish depends on details of how density regulation works in the population. To see the potential for effects, however, we can use the empirical observation that fecundity in fishes is roughly proportional to adult weight. The evolutionary decreases in size seen in the simulation would result in a 6% reduction in fecundity after 5 generations and a 25% reduction after 25 generations, as shown in Figure 5.

DISCUSSION

A clear conclusion from the simulation is that a rapid evolutionary response in growth and size can be generated from size-selective harvesting pressure. The evolutionary rates that will be seen in any real population will of course depend on the harvesting methods, the patterns of genetic variation, and other forces of selection. But there is no reason to think that these factors will prevent substantial genetic changes from occurring in only a few generations under the strong selection pressures generated, for example, by the standard commercial gear widely used in marine fisheries.

Although this may surprise those resource managers who have not considered the genetic dimension of human impact on exploited natural populations, there is every reason to think the conclusion is robust. Size and growth (as well as many other traits) respond rapidly to artificial selection pressures in virtually all domesticated populations. The evolutionary rates seen in the simulations presented above are by no means exceptional when compared with results from selection experiments (e.g. Dudley 1977). Natural populations have also shown substantial genetic responses to human exploitation over short time spans, and several workers have emphasized in particular that size-selective fishing practices can generate strong selective forces on the growth and life histories of exploited populations (see Ricker 1981, Beverton et al. 1984, Nelson and Soulé 1987). Rational management of harvested natural populations must take the genetic impact of harvesting into account. Most management attention to date has focused only on population sizes, but there is a growing appreciation of the genetic implications of harvesting.

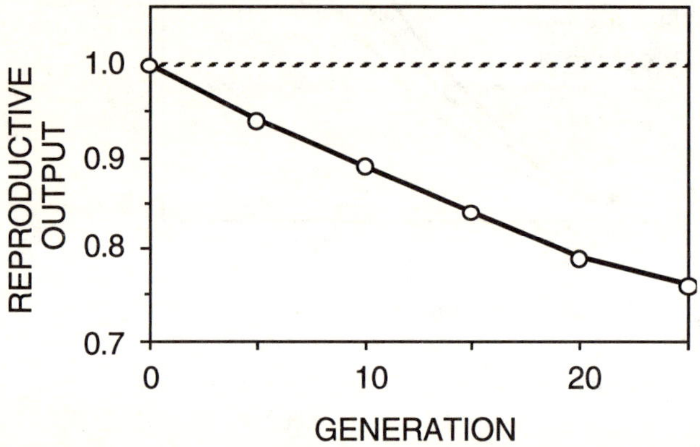

Figure 5. Decline of population's reproductive output caused by the evolutionary decrease in final adult size shown in Figure 4.

Two approaches are now being developed that incorporate an evolutionary dimension into models of harvesting. Optimality models are being developed to predict the ultimate evolutionary outcome in a population subjected to harvesting (e.g. Blythe and Stokes this volume). These models are used to examine how the evolutionary outcome varies as a function of harvesting methods. Ideally, harvesting methods would then be adjusted to maximize the economic yield and probability of survival of the population. An advantage of this approach is that results can be obtained even when many factors (such as growth rate, fecundity and mortality schedules, and density-dependent responses) are included in the model.

Law (1991 and this volume) and others have suggested an alternative approach accounting for the genetic impact of harvesting that is the basis for this paper. Using quantitative genetics, one can develop a model that predicts the population's evolutionary dynamics. This allows us to assess the short-term consequences of genetic changes as well as its ultimate fate. If the predictions from an optimality model are only reached after long time periods, then a quantitative genetic model would be preferable for short-term predictions and management. This advantage over the optimality approach is offset by the cost of more information: genetic data on the characters is needed. It may not, however, be necessary to develop genetic data for every population. If, for example, patterns of genetic variation for growth are similar among salmonid species, then data from a single species in the family could be used as the basis for models of other species.

The simulation results described above are intended only for illustration because of the simplifying assumptions used in the model and limits in the accuracy of the genetic measurements. The simplifying assumptions are not imposed by the quantitative-genetic approach, but rather by the lack of pertinent data. In particular, it seems important to add reproduction at multiple ages (see Law 1991) and density-dependance of growth rates.

Ultimately, any genetic model will need to be wed to a demographic model. Only by understanding the effects of harvesting on both population size and genetic composition can we get an comprehensive picture on which to base rational management decisions.

ACKNOWLEDGEMENTS

I am in debt to Dr. L.R. McKay, who generously provided additional data to supplement those in her papers on rainbow trout. Drs. Richard Law and Bruce Walsh made helpful suggestions on the manuscript. This research was supported by NSF grants BSR-8657521 and 9107040 and NIH grant GM-4522601.

REFERENCES

Becker, W.A. 1984. Manual of Quantitative Genetics. Academic Enterprises, Pullman, Washington.

Beverton, R.J.H. and 11 others. 1984. Dynamics of single species. Pp 13-58, in R.M. May ed. Exploitation of Marine Communities. Springer-Verlag, Berlin.

Bulmer, M.G. 1985. The Mathematical Theory of Quantitative Genetics. Clarendon Press, Oxford.

Dudley, J.W. 1977. 76 generations of selection for oil and protein percentage in maize. Pp 459-474, in E. Pollak, O. Kempthorne and T.B. Bailey eds. Proceedings of the International Conference on Quantitative Genetics. Iowa State Univ Press, Ames.

Falconer, D.S. 1981. Introduction to Quantitative Genetics. 2nd ed. Longman, London.

Gomulkiewicz, R. and M. Kirkpatrick 1992. Quantitative genetics and the evolution of reaction norms. Evolution **46**: 390-411.

Kirkpatrick, M. and N. Heckman. 1989. A quantitative-genetic model for growth, shape, reaction norms, and other infinite-dimensional characters. Journal of Mathematical Biology **27**: 429-450.

Kirkpatrick, M., D. Lofsvold and M. Bulmer. 1990. Analysis of the inheritance, selection and evolution of growth trajectories. Genetics **124**: 979-993.

Kirkpatrick, M. and D. Lofsvold. 1992. Measuring selection and constraint in the evolution of growth. Evolution **46**: 954-971.

Lande, R. and S.J. Arnold. 1983. The measurement of selection on correlated characters. Evolution **33**: 402-416.

Law, R. 1991. On the quantitative genetics of correlated characters under directional selection in age-structured populations. Philosophical Transactions of the Royal Society London B **331**: 213-223.

McKay, L.R., P.E. Ihssen and G.W. Friars. 1986a. Genetic parameters of growth in rainbow trout, *Salmo gairdneri*, prior to maturation. Canadian Journal of Genetics and Cytolology **28**: 306-312.

McKay, L.R., P.E. Ihssen and G.W. Friars. 1986b. Genetic parameters of growth in rainbow trout, *Salmo gairdneri*, as a function of age and maturity. Aquaculture **58**: 241-254.

Nelson, K. and M. Soulé. 1987. Genetical conservation of exploited fishes. Pp 345-368, in N. Ryman and F. Utter eds. Population Genetics and Fisheries Management, University of Washington Press, Seattle.

Ricker, W.E. 1981. Changes in the average size and average age of Pacific salmon. Canadian Journal of Fisheries and Aquatic Sciences **38**: 1636-1656.

Turelli, M., and N.H. Barton. 1990. Dynamics of polygenic characters under selection. Theoretical Population Biology **38**: 1-57.

COHORT-STRUCTURED POPULATIONS, SELECTION RESPONSES, AND EXPLOITATION OF THE NORTH SEA COD

Richard Law and Catherine A. Rowell

Department of Biology, University of York, York YO1 5DD, UK.

ABSTRACT

Exploitation of natural populations usually involves an element of selection. The traits under selection are typically quantitative, and it is appropriate to investigate evolutionary change arising from such harvesting by quantitative genetics. The theory of quantitative genetics does not however address the cohort structure and non-equilibrium population dynamics often found in these natural resources. This paper describes how the quantitative genetics of correlated characters with age-specific expression can be embedded in an age-structured model of population dynamics so that some evolutionary effects of exploitation may be investigated. The method is illustrated with data on directional selection for reduced body length of North Sea cod arising from exploitation over the period 1984 to 1990. The calculation suggests that the selection response is small, perhaps of an order 1 cm over a 40 year period for length of 1 year old individuals. The reason for this is that the selection differential caused by fishing is moderate, the mean generation time is quite long (5 years), and it is assumed that variation in breeding values is 0.1 of the phenotypic variation. In view of uncertainties about the data at several stages in this calculation, this result should be treated with caution.

INTRODUCTION

In the exploitation of living resources, it is usually the case that some individuals are worth more to the harvester than others, and methods of exploitation are developed to remove preferentially those of high commercial value. Exploitation of these self-renewing resources is therefore likely to be selective. This may have profound consequences leading to genetic change in the traits which are under selection and in others which are genetically correlated with those being selected. The latter traits could for instance include components of the life history which determine the capacity of the population to renew itself and hence the yield which the resource can sustain (Law and Grey 1989).

The potential for selection generated by exploitation has long been known in fisheries, where it is often observed that those individuals still surviving when old tend to have been slow growing when young. This property, known as Lee's phenomenon (Lee 1912, Ricker 1969, 1975:215), implies greater natural or fishery-imposed mortality on faster-growing individuals. Yet despite their potential importance, there has been remarkably little attempt to estimate how great the selection pressures caused by exploitation are, and what the likely evolutionary consequences would be (Nelson and Soulé 1987). Notable exceptions to this are analyses of change in growth rate of Pacific salmon species (Ricker 1981), and calculations on the selective advantages of early maturation in North East Arctic cod (Borisov 1978). In both these cases the changes observed are consistent with theoretical expectations, although the extent to which they are genetic remains to be shown (Ricker 1981, Jørgensen 1990).

One of the reasons for our current ignorance is that methods for the analysis of such evolution are not well developed. For the most part the traits under selection are quantitative, under the influence of many genes with small effects, and should fall within the scope of evolutionary quantitative genetics (Bulmer 1980, Barton and Turelli 1989, Falconer 1989). But quantitative genetics does not standardly address the complex life histories and demographic features which characterize exploited biological resources, at least in the marine environment. For example, the North Sea cod (*Gadus morhua*) comprises several cohorts at any one time, each of which may be experiencing selection, with the selection pressures changing as the cohort grows older. Each cohort contributes to the recruitment of new cohorts in a manner which depends on its relative abundance in the population. Furthermore, the stock is typically far from a demographic steady state, with as much as a tenfold variation in cohort strength from year to year. Such demographic complexities are typical of marine biological resources and to ignore them would be to miss a major factor affecting the response to selection.

Arguably, we need to embed the quantitative genetics in a model of population dynamics which incorporates age-structure and non-equilibrium demography. This paper describes and illustrates a theoretical basis recently developed for doing this (Law 1991). Under certain simplifying assumptions, it enables us to follow the short-term phenotypic evolution of a suite of correlated quantitative characters expressed at arbitrary ages and undergoing directional selection in a non-equilibrium cohort-structured population. The theory brings together ideas from selection in age-structured populations (Hill 1974, Charlesworth 1980:188 et seq.), and from the evolution of correlated phenotypic traits that are simultaneously expressed (Lande and Arnold 1983).

The theory is illustrated with some calculations on selection on the North Sea cod during the period 1984 to 1990. The traits considered are length-at-age 1, 2 and 3 years, these being likely candidates for selection given the size-selective fishery in operation. It is important to stress at the outset that the data available are seriously deficient in several respects, and the numerical answers need to be treated with caution; these answers suggest that the selection responses are small. Despite the inadequacies of the data, we have thought it worth persisting with the analysis for, as well as illustrating the theory, it highlights the information which would be required from fisheries to turn the estimates of selection responses into a more exact science.

THEORY

Consider the evolution of m quantitative traits in a population comprising n+1 cohorts (0, 1, ..., j, ..., n). Each trait is expressed at one age only and traits are numbered 1, 2, ..., i, ..., m in the order of the age of expression. (The order for traits expressed at the same age is arbitrary.) An individual has from birth onwards a phenotypic value for each trait, although it may of course die before the trait is expressed. The mean phenotypic values within a cohort aged j are given by the column vector $\bar{\mathbf{x}}_j = (\bar{x}_{1j},...,\bar{x}_{mj})^T$. Thus the mean phenotypic values in the whole population at some time t are given by all n+1 of these vectors, which are conveniently collected together into an m by n+1 matrix $\mathbf{X}_t = [\bar{\mathbf{x}}_0,....,\bar{\mathbf{x}}_n]_t$ to retain explicitly the cohort structure of the population.

Our eventual concern lies not with the phenotypic values but with the extent to which they are transmitted to the progeny. While selection alters the phenotypic distributions, it is the inheritance of these changes which is critical for evolution. The genetic architecture which underlies the traits may be complex, and we simplify matters greatly by assuming that an individual's phenotypic value is the sum of two statistically independent components: a breeding value and an environmental deviation. In the language of quantitative genetics, the breeding value is the sum of the average effects of the genes influencing the trait (Falconer 1989:118). For a cohort at age j, the mean breeding values are given by $\bar{\mathbf{a}}_j = (\bar{a}_{1j},...,\bar{a}_{mj})^T$ the vectors for all ages being collected together as before into an m by n+1 matrix $\mathbf{A}_t = [\bar{\mathbf{a}}_0,...,\bar{\mathbf{a}}_m]_t$. By definition, mean phenotypic and breeding values are identical at birth ($\bar{\mathbf{x}}_{0t} = \bar{\mathbf{a}}_{0t}$), but this identity disappears when selection takes place on the cohort as it grows older because the expected value of the environmental deviations is zero. Although it is the changes in **A** which describe the evolutionary response to selection, it is often useful to keep track of **X** as well because this is in part responsible for the selection differentials.

It should be understood that, when selection is applied to a particular trait, this may not be the only one which exhibits changes in mean phenotypic and breeding values, because phenotypic or breeding values (or both) may be correlated with those of other traits (Smith this volume). For instance, individuals which are large at age 1 year may also tend to be large at age 2, in which case selection at age 1 brings about changes in the mean value at age 1 and 2 in this cohort, even though length-at-age 2 is not currently under selection, or even being expressed within the cohort. To keep track of such changes, we need the covariance matrices of phenotypic and breeding values, $\mathbf{P} = [\text{cov}(x_i,x_k)]$ and $\mathbf{G} = [\text{cov}(a_i,a_k)]$ respectively, both matrices being of order m by m. We assume that **P** and **G** are constant over time, attention being focussed on the evolution of the first-order statistics. Results from artificial selection experiments suggest that this assumption is reasonable over several generations (Falconer 1989:191), although in the longer term it is violated by changes in the frequency of genes which underlie the traits. The analysis is therefore concerned only with the short-term evolutionary dynamics. It should also be pointed out that directional selection causes an immediate reduction in the elements of **P** and **G** (Falconer 1989:203); this effect is small as long as the selection differentials and heritabilities of the traits are low, although it could cause more serious errors when selection operates on strongly correlated characters over a series of ages.

The directional selection pressure on trait i is measured as the selection differential (s_{ijt}), the difference between the mean phenotypic value after ($\bar{x}^*_{ijt}$) and before ($\bar{x}_{ijt}$) selection; $s_{ijt} = (\bar{x}^*_{ijt} - \bar{x}_{jt})$. Taking all n+1 cohorts together, we may write the selection differentials as an m by n+1 matrix $\mathbf{S}_t = [s_{ij}]_t$. Clearly a selection differential can only be applied when a trait is being expressed, so most of the elements of S_t are usually zero. But, as we shall see below, traits which are not themselves under selection can nonetheless still have changing mean values by virtue of their covariances with those which are under selection.

To find the changes in mean values of all traits arising from directional selection on a subset of them, we start by writing down a reduced form of the phenotypic covariance matrix $\dot{\mathbf{P}}$. This, like $\mathbf{P}$, is an m by m matrix but it differs in that all covariances for traits expressed at different times are struck out and replaced with zeros; the non-zero elements then constitute blocks along the diagonal of $\dot{\mathbf{P}}$. The regression coefficients for the breeding values of all traits on the phenotypic values of those under selection are given by $\mathbf{G}\dot{\mathbf{P}}^{-1}$, recalling that the environmental deviations are independent of the breeding values. (It must be possible to invert $\dot{\mathbf{P}}$ to do this, but this is unlikely to be a problem in most cases.) Similarly, the regression coefficients for the phenotypic values of all traits are given by $\mathbf{P}\dot{\mathbf{P}}^{-1}$. Assuming that there is a scale on which the breeding and phenotypic values have a joint multivariate normal distribution, these regression coefficients remain unchanged by directional selection, and the changes in mean values are given by $\mathbf{G}\dot{\mathbf{P}}^{-1}\mathbf{S}_t$ and $\mathbf{P}\dot{\mathbf{P}}^{-1}\mathbf{S}_t$ for the breeding and phenotypic values respectively. The terms in $\dot{\mathbf{P}}^{-1}\mathbf{S}_t$ are referred to as the selection gradients (Lande and Arnold 1983).

The step above will tell us how much **A** and **X** change as a result of a single bout of selection, but we have yet to incorporate the population dynamics of the cohorts. (We shall assume that reproduction is seasonal, so that the cohorts are discrete from one another.) There are two processes involved in these dynamics: the generation of mean values for traits in newborn cohorts following each period of reproduction, and the increment in age of cohorts already present while allowing for their gradual decline through selective and non-selective mortality. These processes are described by a matrix $\mathbf{T}_t$ of order $n+1$ by $n+1$, analogous to a Leslie matrix

$$\mathbf{T}_t = \begin{bmatrix} p_0 & 1 & & \\ \cdot & & \cdot & \\ \cdot & & & \cdot \\ p_{n-1} & & & 1 \\ p_n & & & \end{bmatrix}_t$$

(Hill 1974, Charlesworth 1980:91 et seq.). Elements in the first column are essentially weights for the contribution that each cohort makes to the newborn cohort; more precisely, p_{jt} is the probability that a newborn individual inherits a given gene from a parent aged j at time t. The jth element clearly depends on the rate of reproduction at age j (b_j) and on the relative abundance of the jth cohort, which itself depends on the initial strength of this cohort at the time of birth ($n_{0,t-j}$) and the rate of survival from birth to the time of reproduction at age j ($\overline{\omega}_{jt} = \overline{w}_{0,t-j}\, \overline{w}_{1,t-j+1} \ldots \overline{w}_{j,t}$). Note that the $\overline{w}$s include both selective and non-selective mortality. Putting all these components together, gives $p_{jt} = (n_{0,t-j}\, \overline{\omega}_{jt} b_j)/(\Sigma_j\, n_{0,t-j}\, \overline{\omega}_{jt}\, b_j)$. It is evident from this that the elements of $\mathbf{T}_t$ change through time in non-equilibrium populations depending on the abundance of different cohorts; the elements also change with changes in the amount of selective mortality. The line of elements above the diagonal of $\mathbf{T}_t$ increments the age of cohorts by one year at the time of reproduction and discards the cohort aged n years.

The evolutionary dynamics of the mean breeding values are given by the recurrence relation

$$\mathbf{A}_{t+1} = (\mathbf{A}_t + \mathbf{G}\dot{\mathbf{P}}^{-1}\mathbf{S}_t)\mathbf{T}_t \tag{1}$$

which identifies succinctly the three critical components of evolution: the genetic structure ($\mathbf{A}_t$ and $\mathbf{G}$), the selection ($\dot{\mathbf{P}}^{-1}\mathbf{S}_t$), and the population dynamics ($\mathbf{T}_t$). If we wish to keep track of the dynamics of mean phenotypic values, this is done by means of a second recurrence relation coupled to the first

$$\mathbf{X}_{t+1} = (\mathbf{X}_t + \mathbf{P}\dot{\mathbf{P}}^{-1}\mathbf{S}_t)\mathbf{T}_t\mathbf{C} + \mathbf{A}_{t+1}\mathbf{C}' \tag{2}$$

The first part of the right hand side of this describes the dynamics of mean phenotypic values of cohorts already in existence at time t as these cohorts become older, $\mathbf{C}$ being a diagonal matrix with elements unity except for the first which is zero. The second part sets the mean phenotypic values of the newborn cohort to be equal to its mean breeding values, $\mathbf{C}'$ being a square matrix with the first element unity and zeros elsewhere.

While Equations 1 and 2 give the detailed year-to-year dynamics of a non-equilibrium population, it is also possible to derive asymptotic properties of the response to selection. These results make use of the fact that any $\mathbf{T}$ is the transpose of a stochastic matrix (Lancaster 1969:291), so that $\mathbf{T}^{\infty} = \mathbf{u}\,\mathbf{v}\,\tau^{-1}$, where $\mathbf{u}$ is the right dominant eigenvector of $\mathbf{T}$ scaled with the first element of value unity, $\mathbf{v}$ is the left dominant eigenvector scaled so that all elements are unity, and τ is the mean generation time (Hill 1974, Charlesworth 1980:92). If we imagine that $\mathbf{S}$ and $\mathbf{T}$ are constant for all time, then the asymptotic change in mean breeding values from one time to the next is given by

$$\Delta\mathbf{A}_{\infty} = \mathbf{G}\dot{\mathbf{P}}^{-1}\,\mathbf{S}\,\mathbf{u}\,\mathbf{v}\,\tau^{1} \tag{3}$$

(Hill 1974). Although for the reasons discussed earlier it is unrealistic to suppose that $\mathbf{T}$ and the matrices on the right hand side of this equation would remain constant in the long term, the equation can prove useful in examining the consequences of particular patterns of selection without the confounding effects of non-equilibrium demography.

We should emphasize that the theory is based on the simplest of quantitative genetic models; it does not consider non-additive effects of genes within loci (dominance) or between loci (epistasis). Neither does it investigate the effects of interactions between genotype and environment. It assumes the existence of a multivariate normal distribution for breeding and phenotypic values; this can be no better than an approximation because the distribution of breeding values at birth is obtained by combining the distributions of breeding values of parents of different ages and hence, in general, distributions with different means. Although the theory as outlined here assumes that males and females at a given age make the same contribution of genes to newborn cohorts, this assumption is readily relaxed and different selection pressures on the sexes can be allowed (Law 1991). Note also that we are assuming that natural selection is absent, and that there is no genetic correlation between the traits under selection and the life history. It would be interesting to develop methods which allow such correlations, because the traits likely to be under selection in exploited populations, such as length-at-age, are most probably correlated with components of fitness including fecundity. An alternative method to describe the evolution of traits like length which change continuously through time is to convert the covariance matrix of breeding values into a continuous function and to examine its eigenfunctions; the eigenfunctions on which most variation lies indicate the directions in which the trait is most readily 'deformed' by selection (Kirkpatrick 1988, this volume).

EXAMPLE: SELECTION ON LENGTH-AT-AGE IN NORTH SEA COD

BACKGROUND

To illuminate the theory above, there follow some calculations on the selection differentials on length-at-age in the North Sea cod, together with the putative selection responses expected. The cod is chosen as one of the most important commercial stocks in the North Sea, whose continued health is vital for the livelihoods of many of those in the sea fish industry of North West Europe. Since the North Sea represents a large geographical area within which there is some variation in growth and maturation of cod (Daan 1974, Rowell, this volume), the analysis is confined to an area off the East coast of Northern England and Southern Scotland 53° 30' N to 56° N, known as Round Fish Area 4 (RFA 4). The stock is derived at least in part from a spawning ground off Flamborough. Spawning takes place here in February and March, after which the eggs and larvae drift in currents towards the coast (Houghton 1977). Returns of tagged cod (< 70 cm in length) released close to the coast indicate that they mostly remain in RFA 4, although there is substantial movement of larger individuals to and from the area (Houghton 1977).

The characters investigated are body length-at-age 1, 2 and 3 years, over the period 1984 to 1990. We confine attention to changes in these traits in the third quarter of each year (July, August, September) because, for reasons discussed below, it is necessary to make use of the English Groundfish (EGF) survey which takes place in late August and early September each year (Harding, Woolner and Dann 1986). This survey uses a fine-meshed liner in the codend of the trawl, and can be regarded as giving a representative sample of the length frequency distribution down to a body length of 15 cm and perhaps smaller. Being confined to the third quarter, the results are no more than a partial picture of the selection as this is taking place continuously through time. It is however a particularly interesting period because the 1 year cod are partially recruited into the fishery at this time.

Several kinds of fishing gear are used in the exploitation of the stock in RFA 4, including bottom trawls, pair trawls, beam trawls, seine nets, gill nets and lines. These differ in their selectivity; gill nets, for example select larger individuals than do bottom trawls. But, since about 90 per cent of the catch landed at English ports is obtained from bottom and pair trawls (Figure 1), a good indication of the selection is obtained from these two gears alone. The English landings constitute about 75 per cent of the total landings from RFA 4 and are probably representative of the gears used overall (Stokes personal communication). It should be noted that the regulations governing minimum net mesh size and the minimum length of landed cod changed during the period 1984 to 1990; the regulations in force each year are given in Table 1.

Table 1. Data on the North Sea cod: fishery regulations, discards, and estimated mean values of length-at-age of cod in Round Fish Area 4 in the third quarter of each year from 1984 to 1990.

	1984	**1985**	**1986**	**1987**	**1988**	**1989**	**1990**
Regulations							
Minimum mesh (mm)	80	80	80	85	85	90	90
Minimum landed size (cm)	30	30	30	30	30	35	35
Discards (%)	44	3	61	7	15	55	29
Mean length (cm): age 1 year							
Landed catch	34.36	33.25	33.39	34.44	34.14	38.15	38.00
Landed catch + discards	29.19	31.35	28.68	30.91	31.27	31.77	33.60
Stock before fishing	24.39	28.32	25.10	27.72	28.75	27.73	30.27
Stock after fishing	24.15	27.96	24.90	27.50	28.54	27.52	29.97
Selection differential	-0.24	-0.36	-0.20	-0.22	-0.21	-0.21	-0.30
Mean length (cm): age 2 years							
Landed catch	40.58	42.86	36.14	40.95	40.03	45.43	43.72
Landed catch + discards	-	-	-	-	-	45.20	43.36
Stock before fishing	40.41	42.63	35.76	40.20	39.16	44.50	42.68
Stock after fishing	40.36	42.57	35.67	40.01	38.95	44.32	42.51
Selection differential	-0.05	-0.06	-0.09	-0.19	-0.21	-0.18	-0.17

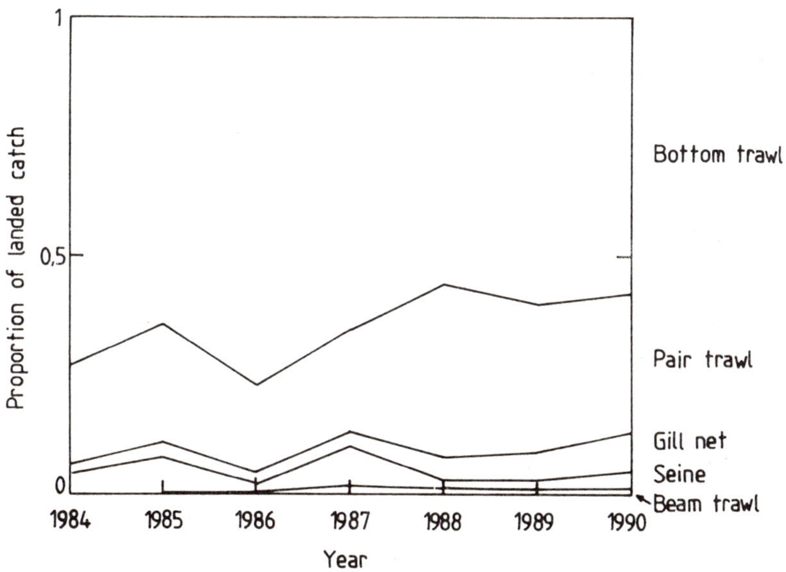

Figure 1. Proportion of cod caught by different gears in Round Fish Area 4 of the North Sea and landed at English ports in the period July to September each year from 1984 to 1990.

SELECTION DIFFERENTIALS

Fisheries are unusual in the context of selection in that detailed information is often available on the fish which die through being caught. For North Sea cod for instance, the catch landed at English ports is monitored by the Fisheries Inspectorate, and the frequency distribution of length-at-age landed by different gears is compiled over three month intervals, age being estimated from banding patterns of otoliths of fishes sampled from the catch. Figure 2 gives the information available for RFA 4 in the third quarter of 1984; these data illustrate the different selectivity of the gears in use, fishes landed from gill nets tending to be larger than those landed from other gears.

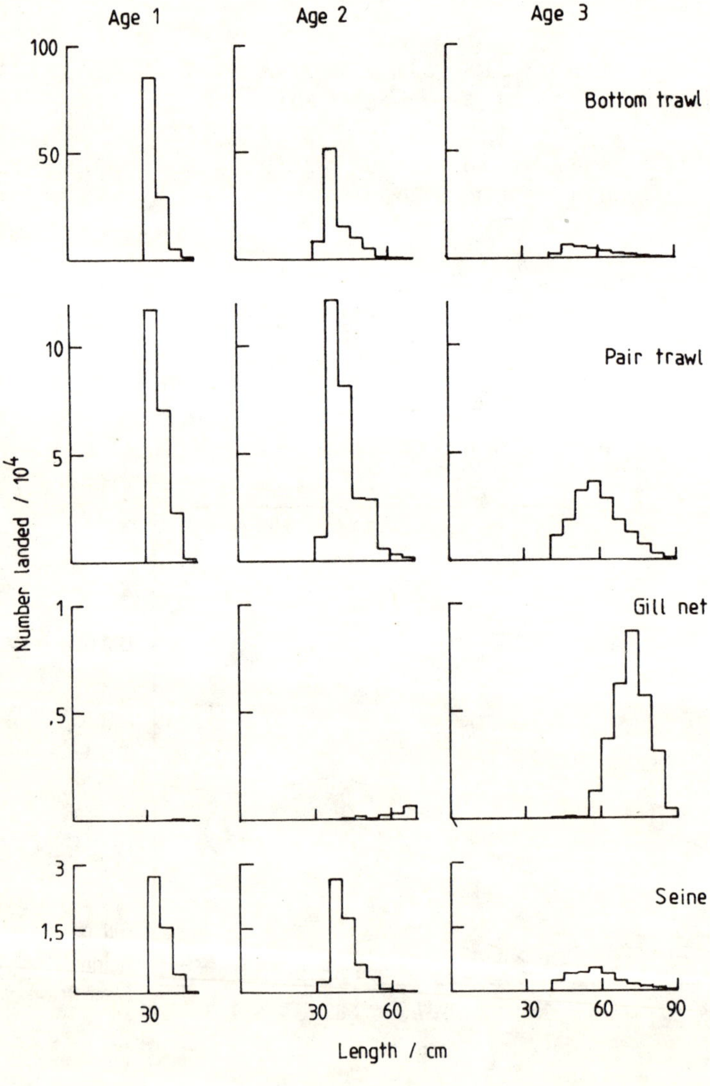

Figure 2. Length distributions of the catch of cod of different ages caught by various gears from Round Fish Area 4 of the North Sea and landed at English ports in the period July to September 1984.

Given detailed information on the landed catch, the approach we have adopted to obtain the selection differentials is to reconstruct from the catch the frequency distribution of length-at-age in the stock before and after the period of fishing in the third quarter. In view of the fact that some growth takes place during this period, it is simplest to envisage that all the fishing takes place at the midpoint and that the selection differential is based on the mean of the stock just before and just after this fishing. Since most of the fishes are caught by bottom and pair trawls, the reconstruction is based on the combined catches of these gears landed at English ports.

Frequency distribution of length before fishing. We start by obtaining the length-frequency distribution of the stock aggregated across all age classes. For length classes at and above the legal limit for landing, the length-frequency distribution of the stock can be reconstructed by making allowance for the size selectivity of the nets being used. To do this rigorously would require information on the range of net mesh sizes in operation and the proportion of the catch which comes from each mesh size. Such information is not available, and we have assumed simply that the average net mesh size is the legal lower limit in operation; this allows for the likelihood that some illegal nets are in use. The proportion of fishes θ_k of length l_k retained by nets of different sizes is then given by

$$\frac{\theta_k}{1-\theta_k} = \exp\{2(al_k+b)\} \tag{4}$$

where the constants are: for 80 mm mesh, a= 0.21, b= 6.1; for 85 mm mesh, a= 0.20, b= 6.0; for 90 mm mesh, a= 0.19, b= 6.1 (Stokes personal communication). Figure 3 gives the length-frequency distributions of the stock reconstructed by this method, together with those from the EGF surveys of RFA 4 carried out at the same time for comparison. Because of the small samples on which the latter are based, the right hand tails have been truncated at the first value at which the sample size is less than 3.

The major snag of using the landed catch is that some of the catch, being below the legal length limit, is discarded prior to landing. This is evident in the catch of 1984 (Figure 2) where, with a minimum landed length of 30 cm in operation, all the frequency distributions are truncated at this point. Since discards are likely to constitute a substantial part of the mortality of 1 year fishes, it is essential to allow for them. Information on discards is usually little more than anecdotal so this is not easy to do; the only data available in this study are the frequencies of small fish in the EGF surveys. We have therefore used the EGF survey data to extend the length-frequency distributions back to 15 cm, rescaling the survey frequencies on to the scale of the catch-based frequencies by means of the frequency of the lowest legal length class for landing. The method introduces a major new source of sampling error because of the small sample sizes of the EGF surveys; the length classes involved are however amongst the most numerous in these surveys. It is notable from these extended length-frequency distributions that discards can constitute a major part of the catch (Figure 3). After allowing for the size selectivity of the fishing gear, we estimate that as many as 50 per cent of the fishes caught by bottom and pair trawls were being discarded in years when the 1 year cohort was particularly abundant (Table 1).

Frequency distribution of length-at-age before fishing. The length-frequency distribution has to be disaggregated into its component age classes to obtain the frequency distribution of length-at-age. For length classes at and above the legal limit, this is done according to the age distribution within each length class of the landed catch. Fishes of lengths in the range 15 to 29 cm, are assumed to be of age 1 year because 2 year olds in this range were rarely found in the EGF surveys. For the 30 to 34 cm length class in 1989 and 1990, the fishes were apportioned to ages on the basis of the age distribution of this length class in the EGF survey in these years.

Frequency distribution of length-at-age after fishing. How much effect size selective mortality has on the stock which remains after fishing depends on the magnitude of fishing mortality; clearly the frequency distribution is little changed if no more than a small proportion of individuals are caught. Recent virtual population analyses (VPAs) of North Sea cod indicate instantaneous fishing mortality rates of approximately 1 year^{-1}, or equivalently 0.25 over 3 month periods, once cohorts have been fully recruited into the fishery (Stokes personal communication). (See Pope (1982) for an introduction to VPA.) Prior to full recruitment, the effect of fishing on the proportion of individuals of age j in length class k (ρ_{jk}) is given by

$$\rho^*_{jk} = \rho_{jk}(1 - \theta_k(1 - e^{0.25})) \tag{5}$$

where ρ_{jk} is the proportion before fishing and θ_k is as before the proportion of fish of length k retained by the net.

Mean lengths and selection differentials. Mean lengths before and after fishing are given in Table 1. To the extent that fishing is size selective, one would expect the mean length-at-age in the catch to be greater than that in the stock. This is evident for the landed catch of 1 year fishes, which have means approximately 8 cm greater than that of the stock. After making allowance for fishes discarded from the catch before landing, this difference becomes substantially smaller, but is still usually greater than 3 cm for 1 year olds. Comparing the mean lengths-at-age of the stock before and after fishing, we get selection differentials on 1 year fishes of approximately 0.2 cm during the third quarter over the years 1984 to 1990 (Table 1). The selection differential on 2 year olds was much smaller when the 80 mm net mesh size was in operation, because very few 2 year fishes could escape from the nets, but as one might expect when the net mesh was increased in 1987 the selection differential on fish of age 2 years became substantially greater. Clearly, the selection differential at ages greater than 2 years was negligible.

PHENOTYPIC COVARIANCE MATRIX AND THE ACCUMULATED CHANGE IN MEAN PHENOTYPIC VALUES

Ideally, the elements of the phenotypic covariance matrix for length-at-age 1, 2 and 3 years would be estimated from measurement of the lengths of individual, tagged fishes in August of their second, third and fourth year of life. Such data are not available for this stock, and an alternative is to use banding of the otoliths of fishes caught at age 3 years as markers of their earlier growth. This technique has been used with success in some other studies on the growth of fishes (van Leeuwen and Rijnsdorp 1986, Rijnsdorp et al. 1990), and it would seem appropriate here because initiation of hyaline bands is thought to take place in the autumn in North Sea cod (Williams and Bedford 1974). We should however point out that for cod the reliability of the technique is open to question, as there is evidence of substantial variation between fishes in the time at which bands are initiated (Rowell 1992). This is a serious weakness of the calculations which could cause estimates of phenotypic variances to have a positive bias; we use the approach simply because there is no alternative available.

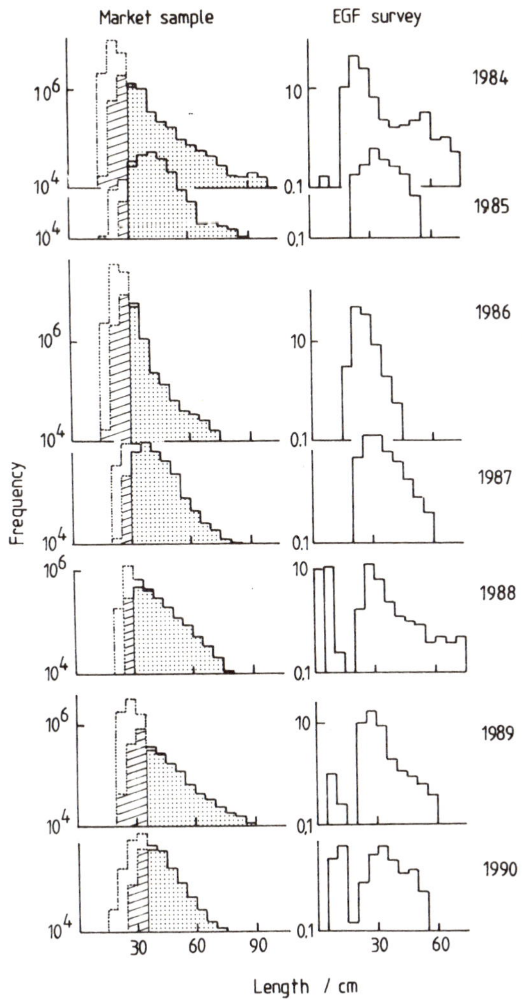

Figure 3. Length-frequency distributions for North Sea cod in Round Fish Area 4 (RFA 4) during the period 1984 to 1990. The shaded histograms give numbers caught. Down to the minimum landed size, these are taken from market samples; below this length (hatching) they are estimated from the English Groundfish (EGF) survey as described in the text. The unshaded parts of the histograms give the length distribution of the stock prior to fishing, where this differs from the catch; below the minimum landed size, this is estimated from the EGF survey (discontinuous lines). For comparison, the length distributions are given for the EGF survey in RFA 4 over the same period. The right-hand tails of the EGF distributions are based on very small numbers of fish and are subject to large sampling variance; the tails have been truncated at the first value at which the sample size falls below 3.

The technique first establishes an allometric relationship between body length and some measure of otolith size. We have done this on all cod caught by the EGF survey in RFA 4 from 1977 to 1990 which were sampled for ageing and found to be of age 1, 2 or 3 years (except 1979, where the data are uninterpretable). The extended period of time back to 1977 is used to increase the size of the sample. On each fish, the radial distance along the wing of the sectioned otolith to the margin was measured (x), and body length from snout to the tip of the tail (y). The relationship can be described by a linear regression of log(y) on log(x) with intercept $\alpha = 0.56$ and gradient $\beta = 1.37$, with residual variance $\sigma^2 = 0.0014$ ($R^2 = 94$ %, $n = 1217$).

Given the relationship between otolith size and body length, an estimate of the length at earlier ages of a fish caught at age 3 years can be obtained from the radial distance from the centre of its sectioned otolith to the inner edge of the appropriate hyaline band. For this calculation, all cod caught and shown to be of age 3 years in the EGF survey from 1977 to 1988 are used. Those caught in 1989 and 1990 were excluded because the sampling method was changed from stratified random to a simple random sample. The logarithm of body length of the kth fish at an earlier age j was taken as a single realization of the distribution:

$$N\{\alpha + \beta \log (x_j(k)) + \rho\varepsilon_3(k) \, , \, \sigma^2(1-\rho^2)\}, \text{ for } j = 1,2 \tag{6}$$

where $x_j(k)$ is the otolith measure for the kth fish at age j, σ^2 is the residual variance from the regression, and $\varepsilon_3(k)$ is the residual for fish k at age 3, after fitting the regression (Grey personal communication). We assume that the parameters α, β and σ^2 are known without error, in view of the large sample on which the regression is based. The residual $\varepsilon_3(k)$ is included for the following reason. It is likely that the deviations from the regression line are not independent; a fish with large otoliths relative to body size at age 3 years is also likely to have had the same property at earlier ages. The coefficient ρ which measures these correlations of residuals is not known, and we have assumed a value 0.5 between age 1 and 3, and 0.75 between age 2 and 3. Tests in which ρ is varied suggest that the phenotypic variances and covariances are relatively insensitive to its value, because the residual variation around the regression line is small.

To estimate the phenotypic variances and covariances, we pool all cod caught and shown to be of age 3 years in the EGF surveys from 1977 to 1988. Allowance is made for the stratified random sample on which the EGF surveys are based, in which a subsample of a preset number of individuals is taken from each length class for age determination. In this way, the phenotypic covariance matrix is estimated as

$$\mathbf{P} = \begin{bmatrix} 26.4 & 23.3 & 23.7 \\ 23.3 & 52.4 & 62.8 \\ 23.7 & 62.8 & 95.2 \end{bmatrix} \text{cm}^2 \tag{7}$$

Since there is only one trait expressed at each age, $\dot{\mathbf{P}}$ is a diagonal matrix, diag(26.4 52.4 95.2), and $\dot{\mathbf{P}}^{-1}$ = diag(0.038 0.019 0.011).

The non-zero phenotypic covariances between traits expressed at different ages have the effect of accumulating changes in a trait's mean phenotypic value even when it is not being expressed. These changes can be followed, using the first part of the right hand side of Equation 2, as shown in Table 2. Consider for example the 1983 yearclass. Fishing in the third quarter of 1984 causes a reduction in mean phenotypic values for length-at-age 2 and 3 years (0.21 cm and 0.22 cm respectively) by virtue of the positive covariance between these traits and length-at-age 1 year (the trait under selection at that time). When the 1983 yearclass reaches the third quarter of 1985, a further reduction in mean length-at-age 2 and 3 years comes about through fishing (0.06 cm and 0.07 cm respectively). So the accumulated reduction in mean phenotypic values at ages 2 and 3 years due to fishing in the third quarters of both 1984 and 1985 are 0.27 cm and 0.29 cm respectively. Bearing in mind that these results refer only to fishing over a quarter of each year, it is likely that the overall accumulated reduction would be of an order of 1 cm within a cohort.

Table 2. The accumulated change in mean phenotypic values for length-at-age 1, 2 and 3 years ($\Delta\bar{x}_1,\Delta\bar{x}_2,\Delta\bar{x}_3$) due to fishing on the North Sea cod in RFA 4 during the third quarter of each year. Yearclasses run diagonally to the right down the table. The trait expressed at each age is shown in italics.

	1984	1985	1986	1987	1988	1989	1990
Age1							
$\Delta\bar{x}_1$	*−0.24*	*−0.36*	*−0.20*	*−0.22*	*−0.21*	*−0.21*	*−0.30*
$\Delta\bar{x}_2$	−0.21	−0.32	−0.18	−0.19	−0.19	−0.19	−0.26
$\Delta\bar{x}_3$	−0.22	−0.32	−0.18	−0.20	−0.19	−0.19	−0.27
Age2							
$\Delta\bar{x}_2$		*−0.27*	*−0.41*	*−0.36*	*−0.40*	*−0.36*	*−0.35*
$\Delta\bar{x}_3$		−0.29	−0.43	−0.41	−0.45	−0.40	−0.39
Age3							
$\Delta\bar{x}_3$			*−0.29*	*−0.43*	*−0.41*	*−0.45*	*−0.40*

SELECTION RESPONSE

How great the selection response is, depends on the covariance matrix of breeding values, **G**. This can only be obtained from controlled breeding programmes or artificial selection experiments. With the exception of a preliminary cross between North East Arctic cod and Norwegian coastal cod, in which no difference in growth rate between the coastal cod and offspring of the cross was found (Godø and Moksness 1987), such experiments have not been carried out. Results of experiments on other species, mostly carried out in connection with the aquaculture industry, clearly indicate the existence of variation in breeding values for production-related traits (Gjedrem 1983, Gjerde 1986). For instance, Gjedrem (1983) quotes heritabilities from sires over a range 0.08 to 0.44 for size-at-age in Atlantic salmon. In the absence of appropriate information on North Sea cod, we have assumed that the elements of **G** are 1/10 of those in **P**. We take this to be a conservative assumption, but note that it could be seriously adrift if there are strong interactions between genotype and environment, or if the effect of the environment on phenotypic variation is substantially greater for cod in RFA 4 than in the experimental systems in which heritabilities have been estimated (Falconer 1989:164).

In view of the uncertainties which have accumulated in the course of the analysis, it would be unrealistic to provide a detailed forecast of evolution from year-to-year in RFA 4 as fishing pressures and cohort strengths fluctuate. Instead, we give a projection (sensu Caswell 1989:20) for the asymptotic rate at which the traits would evolve given the average selection differentials over the years 1987 to 1990, -0.23 cm at age 1 year and -0.19 cm at age 2 years (Equation 3). The change within any one year is given by the matrix product

$$\mathbf{G\dot{P}^{-1}S} = \begin{bmatrix} 0 & -0.023 & -0.008 & 0 & \dots \\ 0 & -0.020 & -0.019 & 0 & \dots \\ 0 & -0.021 & -0.023 & 0 & \dots \end{bmatrix} \qquad (8)$$

in which the rows are traits (length-at-age 1, 2, 3) and there are 15 columns to match the 15 cohorts assumed to exist in the calculations below, **S** being of order 3 by 15. (The older cohorts clearly have very little influence on the population dynamics.) The only non-zero columns in this product correspond to the cohorts at age 1 and 2 years, i.e. the cohorts in which selection is taking place.

To obtain the asymptotic response to selection, some information on the average life history of cod in RFA 4 is needed for entry into the matrix of population dynamics **T**, as described in Table 3. Assuming that the elements of **T** remain unchanged, the asymptotic values $\mathbf{T}^{\infty}$n be written as 15 identical columns given by the right dominant eigenvector of **T** (Table 3), divided by the mean generation time ($\tau = 5.0$). The asymptotic change in mean breeding values from one year to the next is the matrix product

$$\mathbf{G\dot{P}^{-1}ST^{\infty}} = \begin{bmatrix} -0.0062 & -0.0062 & -0.0062 & \dots & \dots \\ -0.0076 & -0.0076 & -0.0076 & \dots & \dots \\ -0.0084 & -0.0084 & -0.0084 & \dots & \dots \end{bmatrix} \text{cm y}^{-1}.$$

This is a 3 by 15 matrix in which the change is the same at all ages, and only the changes at ages 0, 1 and 2 years are shown explicitly above.

It would seem from these calculations that the response to selection is small. This has come about as a consequence of the initially moderate selection differentials having a smaller effect on mean breeding values (through the assumption that **G** = **P**/10), with the asymptotic response being made still smaller by the fairly long mean generation time. Recalling that only selection caused by fishing in the third quarter of each year is being considered, one might expect the overall selection differentials on a cohort to be somewhat greater, perhaps increasing the selection response by a factor of between 2 and 4. It would still be the case however, that the selection responses are small, perhaps of an order of 1 cm over a 40 year period for body length at these early ages.

CONCLUDING COMMENTS

The selection responses arising from exploitation of the North Sea cod calculated above suggest that the evolution of growth of these fishes is slow on the time scale for management. We would however caution against reading too much into the numerical answers because information is missing at at least three stages of the calculations. First there is uncertainty about the frequency distribution of body length in the stock; this could be rectified by increasing the intensity of sampling in fisheries surveys. Secondly, the phenotypic covariance matrix estimated from otoliths of 3 year old cod may well be unreliable, because of the likelihood of variation in the time at which hyaline bands are initiated. Other methods may be needed to estimate this matrix in the case of cod, perhaps involving release and recapture of tagged individuals. The method of back calculation could still prove useful for other species such as plaice, in which the markers of earlier growth appear to be more reliable (Rijnsdorp et al. 1990). Thirdly, we lack information on the genetic architecture of the traits. While it is becoming technically feasible to carry out the necessary experiments to obtain this information on cod (Godø and Moksness 1987), this would be a major undertaking; yet without it predictions about the response to selection inevitably remain highly speculative. It should also be remembered that there are sampling errors associated with the estimates. We have not attempted to calculate these, but they are certainly large relative to the selection response.

With selection responses of the magnitude suggested by this analysis, the absence of changes in growth rate of North Sea cod found earlier in the century is not surprising (Graham 1934, Daan 1974). (These results, being phenotypic values, include both effects of genotype and environment.) It should also be remembered that the fishing mortality rates on which selection differentials depend were substantially smaller earlier in the century; those for ages 1 and 2 years estimated from VPAs have more than doubled since 1960 (Stokes personal communication). Another factor to bear in mind is that variability in the net mesh sizes between fishing boats reduces the overall selectivity of fishing. Moreover, it is possible that the towed nets used by fishermen are in fact less size selective than the mesh selection ogives of fisheries managers would suggest. For instance, diagonal meshes in the codends of trawl nets tend to collapse under heavy loads making escape through them unlikely (Stewart and Robertson 1985); the selectivity of gear also depends on aspects of its overall structure (Armstrong et al. 1989). A further critical factor is how much phenotypic variation is present in size-at-age. This is important both because of its effect on the selection gradients, and also because it influences the period of time over which a cohort is under strong selection while it is being recruited into the fishery. For cod in RFA 4, this period is unlikely to be much more than one year, because the variation in length-at-age is relatively small compared with the rate at which they grow. In other species, such as North Sea plaice, the period may be somewhat greater (Bannister 1977). One should not suppose that a small selection response by one species can be extrapolated to another.

We would argue that, notwithstanding the small selection responses suggested by this analysis, it would be a mistake to ignore the potential for evolution due to exploitation of cod and other marine biological resources. First, the influence of fishermen on these resources as measured by fishing mortality rates is tending to increase, at least in the North Sea. Secondly, although the performance of nets under commercial fishing conditions may have been less selective than thought in the past, there is active interest in making nets more selective, for instance by switching from diagonal to square meshes (Anon 1991). Thirdly, selection generated by the size specificity of nets is but one of the selective pressures generated by exploitation. Simply by imposing extra mortality, exploitation acts as a selection pressure on the life history (Law 1979, Michod 1979), and can generate strong differentials on vital rates such as age and size at sexual maturation (Borisov 1978, Law and Grey 1989, Rowell, this volume). Clearly, on the year to year basis which occupies those concerned with setting catch quotas, these evolutionary changes are unimportant. But if we are concerned with the maintenance of the genetic health of these resources in the longer term, it would be sensible to consider the selection pressures generated by exploitation, and how they are modulated by the rules imposed to regulate this exploitation.

Table 3. Information concerning the matrix of population dynamics (**T**) for North Sea cod: vital rates, matrix elements, and right dominant eigenvector.

Age (j)	Proportion of females mature	Number of eggs per female 10 [6]	M_j	F_j	p_j	Right dominant eigenvector
(1)	(2)	(3)	(4)	(5)	(6)	
0	0.00	0.00	–	0.00	0.00	1.00
1	0.09	0.20	0.69	0.14	0.07	1.00
2	0.17	0.80	0.22	0.97	0.17	0.93
3	0.28	1.80	0.16	0.90	0.21	0.76
4	0.42	3.00	0.11	0.87	0.20	0.55
5	0.58	4.10	0.11	0.83	0.15	0.35
6	0.73	5.20	0.11	0.80	0.10	0.20
7	0.83	6.10	0.11	0.81	0.05	0.10
8	0.91	6.90	0.11	0.77	0.03	0.05
9	0.95	7.60	0.11	0.79	0.01	0.02
10	0.97	8.10	0.11	0.72	0.01	0.01
11	0.99	8.50	0.11	0.81	<0.01	<0.01
12	0.99	8.80	0.11	0.78	<0.01	<0.01
13	1.00	9.10	0.11	0.78	<0.01	<0.01
14	1.00	9.20	0.11	0.78	<0.01	<0.01

Notes for **Table 3**:

1 Age is given at the time of fishing, in the third quarter of the year; reproduction occurs in the following February and March.

2 The proportion mature was based on data provided by ICES from the International Young Fish Survey in RFA 4 which takes place around the time of reproduction. The data include 1980 and 1982 to 1990 (1412 females). A regression of proportion mature by age was carried out on logit transformed values using GLIM, giving an intercept -3.5 and gradient 0.65 for females. Males mature somewhat more gradually than females but, lacking any information on male mating success as a function of age, we have simply assumed that males contribute equally to females at each age (but see Brawn 1961).

3 Egg production was estimated from growth. A von Bertalanffy growth model for length was used with the parameters given by Daan (1974) for the Northern North Sea: l_∞ 118.7 cm, k=0.269, t_0=0.87 y. Length was converted to weight using the condition factor 0.0104 g cm^{-3} (Daan 1975), and this in turn was converted to eggs by the factor 569 g^{-1} (Oosthuizen and Daan 1974).

4 Instantaneous rates of natural mortality (M_j) are not available for RFA 4. The ones given here are preliminary estimates from an international fish stomach sampling survey for the North Sea (Macer and Easey 1988). The rate at age 0 is not known and has no effect on p_j because it is a common factor for all ages.

5 Instantaneous rates of fishing mortality (F_j) are the averages over 1987 to 1989 from a recent VPA for the whole North Sea (Stokes personal communication).

6 The p_js are the elements of the first column of **T**.

ACKNOWLEDGEMENTS

We thank T. K. Stokes and D. R. Grey for advice and discussion at many stages of this work.

REFERENCES

Anonymous 1991. Selectivity in towed fishing gears. Sea Fish Industry Authority, Technical Information Service 1991/1/FG. Sea Fish Industry Authority, St. Andrews Dock, Hull.

Armstrong, D. W., R. J. Fryer, S. A. Reeves and K. A. Coull. 1989. Cod-end selectivity of cod, haddock and whiting by Scottish trawlers and seiners. I.C.E.S. C.M./**B**:55.

Bannister, R. C. A. 1977. North Sea plaice. Pp 243 -282, in J. A. Gulland ed. Fish Population Dynamics. Wiley, London.

Barton N. H. and M. Turelli 1989. Evolutionary quantitative genetics: how little do we know? Annual Review of Genetics **23**: 337-370.

Borisov, V. M. 1978. The selective effect of fishing on the population structure of species with a long life cycle. Journal of Ichthyology **18**: 896-904.

Brawn, V. M. 1961. Reproductive behaviour of the cod (*Gadus callarias* L.). Behaviour **18**: 177-198.

Bulmer, M. G. 1980. The Mathematical Theory of Quantitative Genetics. Clarendon Press, Oxford.

Caswell, H. 1989. Matrix Population Models: Construction, Analysis and Interpretation. Sinauer, Sunderland Massachusetts.

Charlesworth, B. 1980. Evolution in Age-structured Populations. Cambridge University Press, Cambridge.

Daan, N. 1974. Growth of North Sea cod, *Gadus morhua*. Netherlands Journal of Sea Research **8**: 27-48.

Daan, N. 1975. Consumption and production in North Sea cod, *Gadus morhua*: an assessment of the ecological status of the stock. Netherlands Journal of Sea Research **9**: 24-55.

Falconer, D. S. 1989. Introduction to Quantitative Genetics. 3rd edition. Longman, Harlow.

Gjedrem, T. 1983. Genetic variation in quantitative traits and selective breeding in fish and shellfish. Aquaculture **33**: 51-72.

Gjerde, B. 1986. Growth and reproduction in fish and shellfish. Aquaculture **57**: 37-55.

Godø, O. R. and E. Moksness 1987. Growth and maturation of Norwegian Coastal cod and Northeast Arctic cod under different conditions. Fisheries Research **5**: 235-242.

Graham, M. 1934. Report on the North Sea cod. Fisheries Investigations, London **13**: 1-160.

Harding, D., L. Woolner and J. Dann. 1986. The English Groundfish Surveys in the North Sea. I.C.E.S. C.M./**G**:13.

Hill, W. G. 1974. Prediction and evaluation of response to selection with overlapping generations. Animal Production **18**: 117-139.

Houghton, R. G. 1977. Spawning and migration of cod on the North-East coast of England. Fisheries Notice Number **52**, Ministry of Agriculture Fisheries and Food, Lowestoft.

Jørgensen, T. 1990. Long-term changes in age at sexual maturity of Northeast Arctic cod (*Gadus morhua* L.). Journal du. Conseil International pour l' Exploration de la Mer **46**: 235-248.

Kirkpatrick, M. 1988. The evolution of size in size-structured populations. Pp 13-28, in B. Ebenman and L. Persson eds. Size-structured Populations. Springer-Verlag, Berlin.

Lancaster, P. 1969. Theory of matrices. Academic Press, New York.

Lande, R. and S. J. Arnold 1983. The measurement of selection on correlated characters. Evolution **37**: 1210-1226.

Law, R. 1979. Optimal life histories under age-specific predation. American Naturalist **114**: 399-417.

Law, R. 1991. On the quantitative genetics of correlated characters under directional selection in age-structured populations. Philosophical Transactions of the Royal Society London, Series B **331**: 213-223.

Law, R. and D. R. Grey 1989. Evolution of yields from populations with age-specific cropping. Evolutionary Ecology **3**: 343-359.

Lee, R. M. 1912. An investigation into the methods of growth determination in fishes by means of scales. Conseil de l'Exploration de la Mer Publications de Circonstance **63**: 1-34.

Macer, C. T. and M. W. Easey 1988. The North Sea cod and the English fishery. Ministry of Agriculture, Fisheries and Food, Directorate of Fisheries Research, Lowestoft. Laboratory Leaflet number **61**.

Michod, R. E. 1979. Evolution of life histories in response to age-specific mortality factors. American Naturalist **113**: 531-550.

Nelson, K. and M. Soulé. 1987. Genetical conservation of exploited fishes. Pp 345-368 in N. Ryman and F. Utter eds. Population Genetics and Fisheries Management. University of Washington Press, Seattle, Washington.

Oosthuizen, E. and N. Daan 1974. Egg fecundity and maturity of North Sea cod, *Gadus morhua*. Netherlands Journal of Sea Research **3**: 378-397.

Pope, J. G. 1982. Background to scientific advice on fisheries management. Ministry of Agriculture Fisheries and Food, Directorate of Fisheries Research, Lowestoft. Laboratory Leaflet number **54**.

Ricker, W. E. 1969. Effects of size-selective mortality and sampling bias on estimates of growth, mortality, production and yield. Journal of the Fisheries Research Board, Canada **26**: 479-541.

Ricker, W. E. 1975. Computation and interpretation of biological statistics of fish populations. Department of Fisheries and Oceans, Ottawa.

Ricker, W. E. 1981. Changes in the average size and average age of Pacific salmon. Canadian Journal of Fisheries and Aquatic Sciences **38**: 1636-1656.

Rijnsdorp, A. D., P. I. van Leeuwen and T. A. M. Visser. 1990. On the validity and precision of back-calculation of growth from otoliths of the plaice *Pleuronectes platessa* L. Fisheries Research **9**: 97-117.

Rowell, C.A. 1992. Selection differentials imposed by fishing activity on the North Sea cod (*Gadus morhua* L.). D.Phil. Thesis, University of York, York.

Stewart, P.A.M. and J.H.B. Robertson. 1985. Attachments to codends. Department of Agriculture and Fisheries Scotland, Scottish Fisheries Report Number **33**, Aberdeen.

van Leeuwen, P. I. and A. D. Rijnsdorp. 1986. The analysis of the growth of turbot by back calculation of otoliths. I.C.E.S. C.M./**G**:50.

Williams T. and B. C. Bedford. 1974. The use of otoliths for age determination. Pp 114-123, in T. B. Bagenal ed. Ageing of Fish. Unwin, Woking.

PART IV
MANAGEMENT AND EVOLUTION

EVOLUTIONARILY STABLE OPTIMAL HARVESTING STRATEGIES

D. R. Grey

Department of Probability and Statistics, The University of Sheffield, PO Box 597, Sheffield S10 2UN, U.K.

ABSTRACT

The adaptation of optimisation techniques in age-specific harvesting to populations whose life-histories may evolve as a result of harvesting was introduced in Law and Grey (1989). It was found possible to obtain optimal patterns which are more complex than the well-known two-age-class pattern for models without evolution. Here a more general model is proposed and an algorithm described for finding the solution to the optimality problem where density-dependent regulation and competition between phenotypes are both present.

GENERAL PRINCIPLES OF ESOHS

Harvesting a reproducing population according to a fixed pattern may lead to a steady state, *equilibrium*, in which the yield is obtained at a constant rate and the composition of the population remains constant over time.

If the population consists of a single type of individual, fixed in its life history traits (mortality and fecundity as functions of age and population density), then it is often possible to find the harvesting pattern which maximises the yield at equilibrium (the *O*ptimal *H*arvesting *S*trategy).

If, however, the population consists of a variety of different *types* (phenotypes) then the initial effect of a harvesting pattern is to impose differential selection on these types in such a way that only the fittest type will ultimately survive. Thus an equilibrium is achieved in which, typically, only one type (determined by the harvesting pattern) is present. This equilibrium is stable in the sense that other types cannot invade the population.

It may be possible to calculate, for each harvesting pattern, the fittest type and then the equilibrium yield corresponding to this pattern and type. The harvesting pattern which maximises this equilibrium yield is called the ESOHS (*E*volutionarily *S*table *O*ptimal *H*arvesting *S*trategy).

BACKGROUND

AGEING AND REPRODUCTION - LESLIE MATRIX MODEL

Consider first population dynamics without harvesting. Reproduction takes place at a definite time of year. The proportion of new-born individuals surviving to age i is denoted by l_i up to a maximum age of k. An individual surviving to age i produces b_i offspring at that age. If $x_i^{(n)}$ denotes the number of individuals of age i at time n and $\mathbf{x}^{(n)} = (x_0^{(n)},..., x_k^{(n)})^T$, then the dynamics of the population may be written as

$$\mathbf{x}^{(n+1)} = \mathbf{M}\mathbf{x}^{(n)}$$

$$\text{where} \quad \mathbf{M} = \begin{pmatrix} s_1b_1 & s_2b_2 & \cdots & s_kb_k & 0 \\ s_1 & 0 & \cdots & 0 & 0 \\ 0 & s_2 & \cdots & 0 & 0 \\ \cdot & \cdot & & \cdot & \cdot \\ \cdot & \cdot & & \cdot & \cdot \\ \cdot & \cdot & & \cdot & \cdot \\ 0 & 0 & \cdots & s_k & 0 \end{pmatrix}$$

with $s_1 = l_1$ and $s_i = l_i / l_{i-1}$ for $2 \le i \le k$. $\mathbf{M}$ is known as the Leslie matrix. The average number of offspring born to an individual in the course of its lifetime is

$$R_o = \sum_{i=1}^{k} l_i b_i .$$

If the life history parameters $\{l_i\}$ and $\{b_i\}$ are taken as constant then, in the absence of any other factors, a population with $R_o > 1$ will grow indefinitely, a population with $R_o < 1$ will diminish indefinitely, and a population with $R_o = 1$ will approach equilibrium. In general also, a *stable age distribution* will be achieved.

The effect of harvesting will be to reduce the value of R_o in some way. Thus, in order to achieve the equilibrium value $R_o = 1$, we must start with a population with $R_o > 1$; any other population cannot be steadily harvested.

DENSITY-DEPENDENT REGULATION

In reality, a population cannot be expected to grow indefinitely even in the absence of harvesting; the effect of overcrowding will be felt. One way of expressing this is in terms of a *density*

$$d = \sum_{i=0}^{k} a_i x_i \tag{1}$$

where x_i is the number of individuals present of age i and $a_i \geq 0$ is a constant representing the contribution to density from an individual of age i. We then allow the life history parameters to depend upon d, and write

$$R_o(d) = \sum_{i=1}^{k} l_i(d) b_i(d).$$

Typically, l_i and b_i will decrease continuously as d increases, and provided $R_o(d) > 1$ for sufficiently small d and $R_o(d) \to 0$ as $d \to \infty$, there will be a unique value of d such that $R_o(d) = 1$. At this density, equilibrium is possible, with the corresponding stable age distribution yielding the age-class sizes via Equation 1 above. The question of *stability* of this equilibrium is a complicated one. Technically if d^* denotes the equilibrium density, $\mathbf{x}^*$ denotes the corresponding population structure, $\mathbf{a} = (a_o,\ldots,a_k)^T$ and $\mathbf{M}(d)$ denotes the Leslie matrix as a function of density (so that in particular $\mathbf{M}(d^*)\mathbf{x}^* = \mathbf{x}^*$), then for local stability of the equilibrium we require the matrix $\mathbf{M}(d^*) + \mathbf{M}'(d^*)\mathbf{x}^*\mathbf{a}^T$ to have all eigenvalues less than one in absolute value. We shall not pursue the question of stability here, but be content to identify equilibria.

THE EFFECT OF HARVESTING

Harvesting is taken to be age-specific and proportional to population size within each age class. A harvesting pattern is described by a vector $(\eta_1, \eta_2, \ldots.\eta_k) = \eta$ of numbers, where η_i represents the proportion of individuals who are effectively earmarked at birth to be harvested at age i (immediately after reproduction) provided they survive natural mortality to that age. These numbers must satisfy the constraints

$$\eta_i \geq 0 \text{ and } \sum_{i=1}^{k} \eta_i \leq 1.$$

The effect of harvesting is to replace the survival parameter l_i by $l_i\xi_i$, where

$$\xi_i = 1 - \sum_{j=1}^{i-1} \eta_j$$

is the proportion of individuals who, if they survive natural mortality to age i, survive harvesting to that age as well. The average lifetime's offspring born to an individual under harvesting is therefore

$$R = \sum_{i=1}^{k} l_i \xi_i b_i \, .$$

Assuming density-dependence, the effect of a harvesting pattern will be to bring the density d to a level at which $R = 1$ (if such exists), so that equilibrium is achieved. At this density d, a stable age distribution (specific to the harvesting pattern) emerges and may be used to calculate, via Equation 1, the number x_o of new recruits to the population each year. The annual yield to the harvester may then be written as

$$H = x_0 \sum_{i=1}^{k} l_i \eta_i \omega_i$$

where ω_i denotes the value (weight) of an individual harvested at age i.

THE 'CLASSICAL' SOLUTION

The problem is to maximise H above over all feasible harvesting patterns, bearing in mind that each pattern determines a density d upon which the values of x_o and l_i depend. The way forward is as follows. Firstly consider a fixed d, and maximise the harvest H for that d. Then maximise over d.

The 'classical' solution (Reed 1980, Getz 1980) is to formulate the first maximisation as a linear programming problem, after a change of variable to $y_0 = x_0$ and $y_i = x_0 \eta_i$ for $1 \leq i \leq k$. Following Grey (1988), the problem may be written: maximise $\mathbf{w}^T\mathbf{y}$ subject to

$$\mathbf{y} \geq \mathbf{0}$$

$$\mathbf{m}^T\mathbf{y} \geq 0$$

$$\mathbf{a}^T(\mathbf{M} - \mathbf{I})^{-1} \mathbf{y} = d$$

where $\mathbf{m}^T$ is the last row of the matrix $(\mathbf{M} - \mathbf{I})^{-1}$ and the other vectors have obvious meanings. In this problem there are $k + 1$ variables and $k + 3$ constraints operating. The form of solution is therefore as follows: at most two of the η_i will be non-zero, and if there are two non-zero then they add up to 1. So the optimal pattern will be to harvest (partially) at one age and then either to harvest completely at a later age or not to harvest any further.

Since for each fixed d the optimal pattern takes the above form, the pattern which yields maximum harvest over all possible d has the same form. The maximisation over d may not be analytically feasible, but it is a one-dimensional maximisation which may be treated numerically without too much trouble.

To summarise: in principle the problem of finding the OHS may be solved via a linear programming problem followed by a one-dimensional maximisation. The solution takes the form of harvesting (partially) at one age only or of harvesting partially at one age and then completely at a later age.

POPULATIONS WITH SEVERAL TYPES

THE POSSIBILITY OF EVOLUTION

So far we have assumed a single type of individual for which the density function $\Sigma\, a_i x_i$ and the life history parameters $l_i(d)$ and $b_i(d)$ are fixed and known.

We now add the possibility of a number of different types co-existing. To each type corresponds an index parameter p. For the moment we assume no harvesting. The performance of the population is taken to be affected by an overall measure of density

$$d = \sum_p \sum_i a_i(p) x_i(p)$$

to which all types contribute in an additive way. Life history parameters depend on type as well as density; so we may now write

$$R_0(p,d) = \sum_i l_i(p,d) b_i(p,d)$$

for the average number of offspring during the lifetime of a newborn individual of type p, assuming constant density d.

For an equilibrium in which the composition of the population remains constant over time, we require *all* the types present to satisfy $R_0(p,d) = 1$, where d is the density corresponding to this composition.

STABILITY OF EQUILIBRIUM

Without going too deeply into the density-dependent dynamics of a population with several types not in equilibrium, we can argue intuitively that the only density d which can sustain a *stable* equilibrium (one which is resistant to invasion by other types) is the *maximum* value of d which can sustain any kind of equilibrium, i.e. the maximum d for which there exists a type p satisfying $R_0(p,d) = 1$. Call this value d^*.

The argument runs as follows. Suppose we have an equilibrium at some density $d < d^*$. Let p^* be a type satisfying $R_0(p^*,d^*) = 1$. If a small quantity of this type is introduced into the population then it will grow, because $R_0(p^*,d) > 1$, yielding an immediate increase in density which will in turn cause the type(s) p originally present to diminish, since they will now satisfy $R_0(p,d) < 1$. Thus the effect of the disturbance will be magnified and the only way in which a new equilibrium can be reached is when the new type p^* has entirely replaced the old type(s) and a higher density d^* has been established.

This argument suggests that, in the absence of harvesting, only those types (possibly unique) which can maintain equilibrium at the highest density can ultimately be selected. This result has been noted before: see for example Roughgarden (1979: 454 et seq.).

THE EFFECT OF HARVESTING

As before, the effect of a harvesting pattern $\eta = (\eta_1, \eta_2, \ldots, \eta_k)$ is to replace the life history parameters l_i by $l_i \xi_i$ where

$$\xi_i = 1 - \sum_{j=1}^{i-1} \eta_j .$$

Now

$$R(p,d) = \sum_i l_i(p,d) \xi_i b_i(p,d).$$

The effect of prolonged harvesting at a steady rate will be to select a stable equilibrium. By what has been said previously, this means that d is selected as large as possible subject to there being at least one p satisfying $R(p,d) = 1$, and *only* such p will be present in the selected equilibrium. If p is a one-dimensional parameter, typically there will be only one such p. We assume this from now on.

Once this type has been selected, the annual harvest may be written as

$$H = x_0 \sum_{i=1}^{k} l_i \eta_i \omega_i$$

as before, but now x_0, l_i and ω_i depend upon the type selected as well as the density.

The principle of ESOHS is now to find the harvesting pattern which yields the maximum value of H.

AN EARLIER EXAMPLE

In Law and Grey (1989) we looked at a rather more restricted model in which density is measured purely in terms of number of new-borns, and the effect of density is simply to determine the proportion of these new-borns which survive to age one (independently of type). This model is the special case of the present one in which

$$d = x_0 \qquad \text{(i.e. } a_0 = 1, a_2 = \ldots = a_k = 0\text{)}$$

$$b_i(p,d) = b_i(p)$$

$$l_i(p,d) \equiv l_i g(d)$$

where g is the survival function which does not depend upon type, each l_i is a parameter which depends neither on density nor type, and each b_i does not depend upon density.

Within this more restricted model it was found possible to obtain an ESOHS which was intrinsically more complex than the 'classical' solution to the problem without selection. An example was found where the ESOHS was to harvest partially at one age, then partially at a later age, and then leave the remainder of the population unharvested.

FINDING THE ESOHS

THE ALGORITHM

To find the ESOHS in particular examples, it appears that we need an algorithm involving the following steps.

Stage I. Fix the harvesting pattern η .

Stage II. For each density d, find the type p which at density d has the largest R.

Stage III. Find the value of d for which the maximum in Stage II is $R = 1$. Denote this by d^* and the corresponding type by p^*.

Stage IV. Use the stable age distribution for type p^* at density d^* under harvesting pattern η to calculate the value of x_0 ; more specifically, solve the equation

$$d^* = x_0 \sum_{i=0}^{k} a_i(p^*) l_i(p^*,d) \xi_i$$

for x_0,where here $l_0 = 1$, $\xi_0 = 1$.

Stage V. Calculate the corresponding annual harvest

$$H = x_0 \sum_{i=1}^{k} l_i(p^*, d^*)\eta_i \omega_i(p^*).$$

Stage VI. Now vary η and find where H is maximised.

AN EXAMPLE

The programme outlined in the above algorithm is ambitious; it is unlikely that it will be possible to perform all the steps analytically, leading to an explicit formula for the ESOHS. Sooner or later, numerical methods may be needed.

In the example which we consider here, at least Stages I to V may be performed analytically; it is only the final stage at which we resort to computation.

The details of the example are as follows. We suppose density to be measured by the number of newborns: $d = x_0$. This is not an ideal choice but it is made so as to keep calculations tractable as far as possible. There are three age classes ($k = 3$); harvesting is not possible at age 3 ($\omega_3 = 0$) and so a harvesting pattern may be considered as $\eta = (\eta_1, \eta_2)$. Individuals of age 1 do not reproduce ($b_1 = 0$) but fecundity at ages 2 and 3 is proportional to a (positive real) parameter p which is taken as the *type* of individual:

$$b_i(p,d) = \beta_i p \quad \text{for } i = 2,3$$

where β_2, β_3 are positive parameters. To offset the greater fecundity for larger p, we introduce higher mortality for larger p, via the formula

$$l_i(p,d) = (1 - \lambda_i \, dp)^+ \quad \text{for } i = 1,2,3$$

where x^+ denotes x if $x \geq 0$ and 0 if $x < 0$. Here, λ_1, λ_2 and λ_3 are positive parameters satisfying $\lambda_1 < \lambda_2 < \lambda_3$.

Note that $l_i(p,d)$ is a decreasing linear function of density d as long as $1 - \lambda_i dp$ is positive; if it is negative, type p cannot survive to age i at density d.

As a final ingredient in the example, we take weight to be independent of type, i.e. ω_1 and ω_2 are fixed. Again, this is not an ideal choice, but is made for comparative computational simplicity.

Stages I and II. In this example, we have, for a given η

$$R(p,d) = \xi_2(1 - \lambda_2 dp)^+ \beta_2 p + \xi_3(1 - \lambda_3 dp)^+ \beta_3 p.$$

In the range where $1 - \lambda_2 dp$ and $1 - \lambda_3 dp$ are positive, the above expression is a quadratic in p which has a maximum when

$$p = \frac{\xi_2\beta_2 + \xi_3\beta_3}{2d(\xi_2\beta_2\lambda_2 + \xi_3\beta_3\lambda_3)}.$$

It is not hard to show that this yields an overall maximum of R provided $\lambda_3 < 2\lambda_2$, and we assume this as an extra condition on our parameters.

The maximised value of R is found by substituting back, yielding

$$R = \frac{(\xi_2\beta_2 + \xi_3\beta_3)^2}{4d(\xi_2\beta_2\lambda_2 + \xi_3\beta_3\lambda_3)}.$$

Stage III. The above maximised value of R is therefore equal to 1 when

$$d = d^* = \frac{(\xi_2\beta_2 + \xi_3\beta_3)^2}{4(\xi_2\beta_2\lambda_2 + \xi_3\beta_3\lambda_3)}.$$

and, substituting back, it follows that

$$p^* = \frac{2}{\xi_2\beta_2 + \xi_3\beta_3}.$$

Stage IV. This is trivial: $x_o = d^*$.

Stage V.

$$H = d^* \{(1 - \lambda_1 d^*p^*)\, \eta_1\, \omega_1 + (1 - \lambda_2 d^*p^*)\, \eta_2\, \omega_2\}.$$

This can now be expressed entirely in terms of the variables ξ_2 and ξ_3 using the preceding formulae for d^* and p^*, and noting that $\eta_1 = 1 - \xi_2$ and $\eta_2 = \xi_2 - \xi_3$.

Stage VI. The above H needs to be maximised with respect to ξ_2 and ξ_3 subject to the constraints (implied by the constraints on η) $0 \le \xi_3 \le \xi_2 \le 1$. Bearing in mind that the maximum may well lie on the boundary of the feasible region, and recognising the non-linearity of this problem, it seems necessary at this stage to resort to numerical computation, for specific values of the given parameters.

This computation was performed in a fairly crude way, calculating H at each of a lattice of points in the region (described in terms of η) $\eta_1 \geq 0$, $\eta_2 \geq 0$, $\eta_1 + \eta_2 \leq 1$, observing where H appears greatest, laying down a finer lattice and repeating the process until the maximum is located with reasonable accuracy.

Of particular interest are cases where the maximum occurs in the interior of the feasible region, i.e. the inequalities $\eta_1 > 0$, $\eta_2 > 0$ and $\eta_1 + \eta_2 < 1$ are all satisfied. Such a harvesting pattern differs from the 'classical' optimal pattern in a model without selection, being a mixture of three 'pure strategies' (harvest at age 1, harvest at age 2, do not harvest).

This phenomenon was found when the parameters take the following values: $\lambda_1 = 1$, $\lambda_2 = 2$, $\lambda_3 = 3$; $\beta_2 = 2$, $\beta_3 = 3$; $\omega_1 = 5$, $\omega_2 = 4$. In this case, H achieves a maximum value of 0.5186 when $\eta_1 = 0.08$ and $\eta_2 = 0.74$.

SUMMARY AND FURTHER THOUGHTS

We have produced a general model for age-specific harvesting in the presence of density-dependent regulation and the possibility of competition between phenotypes with differing life-history characteristics. The ESOHS, and an algorithm for attempting to find it, have been described. It has been shown by an artificial example that even when many elements of the model are simplified (e.g. only three age classes, density measured by number of new-borns, weights independent of type), the problem of finding the ESOHS is not straightforward.

Nevertheless it is hoped that a model has been produced which will allow more realistic problems to be tackled, if not to find the ESOHS, then at least to compare different harvesting policies for their likely productivity.

What is lacking so far is any general theory arising from the model, e.g. concerning the form taken by the ESOHS in general. It may be that simplifying assumptions will have to be made before it is possible to produce any mathematical proofs.

REFERENCES

Getz, W.M. 1980. Optimal harvesting of structured populations. Mathematical Biosciences **48**: 279-292.

Grey, D.R. 1988. Harvesting under density-dependent mortality and fecundity. Journal of Mathematical Biology **26**: 193-197.

Law, R. and Grey, D.R. 1989. Evolution of yields from populations with age-specific cropping. Evolutionary Ecology **3**: 343-359.

Reed, W.J. 1980. Optimum age-specific harvesting in a non-linear population model. Biometrics **36**: 579-593.

Roughgarden, J. 1979. Theory of Population Genetics and Evolutionary Ecology: an Introduction. Macmillan, New York.

ECOGENETIC ANALYSIS AND EVOLUTIONARY STABLE STRATEGIES IN HARVESTED POPULATIONS

Wayne M. Getz[1] *and Veijo Kaitala*[2]

[1]Department of Entomology, University of California, Berkeley, CA 94720, USA
[2]Systems Analysis Laboratory, Helsinki University of Technology, 02150 Espoo, Finland and Department of Zoology University of Helsinki, 00100 Helsinki, Finland.

ABSTRACT

The demographic and genetic structure of dynamic populations is usually severally impacted by management, especially harvesting, practices. Changes to this structure are often difficult to predict because of nonlinearities inherent in the underlying demographic, ecological, and evolutionary processes. Here we describe a step-by-step procedure for separating out elements of these dynamic processes, thereby enhancing our understanding of the nature of changes that have been observed to occur in exploited populations. The methods outlined are then applied to a discussion of reproductive life-history changes that may occur in harvested salmon populations.

INTRODUCTION

The demographic response and bioeconomic properties of harvested populations have been analyzed in considerable detail over the past twenty years (Clark 1976; Getz and Haight 1989). More recently, attention has been focused on the genetic and/or phenotypic response of populations to harvesting (Bergh and Getz 1989) and evolutionarily stable strategy (ESS) analysis has been employed in this regard (Law and Grey 1989).

The concept of an evolutionarily stable strategy (ESS), as defined by Maynard Smith and Price (1973; for a comprehensive overview see Maynard Smith 1982; Hines 1987), has proven extremely valuable in helping us understand the evolution of certain types of behavioral and life history strategies in plants and animals. Maynard Smith's original ESS formulation was in the context of the deterministic dynamics of frequencies of genes or individuals playing agonistic matrix games. As such, Maynard Smith's formulation fell within the realm of the mathematical theory of discrete games.

Standard ESS analysis is based on the assumptions of infinite population size and asexual inheritance of behavioral or life history traits. As Maynard Smith (1982) points out, the results of this theory usually apply almost as well to large but finite sexually reproducing populations. Maynard Smith does identify some exceptions when the incorporation of genetic structure into a population model is crucial to understanding the evolution of behavioral traits: specifically with regard to traits associated with sexual selection.

Standard ESS analyses, which assume that the fitness of individuals playing particular strategies (i.e. players) is purely a function of the relative frequencies of the various players, are severely limited when ecological factors play a central role in determining the dynamics of population interactions. In such cases, as discussed in Vincent and Brown (1988), it is more appropriate to model the population process using ecological rather than purely frequency dependent fitness constructs. This approach no longer relies on an infinite population size assumption although in the deterministic case we implicitly assume that the population is large enough to effectively smooth out stochastic birth and death events at the level of the individual. Stochastic systems are much more difficult to analyze but some problems have been addressed which relate to populations growing in stochastic environments (Ellner 1985a, b; Kaitala, Kaitala, and Getz 1989).

Over the past decade, Vincent, Brown and others (see the review by Vincent and Brown [1988] and references therein) have extended the ESS approach to a more general ecological setting where both frequencies and densities, rather than just frequencies, of groups with different life-history or behavior patterns (i.e. different strategies) influence the dynamics of the population process (i.e. of the game). The focus of this extension is the development of a comprehensive dynamic framework which can be used to identify and construct ESS solutions. The derivation of ESS solutions, however, like the derivation of optimal solutions to optimization problems, is only a small part of the story. A comprehensive understanding of evolutionary games requires some knowledge of the dynamic properties of the underlying populations process. Only with this knowledge can we begin to anticipate how the ESS might respond to changing environmental circumstances, including anthropogenic factors such as harvesting. Thus, for example, to obtain a better understanding of the relationship between ecological and evolutionary factors in our models we explicitly identify a set of life history parameters that pertain to the growth of each strategy-homogeneous group within a population and identify a set of interaction parameters which define how different strategy-homogeneous groups within the population impact on one another.

One issue that we have little insight into at this time is how the genetic structure of populations responds in a changing environment and how much explicit genetic structure is required in our models to address these types of evolutionary ecology questions. Since the direct evaluation of ESS solutions does not provide sufficient insight into both the ecological and genetic process at work, we propose that the problem of determining how the genetic structure of a population might respond to harvesting be approached through a sequence of steps, where the analysis at each step builds on the understanding gained from the previous step. One such sequence is to: 1) characterize the ecological dynamics of a strategy-homogeneous population under various harvesting regimes, 2) study the competitive dynamics of a strategy-stratified population under those same harvesting regimes, 3) use a game theoretic setting to identify the how ESS solutions may change in response to harvesting; and, if applicable, 4) incorporate genetic structure to obtain a deeper insight into the evolutionary process under harvesting.

As we discuss below, step 1 is often straightforward and a large literature exists on how to approach this analysis in discrete and continuous time single species models (Clark, 1976, 1985; Haight and Getz 1989). Step 2 is more complicated because the number of parameters in the model can become unmanageable even when competition between only two populations is considered (Bergh and Getz 1989). The search for an ESS in step 3, given a particular harvesting policy, usually requires a large number of numerical simulations (Hines, 1987). Hence to characterize the ESS as a function of different harvesting policies can present difficulties. Further, it may not be clear *a priori* whether an ecological model without explicit genetic structure is adequate for characterizing the ESS strategy or whether a genetically detailed model is required for this purpose, as mentioned in step 4. As we discuss later, in some cases explicit genetic models might provide a means for by passing step 3 to obtain the ESS life-history of a population under various harvesting strategies; although we do not yet understand how to characterize the conditions when this approach applies.

MODELING FRAMEWORK

In this section we discuss the implementation of steps 1 to 3 above in a general discrete time modelling framework. Our approach applies equally well to continuous time models. To keep the notation simple in our presentation of a general modelling approach, we assume that the population densities of all groups of players can be characterized by a non-negative scalar variable, although the approach holds (as will be illustrated in the case of specific examples) for structured groups of players (e.g., structured into age-classes, spatial regions, etc.).

HOMOGENEOUS POPULATION DYNAMICS

The first step in the analysis is to obtain understanding of underlying dynamic processes in terms of existence of equilibria and their stability properties under harvesting. It is important to understand how behavioral and/or life-history strategies influence the value and properties of the equilibria and whether the population dynamics exhibit any notable properties such as cyclic or chaotic trajectories.

More specifically, let $x(t)$ represent the density of individuals (numbers or biomass) in a population at time t. If the population has age, stage, or spatial structure, we can regard x as a vector of appropriate population classes $\boldsymbol{x}(t) = (x_1, \ldots, x_n)'$. For the scalar case, $x(t)$, we assume that we are able to characterize the transition of the population from one time period to the next by the equation

$$x(t+1) = x(t)f(x(t), \boldsymbol{\alpha}, v(t)), \tag{1}$$

where $f(\cdot)$ is a per-capita "growth" function, $\boldsymbol{\alpha} = (\alpha_1, \ldots, \alpha_r)'$ is a vector of r parameters appearing in the function f, and v is a harvesting intensity variable. (In the vector case $\boldsymbol{x}$, we have a vector valued function $\boldsymbol{f}$ that has both growth and transition components. In this case, equation (1) generalizes to $x_i(t+1) = x_i(t)f_i(\boldsymbol{x}(t), \boldsymbol{\alpha}, v(t)), i = 1, \ldots, n$). The parameters $\boldsymbol{\alpha}$ are thought of as characterizing the strategy "$\boldsymbol{\alpha}$" of individuals within this strategy-homogeneous population.

Note that it is problematic to include harvesting as follows,

$$x(t+1) = x(t)f(x(t), \boldsymbol{\alpha}, v(t)) - h(x(t)),$$

($h(\cdot)$ is a "harvest" which specifies how much biomass has been extracted from the population during time period $(t, t+1]$), since we have to pay particular attention to the simultaneous dependence of f and h on $x(t)$. For example, if harvesting occurs at the beginning of the interval $(t, t+1]$, then the growth may be more appropriately evaluated by replacing $x(t)$ in the right-hand-side of equation (1) with the residual quantity $x(t) - h(x(t))$ (i.e., r.h.s. becomes $(x(t) - h(x(t)))f(x(t) - h(x(t)), \boldsymbol{\alpha})$. On the other hand, if harvesting occurs at the end of the interval $(t, t+1]$, then it might be more appropriate to determine the harvested amount in terms of $h(f(x(t), \boldsymbol{\alpha}))$. Finally, in many resource systems, especially fisheries, harvesting takes place over an interval of time, rather than at a specific point in time (see Getz and Haight [1989] for further consideration of this issue)

The analysis of the dynamic properties of equation (1) follows standard procedures (Murray

1989) and begins by identifying the set of equilibrium solutions $x_j^\alpha(v)$ which satisfy

$$x_j^\alpha(v) = f(x_j^\alpha, \boldsymbol{\alpha}, v). \tag{2}$$

Because equation (1) represents a biological process $x_0^\alpha(v) \equiv 0$ for all v is a solution of equation (2). In viable populations—that is populations able to increase at low densities—0 is often unstable (populations grow away from it) and a second equilibrium $x_1^\alpha(v) > 0$ exists. This equilibrium may be stable or unstable. If both 0 and $x_1^\alpha(v)$ are unstable then the population approaches cyclic or exhibits chaotic dynamics (May and Oster 1976).

As an illustration, consider the model (Getz and Kaitala 1989)

$$f(x, \boldsymbol{\alpha}, v) = \frac{\alpha_1 \alpha_2(v)}{1 + (x/\alpha_3)^{\alpha_4}}, \tag{3}$$

where $\alpha_1 > 0$ is a birth rate, $0 < \alpha_2(v) < 1$ is a survival rate which depends on the harvesting intensity v, and $\alpha_3 > 0$ and $\alpha_4 > 1$ are parameters associated with density dependent effects. (The detailed properties of this model—which, for different parameter values, exhibits extinction, stable, oscillatory, and chaotic behavior—are discussed in Getz and Kaitala, 1989.) Given that an organism has limited amounts of energy, the more it invests in increasing its birth rate α_1 the less energy it has to invest in increasing its survival rate by, for example, evolving mechanisms for coping with parasites, avoiding predators, or coping with climatic extremes. Thus, when we think of groups playing contrasting strategies, we should think in terms of, say, some individuals having large α_1 (high birth rates) and small α_3 (low survival at high densities) versus small α_1 and large α_3. Consequently, whenever we formulate a problem, we need to consider how the parameters α_i are to be constrained: that is, we identify a constraint set A such that $\boldsymbol{\alpha} \in A$.

Although considerable controversy still surrounds the question of which models are most appropriate to use in particular ecological situations (e.g. Oksanen 1991), a number of different models, of which equation (3) is just one, have been proposed, particularly for populations with discrete synchronized generations (May 1981; Getz and Kaitala 1989). Populations exhibiting depensatory effects at low densities (the so-called Allee effect in which populations are not viable at low densities because, for example, they are highly social or have difficulty finding mates) typically have 3 equilibria when $v = 0$: namely, 0 which is stable, $x_1^\alpha(0) > 0$ which is unstable and $x_2^\alpha(0) > x_1^\alpha(0)$ which may be stable or unstable. As the harvesting intensity is increased, the two larger equilibria may disappear leaving 0 as the only stable equilibrium in the population (Clark 1976).

Before we go on to explore the dynamics of populations structured into groups playing different strategies (say e.g. two groups playing strategies $\boldsymbol{\alpha}$ and $\boldsymbol{\beta}$) we need to understand the properties of the equilibria and the nature of the dynamic solutions. In addition we should characterize the set of viable strategies (i.e., those that do not cause the population to go extinct) and identify which strategies in the constraint set A maximize population growth or density in some sense (e.g., if a single stable equilibrium is associated with a set of strategies, then which strategy provides for the maximum equilibrium value).

COMPETION BETWEEN NONINTERBREEDING POPULATIONS PLAYING DIFFERENT STRATEGIES

In the simplest case we can consider the interactions between two or more groups in a population that differ only in terms of the particular strategies they play. An analysis of the dynamics of this competitive interaction will often reveal that one group is superior to the others in the sense that the inferior groups eventually go extinct (or asymptotically approach zero). This analysis falls under the rubric of the theory of competitive exclusion, which is well known in the context of the Gause-Lotka-Volterra continuous time model (Yodzis 1989), but has also been studied in the context of more general continuous time models (Levin 1970).

Although, as discussed later, we have some insights into competition of discrete populations with and without age-structure, we still lack a good appreciation of the complexities of competing populations in a discrete time setting. Let $x_i(t)$, $i = 1, \ldots, n$, represent density of n groups in the population, respectively playing strategies $\boldsymbol{\alpha}_i$. To keep the parameteric structure of the models transparent, we have departed from the notation of Vincent and Brown (1988) by regarding $\boldsymbol{\alpha}_i$ as a vector of parameters $\boldsymbol{\alpha}_i = (\alpha_{i1}, \ldots, \alpha_{ir})'$ that appear in the model. (For game theoretic reasons, Vincent and Brown identify a set of strategies $\boldsymbol{u}^i$ and then make their model parameters functions of these strategies. The difference between the two approaches is purely technical.) We also explicitly identify a set of interaction parameters $\boldsymbol{\beta}_i = (\beta_{i1}, \ldots, \beta_{is})'$ which are constrained to lie in the set B, which characterizes the effect that the groups have on one another's growth rate. To simplify our presentation, we assume that the interaction parameters $\boldsymbol{\beta}_i$ are independent of the strategies $\boldsymbol{\alpha}_i$ played by the various groups of players. (A more general treatment would be to define group strategies $\boldsymbol{u}_i$, $i = 1, \ldots, n$, and game strategy vectors $U = (\boldsymbol{u}_1, \ldots, \boldsymbol{u}_n)'$, and then to assume that each parameter $\boldsymbol{\alpha}_i$ is a function of $\boldsymbol{u}_i$ and all parameters $\boldsymbol{\beta}_i$ are functions of U—see Vincent and Brown 1988). In our more restrictive setting, the formulation is sufficient to address questions relating to the evolution of ecological parameter values—the focus of this presentation, but generally will be inadequate to address questions pertaining to the evolution of agonistic forms of behavior. (The extension is straightforward but notationally more cumbersome). We should expect, however, that the ecological dynamics of the ith group will be impacted by the density of the remaining groups which we denote by $\boldsymbol{x}^i = (x_1, \ldots, x_{i-1}, x_{i+1}, \ldots, x_n)$.

Under the above assumptions, a discrete time model of n interacting groups playing strategies $\boldsymbol{\alpha}_i \in A$ and $\boldsymbol{\beta}_i \in B$, $i = 1, \ldots, n$, has the general form

$$x_i(t+1) = x_i(t) g(x_i(t), \boldsymbol{\alpha}_i, v_i(t); \boldsymbol{x}^i(t), \boldsymbol{\beta}_i) \qquad i = 1, \ldots, n, \tag{4}$$

where $v_i(t)$ is the intensity of harvesting on the ith strategy group. Note that the effects of competition typically imply that $g(x_i, \boldsymbol{\alpha}_i, v_i; \boldsymbol{x}^i, \boldsymbol{\beta}_i)$ is a decreasing function of x_j for $j \neq i$.

To ensure consistency, equations (4) are related to equation (1) by the condition that

$$g(\cdot, \cdot, \cdot; 0, \boldsymbol{\beta}_i) = f(\cdot, \cdot, \cdot) \tag{5}$$

for all $\boldsymbol{\beta}_i \in B$. From these consistency conditions it immediately follows that if $x^\alpha(v)$ is an equilibrium solution to equation (1), then the vector $(0, \ldots, 0, x^\alpha(v), 0, \ldots)'$ (where $x^\alpha(v)$ appears in the jth position) is the so-called isolation equilibrium for the jth group with regard to the population process modeled by equations (4).

Following Vincent and Brown (1988), we can reformulate equations (4) in terms of total population density

$$N(t) = \sum_{i=1}^{n} x_i(t) \tag{6}$$

and group frequencies

$$p_i(t) = x_i(t)/N(t) \tag{7}$$

If we now define $\boldsymbol{p} = (p_1, \ldots, p_n)'$, $\boldsymbol{p}^i = (p_1, \ldots, p_{i-1}, p_{i+1}, \ldots, p_n)'$ and $\boldsymbol{v} = (v_1, \ldots, v_n)'$ $i = 1, \ldots, n$, we can use identities (6) and (7) to transform variables employing the following notation:
$G(N(t), p_i(t), \boldsymbol{p}^i(t), \boldsymbol{\alpha}_i, \boldsymbol{\beta}_i, v_i) \equiv g(x_i(t), \boldsymbol{\alpha}_i, v_i; \boldsymbol{x}^i(t), \boldsymbol{\beta}_i)$, $\mathcal{A} = (\boldsymbol{\alpha}_1, \ldots, \boldsymbol{\alpha}_n)'$ and $\mathcal{B} =$

$(\beta_1, \ldots, \beta_n)'$. With this notation, and the identity

$$\mathcal{G}(N(t), \boldsymbol{p}(t), \mathcal{A}, \mathcal{B}, \boldsymbol{v}) = \sum_{i=1}^{n} p_i G(N(t), p_i(t), \boldsymbol{p}^i(t), \boldsymbol{\alpha}_i, \boldsymbol{\beta}_i, v_i), \tag{8}$$

it can be shown that equations (4) transform to (Vincent and Brown, 1988)

$$\begin{aligned} N(t+1) &= N(t)\mathcal{G}(N(t), \boldsymbol{p}(t), \mathcal{A}, \mathcal{B}, \boldsymbol{v}) \\ p_i(t+1) &= p_i(t)\frac{G(N(t), p_i(t), \boldsymbol{p}^i(t), \boldsymbol{\alpha}_i, \boldsymbol{\beta}_i, v_i)}{\mathcal{G}(N(t), \boldsymbol{p}(t), \mathcal{A}, \mathcal{B}, \boldsymbol{v})} \qquad i = 1, \ldots, n-1. \end{aligned} \tag{9}$$

Note that an equation for p_n is redundant since the relationship $\sum_{i=1}^{n} p_i = 1$ (system (9) consists of n equations as it stands, as does system (4)).

The function $G(N(t), p_i(t), \boldsymbol{p}^i(t), \boldsymbol{\alpha}_i, \boldsymbol{\beta}_i, v_i(t))$ is the so-called fitness-generating or G-function of Vincent and Brown (1988) and $\mathcal{G}(N(t), \boldsymbol{p}(t), \mathcal{A}, \mathcal{B}, \boldsymbol{v}(t))$ is the average of these functions at time t. In comparing equations (4) with equations (9), it also follows that, for given $\boldsymbol{v}$, these G-functions are just growth functions which at an equilibrium $(\hat{N}(\boldsymbol{v}), \hat{\boldsymbol{p}}(\boldsymbol{v}))$, corresponding to parameters $\hat{\boldsymbol{\alpha}}_i$ and $\hat{\boldsymbol{\beta}}_i$, must clearly satisfy $G(\hat{N}(\boldsymbol{v}), \hat{p}_i(\boldsymbol{v}), \hat{\boldsymbol{p}}^i(\boldsymbol{v}), \hat{\boldsymbol{\alpha}}_i, \hat{\boldsymbol{\beta}}_i, v_i) = 1$.

A first step in understanding the dynamics of system (4), or equivalently system (9), is to explore the conditions (i.e., the range of values of the life-history parameters $\boldsymbol{\alpha}_1$ and $\boldsymbol{\alpha}_2$, interaction parameters $\boldsymbol{\beta}_1$ and $\boldsymbol{\beta}_2$, and harvesting intensities v_1 and v_2) under which coexistence of two groups and, conversely, competitive exclusion of one group by the other takes place.

More concretely, consider the following G-function related to equation (3),

$$G(N(t), p_i(t), \boldsymbol{p}^i(t), \boldsymbol{\alpha}_i, \boldsymbol{\beta}_i, v_i) = \frac{\alpha_{1i}\alpha_{2i}(v_i)}{1 + (\sum_{i=1}^{n} \beta_{ij} p_j N/\alpha_{3i})^{\alpha_{4i}}}, \tag{10}$$

which satisfies compatible condition (5) provided $\beta_{ii} = 1$, $i = 1, \ldots, n$. For $n = 2$, conditions under which model (10) permits two groups to coexist are discussed in detail in Getz and Kaitala (1989). Their analysis raises a number of points which are germane to the issues considered here.

For example, suppose that $\hat{x}_1$ and $\hat{x}_2$ are unique positive equilibria associated with equations (1) and (3) when the strategies respectively are $\hat{\boldsymbol{\alpha}}_1$ and $\hat{\boldsymbol{\alpha}}_2$ and the harvesting levels respectively are $\hat{v}_1$ and $\hat{v}_2$. Further, suppose that there exists a unique interior equilibrium $(\bar{x}_1, \bar{x}_2)$ of equations (9) and (10) corresponding to strategy parameters $\hat{\boldsymbol{\alpha}}_1$, $\hat{\boldsymbol{\alpha}}_2$, $\hat{\boldsymbol{\beta}}_1$ and $\hat{\boldsymbol{\beta}}_2$ and harvesting levels $\hat{v}_1$ and $\hat{v}_2$. In this case, the internal equilibrium $(\bar{x}_1, \bar{x}_2)$, the isolation equilibria $(\hat{x}_1, 0)$ and $(0, \hat{x}_2)$, and the origin or trivial equilibrium $(0, 0)$ are the only equilibria associated with equations (1) and (3). As illustrated in Getz and Kaitala (1989), the behavior of this system is extremely complicated. For example, coexistence chaotic behavior occurs under conditions where the coexistence and isolation equilibria are themselves unstable.

An important evolutionary issue is to determine whether a particular strategy exists that is superior in some sense to all other available strategies. For example, for given harvesting intensity v, does there exist a strategy $\boldsymbol{\alpha}^*(v)$ that will always competitively exclude all other strategies? Such a strategy automatically satisfies the definition of an ESS since it cannot be invaded by any other strategy. If such a strategy exists, one might ask whether it also is superior in some sense in the context of the homogeneous population dynamics modeled by equation (1). For the general formulation, the answer to this question is open, but answers have been obtained for specific problems.

For example, Bergh and Getz (1989) analysed a discrete Leslie matrix type model of an age-structured population with a density dependent (specifically compensatory) birth function. They defined the strategy of each group to be the set of its life-history parameters $\boldsymbol{\alpha}$ responsible for determining the value $\boldsymbol{x}^{\alpha}(v)$ of that groups isolation equilibrium when the harvesting intensity is v. In a competitive situation, Bergh and Getz showed that as long as a coexistence equilibrium is never possible, the group that wins is the group with the largest isolation equilibrium $\boldsymbol{x}^{\alpha}(\boldsymbol{v})$.

EVOLUTIONARY STABLE STRATEGIES

The simplest notion of an ESS strategy is that a population playing an ESS strategy cannot be invaded by a low frequency (i.e. mutant) group playing any other strategy. As pointed out by Brown and Vincent (1987), the notion is more complicated since the ESS can represent a mix of strategies played by coalitions of groups able to resist invasion by groups playing mutant strategies.

In the case that an ESS is given by the proportions $p_i^*(\boldsymbol{v}) > 0,\ i = 1,\ldots,r < n$ and $p_i^*(\boldsymbol{v}) = 0,\ i = r,\ldots,n$, given harvesting intensity v and strategy types playing strategies $\boldsymbol{\alpha}_i^*(\boldsymbol{v}) \in A$ and $\boldsymbol{\beta}_i^*(\boldsymbol{v}) \in B,\ i = 1\ldots,r$ such that $(N^*(\boldsymbol{v}), \boldsymbol{p}^*(\boldsymbol{v}))$ is an equilibrium of system (9), then it follows that the isolation equilibrium $(N^*(\boldsymbol{v}), \boldsymbol{p}^*(\boldsymbol{v}))$ (located in the subspace corresponding to $p_i^*(\boldsymbol{v}) = 0,\ i = r,\ldots,n$) is locally stable. This stability criterion implies a set of algebraic conditions corresponding to maximizing $G(N^*(\boldsymbol{v}), p_i^*(\boldsymbol{v}), \boldsymbol{p}^{i*}(\boldsymbol{v}), \boldsymbol{\alpha}_i, \boldsymbol{\beta}_i, v_i)$ with respect to $\boldsymbol{\alpha}_i^*(\boldsymbol{v}) \in A$ and $\boldsymbol{\beta}_i^*(\boldsymbol{v}) \in B,\ i = 1\ldots,r$ at the equilibrium point $(N^*(\boldsymbol{v}), \boldsymbol{p}^*(\boldsymbol{v}))$ (Brown and Vincent 1987): that is, when the G-functions are differentiable, at an ESS it is necessary that

$$\frac{\partial G(N^*(\boldsymbol{v}), p_i^*(\boldsymbol{v}), \boldsymbol{p}^{i*}(\boldsymbol{v}), \boldsymbol{\alpha}_i^*(\boldsymbol{v}), \boldsymbol{\beta}_i^*(\boldsymbol{v}), v_i)}{\partial \boldsymbol{\alpha}_i} = 0$$

$$\text{and} \quad \frac{\partial G(N^*(\boldsymbol{v}), p_i^*(\boldsymbol{v}), \boldsymbol{p}^{i*}(\boldsymbol{v}), \boldsymbol{\alpha}_i^*(\boldsymbol{v}), \boldsymbol{\beta}_i^*(\boldsymbol{v}), v_i)}{\partial \boldsymbol{\beta}_i} = 0 \quad i = 1,\ldots,r. \tag{11}$$

Alternatively, consider the case where each $\boldsymbol{\alpha}_i \in [\underline{\boldsymbol{\alpha}}, \overline{\boldsymbol{\alpha}}]$ and $\boldsymbol{\beta}_i \in [\underline{\boldsymbol{\beta}}, \overline{\boldsymbol{\beta}}]$ and we happen to know that the ESS lies at a corner of the boundary of the set of admissible strategies (taking into account that $\boldsymbol{\alpha}$ and $\boldsymbol{\beta}$ may be vectors so that each of the elements in these vectors increases the dimension of the simplex involved). Suppose n is sufficiently large so that all combinations of corner strategies can be played by a set of strategy types in the population. Then one way to search for the ESSs is to locate the stable isolation equilibria of the system (9). This approach is only feasible if the number of strategy parameters is not too large and we know in advance that the ESS does lie at the corner of the admissible strategy set. We will demonstrate this approach in Section 3.

ESS theory breaks down in systems that are inherently unstable. Vincent and Brown (1987) extend ESS theory for discrete systems with stable cycles. But this analysis is equivalent to the equilibrium analysis in a higher dimensional system corresponding to the periods of the cycles involved. Again we illustrate in Section 3 the difficulties of applying ESS analysis to nonequilibrium systems.

Another technique to obtain ESSs, for given harvesting intensities v, is to augment system (9) with a set of equations of the type (Vincent 1990)

$$\alpha_i(t+1) = \alpha_i(t) + \mu \frac{\partial G(N^*(\boldsymbol{v}), p_i^*(\boldsymbol{v}), \boldsymbol{p}^{i*}(\boldsymbol{v}), \boldsymbol{\alpha}_i, \boldsymbol{\beta}_i, v_i)}{\partial \boldsymbol{\alpha}_i}$$

$$\beta_i(t+1) = \beta_i(t) + \mu \frac{\partial G(N^*(\boldsymbol{v}), p_i^*(\boldsymbol{v}), \boldsymbol{p}^{i*}(\boldsymbol{v}), \boldsymbol{\alpha}_i, \boldsymbol{\beta}_i, v_i)}{\partial \boldsymbol{\beta}_i} \quad i = 1, \ldots, r. \tag{12}$$

The parameter μ has the interpretation of a mutation rate, and it follows that the equilibrium $(N^*(\boldsymbol{v}), p_i^*(\boldsymbol{v}), \boldsymbol{p}^{i*}(\boldsymbol{v}), \boldsymbol{\alpha}_i^*(\boldsymbol{v}), \boldsymbol{\beta}_i^*(\boldsymbol{v}))$ of the augmented system (9) and (12) satisfies the ESS necessary conditions (11). Whether the augmented system converges to an ESS solution, however, depends on the initial conditions and the value of the mutation parameter μ.

INCORPORATION OF GENETIC STRUCTURE

ESS analysis, as originally formulated by Maynard Smith and others, assumes that strategy subgroups within populations represent non interbreeding or haploid phenotypes. Most populations, however, consists of interbreeding strategy phenotypes (diploids, haplodiploids, nonclonal organisms, etc.), so it is important to understand when it is necessary to explicitly incorporate genetic considerations into ESS analyses. For example, as Maynard Smith (1982) discusses in some detail, an adequate treatment of behavioral traits under the influence of sexual selection requires explicit consideration of genetic issues.

The issue of explicit genetic structure in frequency-dependent, but not density-dependent, models has been extensively discussed by Thomas (1985a,b) and reviewed by Hines (1987). As pointed out by Hines, genetic constraints might adversely affect the dynamic stability of an ESS solution. However, Hines (1987) also concluded that with the addition of explicit diploid structure we might generally expect the average strategy (across the different genotypes) in the population to evolve towards the ESS solution.

The usual approach to incorporating genetic consideration is to assume that alleles at one or more particular loci influence the expression of the strategy and then to model the fate of particular alleles as the population responds to selective pressures. In the simplest case, we assume that the strategy is governed by a one-locus two-allele diploid system and define $p_1(t)$, $p_2(t)$, and $p_3(t)$ to be the proportions of three different adult strategy phenotypes at time t respectively corresponding to diploid genotypes AA, Aa, and aa. We might then assume that these adult phenotypes give rise to progeny phenotypes at time $t' \in (t, t+1)$ in proportions $p_1(t')$, $p_2(t')$, and $p_3(t')$. These proportions will depend on the adult proportions $(p_1(t), p_2(t), p_3(t))$, the mating structure (random, positively or negatively assortative), and, in some cases, even the adult population density $N(t)$.

In general, for multi-locus multi-allele systems, with corresponding genotype frequency vector $\boldsymbol{p}$, we can model the mating process in terms of a mapping M from the population $frequency \times density$ set to the $frequency$ set—that is:

$$\boldsymbol{p}(t') = M(N(t), \boldsymbol{p}(t)). \tag{13}$$

This equation can be linked to the population dynamics equations (9) by simply replacing $\boldsymbol{p}(t)$ with $\boldsymbol{p}(t')$—that is:

$$\begin{aligned} N(t+1) &= N(t)\mathcal{G}(N(t), \boldsymbol{p}(t'), \mathcal{A}, \mathcal{B}, \boldsymbol{v}) \\ p_i(t+1) &= p_i(t') \frac{G(N(t), p_i(t'), \boldsymbol{p}^i(t'), \boldsymbol{\alpha}_i, \boldsymbol{\beta}_i, v_i)}{\mathcal{G}(N(t), \boldsymbol{p}(t'), \mathcal{A}, \mathcal{B}, \boldsymbol{v})} \qquad i = 1, \ldots, n-1. \end{aligned} \tag{14}$$

In the absence of mutations, all corners of the frequency set—that is all n points of the form $\boldsymbol{p} = (0, \ldots, 0, 1, 0, \ldots, 0)$—are fixed points of the mapping M for all values of N. Specifically, if $\hat{\boldsymbol{p}}$ is any fixed point (not necessarily a corner point) of M, for $N = N^*$ say, then

$$\hat{\boldsymbol{p}}(\boldsymbol{v}) = M(N^*(\boldsymbol{v}), \hat{\boldsymbol{p}}(\boldsymbol{v})). \tag{15}$$

Hence if system (9) has an ESS $(\boldsymbol{\alpha}^*(\boldsymbol{v}), \boldsymbol{\beta}^*(\boldsymbol{v}))$ for $p_1^*(\boldsymbol{v}) = 1$, then system (13) and (14) has the same ESS (i.e. allele A is at fixation and genotype AA plays $(\boldsymbol{\alpha}^*(\boldsymbol{v}), \boldsymbol{\beta}^*(\boldsymbol{v}))$).

Further, for cases (Hines, 1987) where it has been demonstrated that the genetic ESS solution evolves towards a polymorphic frequency distribution $\hat{\boldsymbol{p}}(\boldsymbol{v})$, say, when the strategy genotypes are *a priori* fixed at $(\hat{\boldsymbol{\alpha}}_i(\boldsymbol{v}), \hat{\boldsymbol{\beta}}_i(\boldsymbol{v}))$ such that the average population strategy is $(\boldsymbol{\alpha}^*(\boldsymbol{v}), \boldsymbol{\beta}^*(\boldsymbol{v}))$ –that is:

$$\sum_{i=1}^{n} \hat{p}_i(\boldsymbol{v})\hat{\boldsymbol{\alpha}}_i = \boldsymbol{\alpha}^*(\boldsymbol{v}) \quad \text{and} \quad \sum_{i=1}^{n} \hat{p}_i(\boldsymbol{v})\hat{\boldsymbol{\beta}}_i = \boldsymbol{\beta}^*(\boldsymbol{v}), \tag{16}$$

then it follows that equation (15) must hold (i.e., $\boldsymbol{p} = \hat{\boldsymbol{p}}(\boldsymbol{v})$ is a fixed point of the mapping $M(N^*(\boldsymbol{v}), \boldsymbol{p})$) and that

$$G(N^*(\boldsymbol{v}), \hat{p}_i(\boldsymbol{v}), \hat{\boldsymbol{p}}^i(\boldsymbol{v}), \hat{\boldsymbol{\alpha}}_i(\boldsymbol{v}), \hat{\boldsymbol{\beta}}_i(\boldsymbol{v}), \boldsymbol{v}) = 1. \tag{17}$$

Insights into the class of mating structures and G-functions for which it is possible to find a frequency vector $\hat{\boldsymbol{p}}(\boldsymbol{v})$ which satisfies equations (15), (16) and (17) will help us to understand when the interbreeding genetic and noninterbreeding phenotypic formulations provide equivalent solutions. For now this problem remains an open question.

SALMON CASE STUDY

The reproductive life-history of an organism can be regarded as an evolutionary strategy designed to maximize its long term reproductive fitness. At one extreme, some organisms have opted for semelparity: the organism reproduces once and then dies. This is a strategy adopted by many plants and invertebrates and, among the vertebrates, several species of Pacific salmon provide a notable example (Waples and Teel 1990).

In addition to semelparity as an evolutionary strategy in itself, the timing of the reproductive event can also be regarded as an evolutionary strategy (Myers 1986). We recently (Kaitala and Getz 1992) undertook an extensive study of the dynamics of a semelparous population process that has application to the spawning age response of Pacific salmon under harvesting.

THE MODEL

The model we developed (Kaitala and Getz 1992) is defined in terms of the following variables and strategy parameters. The variable $x_0(t)$ represents the number of individuals spawned at time t of which a proportion s_0 survive to enter the ocean at time $t+1$. The variables $x_i(t)$, $i = 1, 2, 3, \ldots$, are the number of individuals of age i at time t in the ocean that will remain in the ocean for another year, and $y_i(t)$ are the number of individuals of age i at time t that return up river to spawn. We assume that no 1 or 2-year-olds, a proportion $\alpha \in [0,1]$ of 3-year-olds and all remaining 4-year-olds return up river to spawn. We regarded the proportion α as a strategy parameter.

We used the following parameters to specify the structure of the model (we do not regard these parameters as strategy parameters and hence do not use the notation α_i to describe them). The parameters s_i, $i = 1, 2, 3, \ldots$, are the proportions of individuals that survive their ith year of life in the ocean (i.e. period $[i, i+1)$) and σ_i, $i = 3, 4$, are the proportions of i-year-old returning individuals that will make it up river to spawn. We assumed that the marine survival parameters s_i ($i = 2, 3, 4 \ldots,$) have the form

$$s_i = e^{-(\mu_i + q_i v)}, \qquad i = 2, 3, 4, \ldots, \tag{18}$$

where μ_i is a natural mortality rate, q_i is a catchability coefficient, and v is a fishing intensity variable. If the returning adults are harvested as they make their way up river, then the parameters σ_i can also be partitioned into natural and harvest mortality components, although we did not consider this complication. We used the parameters w_i to denote the average biomass of an individual of age i, so that the biomass $B(t)$ of the adult spawning population at time t is given by

$$B(t) = w_3 y_3(t) + w_4 y_4(t). \tag{19}$$

We assumed that the number of individuals spawned was related to the spawning stock biomass through a Ricker function with intrinsic fecundity parameter ρ and biomass scaling parameter γ—that is

$$x_0(t) = B(t)e^{\rho-\gamma B(t)}. \tag{20}$$

Using the above variables and parameters, our salmon population model has the form

$$\begin{pmatrix} x_1(t+1) \\ x_2(t+1) \\ y_3(t+1) \\ x_3(t+1) \\ y_4(t+1) \end{pmatrix} = \begin{pmatrix} 0 & 0 & 0 & 0 & 0 \\ s_2 & 0 & 0 & 0 & 0 \\ 0 & s_3\alpha\sigma_3 & 0 & 0 & 0 \\ 0 & s_3(1-\alpha) & 0 & 0 & 0 \\ 0 & 0 & 0 & s_4\sigma_4 & 0 \end{pmatrix} \begin{pmatrix} x_1(t) \\ x_2(t) \\ y_3(t) \\ x_3(t) \\ y_4(t) \end{pmatrix} + \begin{pmatrix} s_1 B(t)e^{\rho-\gamma B(t)} \\ 0 \\ 0 \\ 0 \\ 0 \end{pmatrix}. \tag{21}$$

We demonstrated (Kaitala and Getz, ms.) that the existence of a positive equilibrium and its stability properties depends on the value of the quantity

$$K(v) = \rho + ln\, C(v). \tag{22}$$

where

$$C(v) = s_1 s_2(v) s_3(v)(s_4(v)\sigma_4 w_4 + \alpha(\sigma_3 w_3 - s_4(v)\sigma_4 w_4)). \tag{23}$$

In particular, a positive equilibrium exists if and only if $K(v) > 0$ and is stable for at least $K(v) \leq 2$. Further, we showed that when $\alpha = 0$ or 1 that equilibrium is unstable for $K(v) > 2$. We also showed that the population strategy which maximizes recruitment and hence overall productivity of the population is given by:

1. the spawning age polymorphism

$$\alpha^r = \frac{\frac{e^{1-\rho}}{s_1 s_2 s_3} - s_4 w_4 \sigma_4}{w_3\sigma_3 - s_4 w_4 \sigma_4}, \tag{24}$$

   whenever the value of $\frac{e^{1-\rho}}{s_1 s_2 s_3}$ lies between the values of $w_3\sigma_3$ and $s_4 w_4 \sigma_4$;

2. early spawning (i.e., $\alpha^r = 0$) whenever $s_4 w_4 \sigma_4$ lies between $w_3\sigma_3$ and $\frac{e^{1-\rho}}{s_1 s_2 s_3}$;

3. late spawning (i.e., $\alpha^r = 1$) whenever $w_3\sigma_3$ lies between $s_4 w_4 \sigma_4$ and $\frac{e^{1-\rho}}{s_1 s_2 s_3}$.

There is no *a priori* reason to expect that the strategy which maximizes the productivity of a population should be an ESS solution. Since $C(v)$ represents an 'expected spawning biomass trade-off' between returning to spawn at age 3 with survival probability σ_3 and spawning fecundity proportional to w_3 and returning to spawn at age 4 with survival probability $s_4\sigma_4$ and spawning fecundity proportional to w_4. If the former has a higher expected value than the latter—i.e., $\sigma_3 w_3 > s_4(v)\sigma_4 w_4$ then an individual salmon can maximize its contribution to the spawning stock by selecting its own α to be

$$\alpha^*(v) = 1(0) \quad \text{whenever} \quad w_3\sigma_3 > (<) w_4\sigma_4 s_4(v). \tag{25}$$

Note that it can be shown (Getz and Kaitala 1992) that $C(v)$ represents the per-capita contribution of an α strategy-homogeneous population to the reproductive biomass $B(v)$ of the population at the nontrivial equilibrium corresponding to harvesting intensity v. Thus our result is in line with Law and Grey (1989) who obtain the same interpretation for an ESS in the context of the more general iteroparous case.

TWO TYPE ANALYSIS

A limited ESS analysis can be carried out by considering a mutant-strategy type playing α^μ under the assumptions that (1) the mutant group is otherwise subject to the same set of population parameters as the resident group and (2) the density-dependent effects on recruitment are calculated in term of their combined biomass as follows:

$$x_0(j) = B(j)e^{\rho-\gamma(B(j)+B^\mu(j))} \tag{26}$$

and

$$x_0^\mu(j) = B^\mu(j)e^{\rho-\gamma(B(j)+B^\mu(j))}, \tag{27}$$

where the μ superscript pertains to the mutant population. We have shown (Getz and Kaitala 1992) that $\alpha^*(v)$, defined by relationship (25), is a unique ESS provided the harvesting intensity v is not too large to violate $K(v) > 0$. If for small v, $w_3\sigma_3 > w_4\sigma_4 s_4(v)$ so that $\alpha^*(v) = 0$ (i.e., individuals return to spawn at age 4), then it follows from identity (18) that s_4 is a decreasing function of v so that increasing v could well cause the ESS to switch to $\alpha^*(v) = 1$ (i.e., individuals return to spawn at age 3).

A more comprehensive ESS analysis can be carried out by assuming that more than one spawning ground exists. For example, suppose two spawning regions exist (perhaps one upstream of the other), each characterized by its own set of spawning parameters (ρ_1, γ_1) and (ρ_2, γ_2) respectively. Further suppose proportions $\beta \in [0,1]$ of the resident population and $\beta^\mu \in [0,1]$ of the mutant populations migrate to spawning ground 1, with the remaining adults migrating to spawning ground 2, then equations (26) and (27) can be extended as follows:

$$x_0(j) = \beta B(j)e^{\rho_1-\gamma_1(\beta B(j)+\beta^\mu B^\mu(j))} + (1-\beta)B(j)e^{\rho_2-\gamma_2((1-\beta)B(j)+(1-\beta^\mu)B^\mu(j))} \tag{28}$$

and

$$x_0^\mu(j) = \beta^\mu B^\mu e^{\rho_1-\gamma_1(\beta B(j)+\beta^\mu B^\mu(j))} + (1-\beta^\mu)B^\mu(j)e^{\rho_2-\gamma_2((1-\beta)B(j)+(1-\beta^\mu)B^\mu(j))}. \tag{29}$$

Note that the parameters β and β^μ are a measure of the degree to which the two strategy types use the same spawning grounds. Obviously, $\beta = 1$ and $\beta^\mu = 0$ implies that the two populations are total independent of one another. Also note that equations (28) and (29) hold both when the mutant is rare (invasion analysis) and when it is common (coexistence analysis).

For the single spawning ground case modeled by equations (26) and (27), the ESS analysis is straightforward provided the system (21) is stable—that is, $K(v)$ defined by identity (22) is not much larger than 2 (Kaitala and Getz 1992). If ρ is relatively large, however, or $\ln C(v)$ is negative (i.e., $C(v) < 1$) then $K(v)$ might be large enough for the system to exhibit oscillatory and even chaotic behavior. In the latter case, the theory for locating ESSs in ecological models—as derived by Brown and Vincent (1987) and discussed in Section 2.3 above—no longer applies.

CASE-STUDY PARAMETERS

Non-equilibrium systems can exhibit surprising behavior, as illustrated by the dynamics of the following imaginary salmon population. We assume that the system is deterministic with survival parameters $s_1 = s_2 = 0.3$, $s_3 = s_4 = 0.5e^{-v}$, and $\sigma_3 = \sigma_4 = 1.0$: that is, only marine stocks over 2-years-old are subject to fishing and all returning spawners return up river to spawn. We assume the weight parameters are $w_3 = 1$, $w_4 = 2.4$ (i.e., biomass units are standardized with respect to the weight of a three year old), and the stock-recruitment biomass scaling parameter is $\gamma = 0.003$. (Note that $K(v)$ defined by expressions (22) and (23) does not depend on γ and hence the existence and stability properties of a nontrivial equilibrium do not depend on γ—Kaitala and Getz 1992). We explore the behavior of this system in a two-strategy type setting (i.e. each strategy satisfies system (20), except the first element in the right-hand-side vector is replaced by (26) for one of the strategy types and (27) for the other of the strategy types) and in the context of how the system is affected by the stock-recruitment intrinsic fecundity parameter ρ, fishing intensity v and adult spawning strategies α and α^{μ}.

Substituting the above set of parameters in identity (25) implies that in the context of an invading mutant type the ESS is (assuming stability of the isolation equilibrium):

$$\begin{aligned} \alpha^* &= 0 \text{ (late spawner)} \quad \text{when} \quad v < 0.18 \\ \alpha^* &= 1 \text{ (early spawner)} \quad \text{when} \quad v > 0.19. \end{aligned} \tag{30}$$

Further, it follows from identities (22) and (23) (and comments below (23)) that our system has a stable positive equilibrium for the case $v = 0$ when

$$\begin{aligned} \alpha = 0 \quad &\text{and} \quad 2.9188 < \rho < 4.1988 \\ \alpha = 1 \quad &\text{and} \quad 3.1101 < \rho < 5.1101, \end{aligned} \tag{31}$$

and has an unstable positive equilibrium when the right-hand inequalities are violated.

NUMERICAL SIMULATIONS

The analytical results represented by expressions (30) pertain to an invasion analysis in the case where the isolation equilibria are stable: that is ρ satisfy inequalities (31). A coexistence analysis (i.e., both strategy types are initially at substantial population levels) reveals for the case $\rho = 4$ and $v = 0.0$ (Figure 1A) that the ESS, $\alpha^* = 0$ (see expressions (30) and (31)), begins to dominate a mutant playing a similar strategy $\alpha = 0.05$, but rapidly excludes a mutant playing the opposite strategy $\alpha = 1.0$ (Figure 2A). From expression (30) $\alpha = 0$ is no longer an ESS for $v = 0.2$ and, not surprisingly, is no longer able to dominate strategy type $\alpha = 0.05$ (Fig. 1B). In fact, the latter strategy type becomes more dominant as v is increased to 0.3 (Figure 1C).

When the value of ρ is increased from 4 to 5, it follows from expressions (31) that the isolation equilibrium of strategy type $\alpha = 0$ is no longer stable. Nevertheless, when $v = 0$, strategy type $\alpha = 0$ is still able to dominate strategy type $\alpha = 0.05$ (Figure 1D), although strategy type $\alpha = 0$ is no longer an ESS (simulations indicated that strategy type $\alpha = 0.05$ can invade strategy type $\alpha = 0$ but cannot penetrate beyond oscillating between 0.1 and 0.2 biomass units while strategy type $\alpha = 0.05$ continues to oscillate between 460 and 930 biomass units) . However, the situation changes dramatically when ρ is increased from 5 to 6. The cyclic behavior under $\rho = 5$ becomes chaotic; but, suprisingly, strategy type $\alpha = 0.05$ now rapidly dominates and then completely excludes strategy type $\alpha = 0$ (Figure. 1E). Strategy type $\alpha = 0.05$ continues to dominate when $v = 0.03$ (Figure 1F), although the chaoatic behavior has been tempered to cyclic behavior (increasing v leads to decreasing $C(v)$ and hence $K(v)$ which tends to restabilize the population — see identities (22) and (23) and the paragraph following these identities) and complete exclusion does not occur within the 500-year simulation.

Let us return to the case $\rho = 4$ and $v = 0$ and a comparison of strategy types $\alpha^* = 0$ and $\alpha = 1$. We know from our analysis (and numerical verification, although the results are not presented here) that strategy type $\alpha = 1$ cannot invade strategy type $\alpha^* = 0$ and also from simulations (Fig. 2A) that strategy type $\alpha^* = 0$ rapidly outcompetes and excludes strategy type $\alpha = 1$. Thus one might be tempted to conclude that strategy type $\alpha^* = 0$ can invade and exclude strategy type $\alpha = 1$. Interestingly, strategy type $\alpha^* = 0$ is able to invade strategy type $\alpha = 1$, but it is not able to make much headway after establishing itself at a relatively low oscillatory level (Figure 2B). Further, a coexistence analysis of these two strategy types for the case $\rho = 6$, $v = 0.3$ indicates that strategy type $\alpha = 1$ dominates strategy type $\alpha = 0$, as would be expected from expressions (30). But unlike the case of strategy type $\alpha = 0.05$ versus strategy type $\alpha = 0.0$ (Figure 1F), the behavior is chaotic and strategy type $\alpha = 0$ shows a virtual-exclusion–reinvasion pattern of behavior. In reality, however, the dynamics of exclusion and reinvasion require a stochastic analysis of extinction probabilities of populations at very low levels and mutation rates for reinvasion analysis.

CONCLUSION

The analysis of the salmon model presented in the previous section can be greatly extended to include explicit genetic structure in the model and the effects of stochastic events as they may relate to mutation rates and extinction probabilities. An initial study (Kaitala and Getz 1992.) indicates that explicit genetic structure does not alter the conclusions relating to which strategy type is an ESS, provided that the ESS proportion can be generated polymorphically, as described in equations (16).

Stochastic effects obviously play an important role at low population levels. This is an aspect that we have not considered here, and should be added as a fifth step to the sequence of analyses outlined in Section 2.

ESS analysis provides a powerful tool for enhancing our understanding of evolutionary processes. It is clear from our discussion, however, that classical ESS analysis is only a small part of the whole dynamic story. A comprehensive understanding of demographic evolutionary processes requires that we understand the underlying population dynamics, coexistence and invasion dynamics, the constraints imposed by genetic and mating structures, and the consequences of stochastic effects when population densities are relatively low.

FIGURE 1. Trajectories of populations 1 and 2 respectively playing strategies $\alpha = 0$ and $\alpha = 0.05$ over the abscissa time interval [0,500]. The population biomass units on the ordinate scale are relative to the weight of a three year old individual ($w_3 = 1$). The initial conditions for both populations are $\boldsymbol{x}(0) = (1000, 500, 100, 100, 50)'$ (coexistence analysis). The population parameters are given in the text except as follows: A. $\rho = 4,\ v = 0$; B. $\rho = 4,\ v = 0.2$; C. $\rho = 4,\ v = 0.3$; D. $\rho = 5,\ v = 0$; E. $\rho = 6,\ v = 0$; and F. $\rho = 6,\ v = 0.3$.

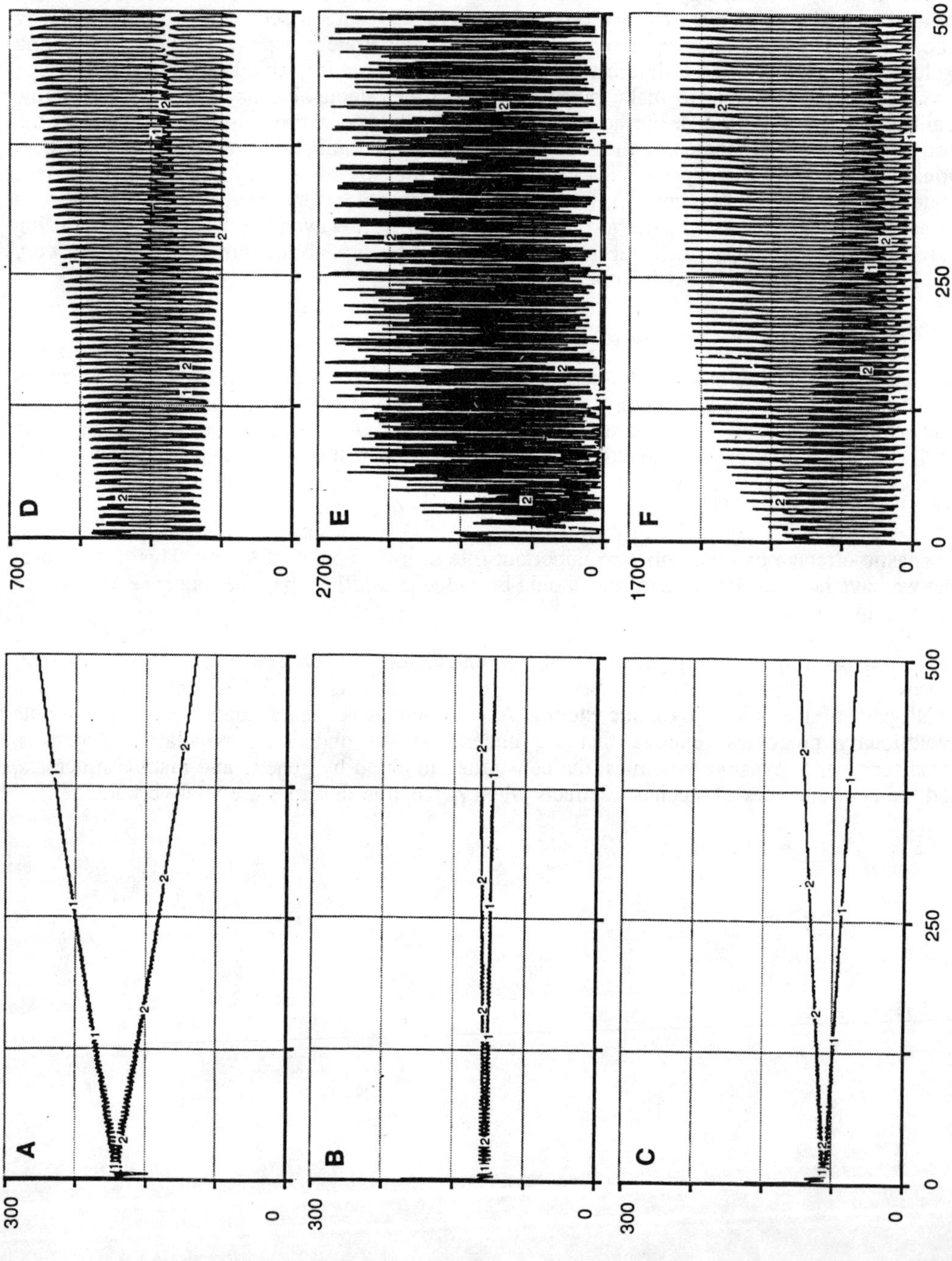

FIGURE 2. Trajectories of populations 1 and 2 respectively playing strategies $\alpha = 0$ and $\alpha = 1$ over the abscissa time interval [0,500]. The population biomass units on the ordinate scale are relative to the weight of a three year old individual ($w_3 = 1$). The initial conditions for both populations in A. and C. (coexistence analysis) and population 2 in B. are $\boldsymbol{x}(0) = (1000, 500, 100, 100, 50)'$ while the initial condition for population 1 in B. is $\boldsymbol{x}(0) = (1, 0, 0, 0, 0)'$ (invasion analysis). The population parameters are given in the text except as follows: A. $\rho = 4,\ v = 0$; B. $\rho = 4,\ v = 0.3$; and C. $\rho = 6,\ v = 0.3$.

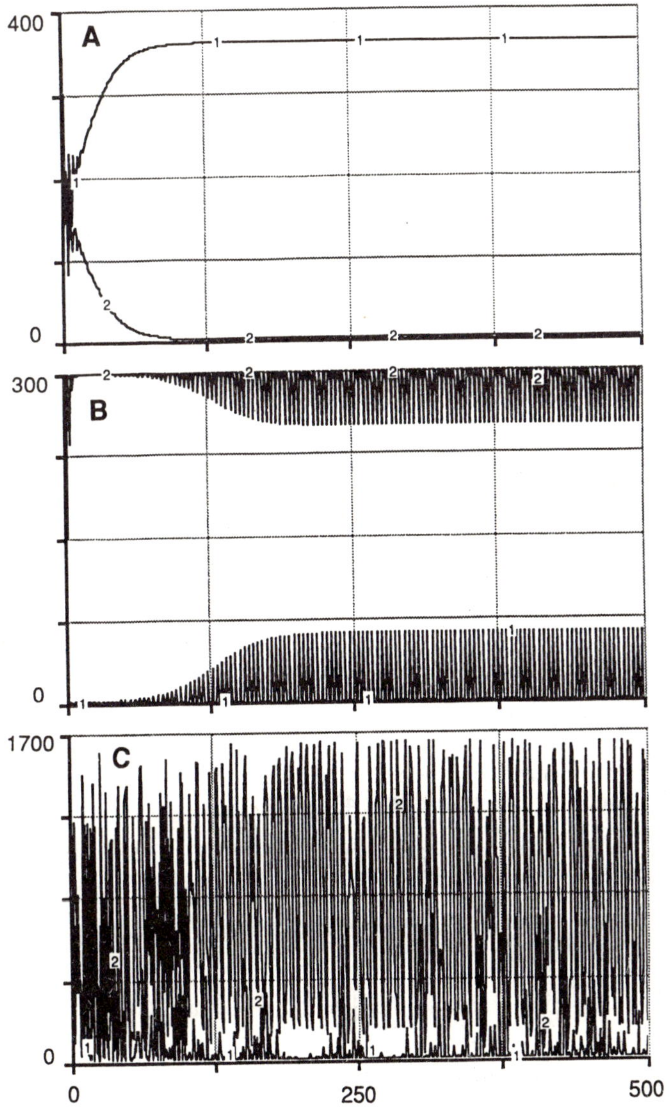

REFERENCES

Brown, J. S. and T. L. Vincent 1987. A theory for the evolutionary game. Theoretical Population Biology **31**: 140–166.

Bergh, M. O. and W. M. Getz 1989. Stability and harvesting of competing population with genetic variation in life history strategies. Theoretical Population Biology **36**: 77–124.

Clark, C. W. 1976. Mathematical Bioeconomics: the Optimal Management of Renewable Resources, Wiley–Interscience, New York.

Clark, C. W. 1985. Bioeconomic Modelling and Fisheries Management Wiley–Interscience, New York.

Ellner, S. 1985a. ESS germination rates in randomly varying environments. I. Logistic-type models. Theoretical Population Biology **28**: 50–79.

Ellner, S. 1985b. ESS germination rates in randomly varying environments. II. Reciprocal yield-law models. Theoretical Population Biology **28**:80–116.

Getz, W. M., and R. G. Haight 1989. Population Harvesting: Demographic Models of Fish, Forests and Animal Resources. Princeton Monographs in Population Biology, Princeton University Press.

Getz, W. M. and V. Kaitala 1989. Ecogenetic models, competition, and heteropatry. Theoretical Population Biology **36**: 34–58.

Hines, W. G. S. 1987. Evolutionarily stable strategies: a review of basic theory. Theoretical Popuation Biolology **31**: 195–272.

Kaitala, V., and W. M. Getz 1992. Harvesting semelparous populations and variability in the life history of salmon. Beijer Discussion Paper Series No. 7, Beijer International Institute of Ecological Economics, Stockholm, Sweden.

Kaitala, V., A. Kaitala and W. M. Getz 1989. Evolutionarily stable dispersal of a waterstrider in a temporally and spatially heterogeneous environment, Evolutionarily Ecology **3**: 283–298.

Law, R. and D. R. Grey 1989. Evolution of yields from populations with age-specific cropping. Evolutionarily Ecology **3**: 343–359.

Levin, S. A. 1970. Community equilibria and stability, and an extension of the competitive exclusion principle. American Naturalist **104**: 413–423.

May, R. M. 1981. Models for single populations. Pp 5 - 29, in Theoretical Ecology, (2nd edition), Sinauer Associates, Sunderland, MA.

May, R. M. and G. F. Oster 1976. Bifurcations and dynamic complexity in simple ecological models. American Naturalist **110**: 573–590.

Maynard Smith, J. 1982. Evolution and the Theory of Games, Cambridge University Press, Cambridge.

Maynard Smith, J. and G. R. Price 1973. The logic of animal conflict. Nature (London) **246**:15–18.

Murray, J. D. 1989. Mathematical Biology. Springer-Verlag, Berlin, pp. 767.

Myers, R. A. 1986. Game theory and the evolutions of Atlantic salmon (*Salmo salar*) age at maturation. Canadian Special Publications in Fisheries and Aquatic Sciences **89**:53–61.

Oksanen, L. 1991. Trophic levels and trophic dynamics: a consensus emerging? TREE **6**:58–60.

Thomas, B. 1985a. Genetical ESS–models. I. Concepts and basic model Theoretical Population Biology **28**:18–32.

Thomas, B. 1985b. Genetical ESS–models. II. Multi-strategy models and multiple alleles. Theoretical Population Biology **28**:33–49.

Vincent, T. L. and J. S. Brown 1985. An evolutionary response to harvesting. Pp. 80 - 97, in T. L. Vincent, Y. Cohen, W. J. Grantham, G. P. Kirkwood, and J. M. Skowronski eds., Modeling and Management of Resources under Uncertainty, Springer-Verlag, Berlin.

Vincent, T. L. and J. S. Brown 1987. Evolution under nonequilibrium dynamics. Mathematical Modelling **8**:766–71.

Vincent, T. L. and J. S. Brown 1988. The evolution of ESS theory. Annual Review of Ecology and Systematics **19** :423–43.

Vincent, T. L. 1990. Strategy dynamics and the ESS. Pp. 236 - 262, in T. L. Vincent, A. I. Mees, and L. S. Jennings eds., Dynamics of Complex Interconnected Biological Systems, Birkhäuser, Boston.

Waples, R.S., and Teel. D.J. 1990. Conservation genetics of Pacific salmon. I. Temporal changes in allele frequency. Conservation Biology **4**:144-156.

Yodzis, P. 1989. Introduction to Theoretical Ecology, Harper and Row, New York.

CATASTROPHE-TYPE REGULATION OF PELAGIC FISH STOCKS : ADAPTIVE MANAGEMENT FOR EVOLVING RESOURCES

Philippe Cury

ORSTOM, Département TOA, 213 Rue LaFayette 75480 Paris, cedex 10. Present address: PFEG/NMFS, P.O. Box 831, Monterey, CA 93942, USA.

ABSTRACT

The temporary extinction of coastal pelagic marine fish populations during unpredictable time periods appears to be a natural phenomenon. For these short life cycle species (life span of 4 to 8 years) the appearance of quick and sustained recoveries after extinction that lasts decades to centuries is common. Considering previous results in terrestrial evolutionary ecology, we hypothesize that the duration of low biomass levels are the result of strong interspecific selective pressures and poor population fitness, with only minor impacts of climatic fluctuations. The recovery of depleted populations could be the result of an adaptive process. A model using catastrophe theory illustrates long term regulations of these ecological systems. Temporary extinctions could appear as a regulatory process of the population adaptability to the environment. Some improvement should be expected in the management of these evolving resources by using a 'groping strategy'.

TEMPORARY EXTINCTION OF STOCKS AS A NATURAL AND UNPREDICTABLE PHENOMENON

Three of the most important pelagic fish stocks in the world are well known for their long term instability: the Japanese sardine (*Sardinops melanosticta*), the Pacific sardine (*Sardinops sagax*) and the Peruvian and Chilean anchoveta (*Engraulis ringens*). Recent fishery statistics (MacCall 1983, Zuta et al. 1983, Kondô 1991), information available in past literature (Tsuboï 1984; Hiramoto 1991), as well as paleoecological studies (Soutar and Isaacs 1969; De Vries and Pearcy,1982) show evidence of drastic population fluctuations in the immediate and remote past (Figure 1). Temporary extinctions of these short life cycle species (life span of 4 to 8 years) may last for decades or centuries without any apparent regularity.

Some authors (Kawasaki 1983, Shannon et al. 1984, Crawford et al. 1991) suggest that large scale environmental phenomena may influence fish stock dynamics, other authors propose that small or gradual changes in the environment are producing sudden and dramatic consequences (Steele and Henderson 1984). It is now recognized that high fishing exploitation rates lead to more frequent collapses (Beverton 1990), however no scientific analysis is actually able to predict when the recovery of a depleted stock will occur. The relations between fish population dynamics and variations in climate or fishing exploitation rates are often questionable when analyzing long term changes (Watanabe 1983, Kawai and Isibasi 1983, Cury 1988) and the temporary extinction of pelagic stocks appears to be a natural and unpredictable phenomenon.

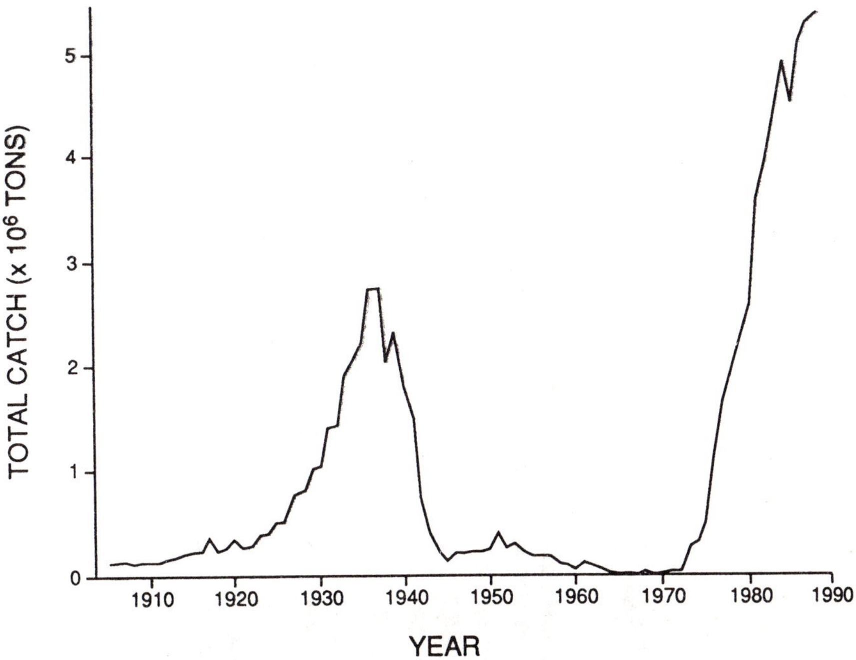

Figure 1. Examples of drastic fluctuations in pelagic fish stock abundance: **a** Sardine catch around Japan between 1905 and 1988: after an increase in the fleet activity between 1920 and 1935 the catch reached 2.5 million metric tons; the stock collapsed between 1940 and 1945 and the stock remained at a very low level for thirty years. After 1975, the sardine reappeared in large abundance (5 million metric tons were landed between 1985 and 1988) (after Kondô 1991).

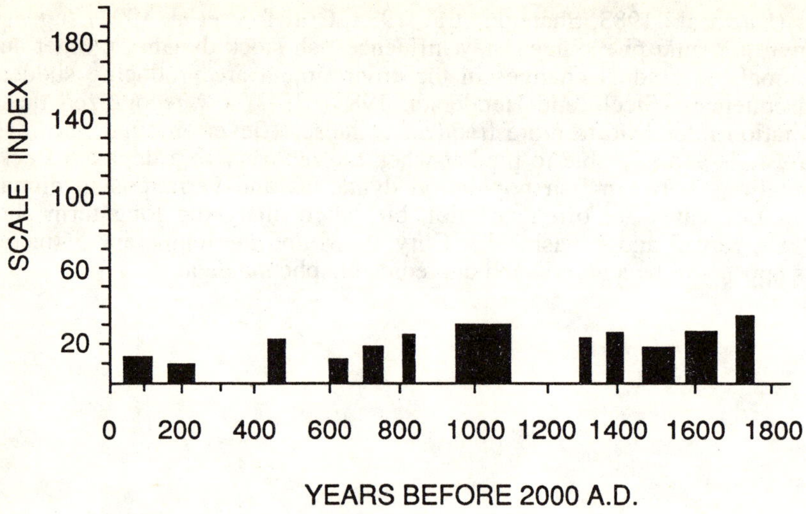

Figure 1b. Population index for Pacific sardine derived from the number of scales found in anaerobic sediments from the Santa Barbara Basin, California: the deposition rates make it possible to reconstruct the prosperous and depleted phases of the population (after Soutar and Isaacs 1969).

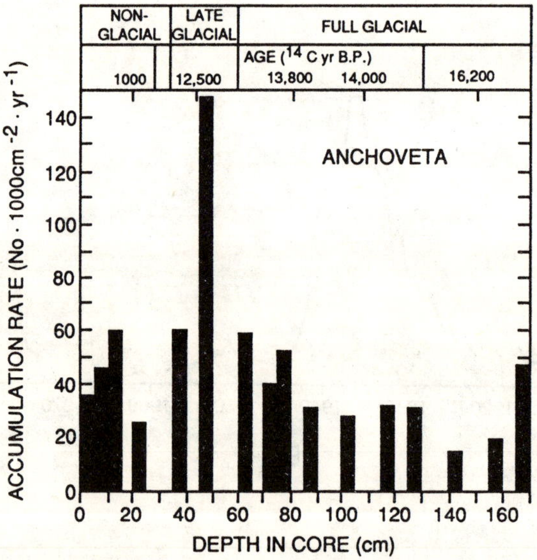

Figure 1c Accumulation rates of anchoveta scales off the Peruvian coast (number per year per 0.1 n). Ages are based on ^{14}C dates and the interpretation of sedimentary structures (after De Vries and Pearcy 1982).

HOW TO RECOVER FROM A DEPLETED BIOMASS LEVEL FUNCTIONING AS A FEED-BACK TRAP ?

In modeling the spasmodic behavior of pelagic fish stocks a self-regenerating stochastic model can be used. It describes the evolution of the recruited population by using the equation $N_{t+1}=N_t exp^{(-F-M)}$ combined with a Ricker stock-recruitment relationships affected by a white noise (see appendix for details). An example of this simulation during one thousand years is presented in Figure 2. Abundant population followed by temporal extinction that lasts during time-periods much longer than the life cycle duration can be described using this model. This reveals the existence of regime with different states, however the understanding of these unstable dynamics is trite in a sense that high recruitments are needed to recover from depleted biomasses which represent a feed-back trap. And the question 'when are important recruitments observed with depleted biomass?' remains unanswered.

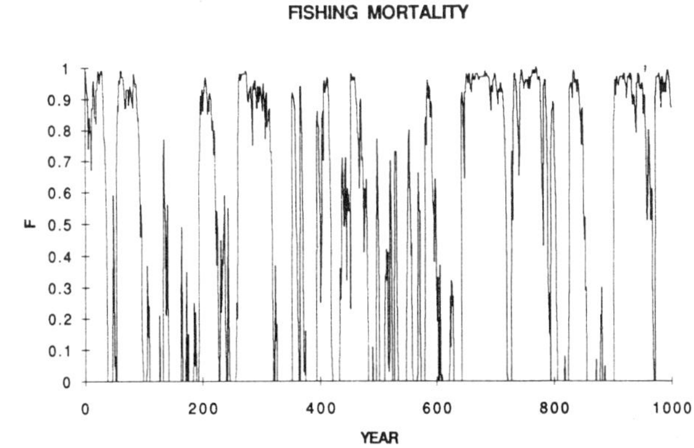

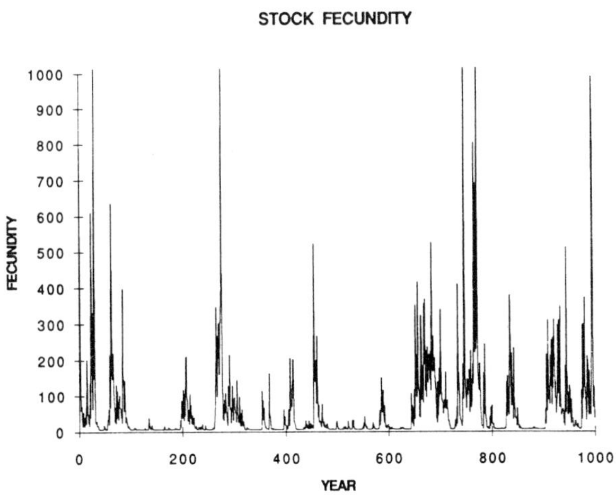

Figure 2. Simulated fishing mortality, stock fecundity, recruitment and catch during 1000 years using a self regenerating stochastic model. (The minimal recruitment is assumed to be not lower than 5‰ of the equilibrium recruitment)

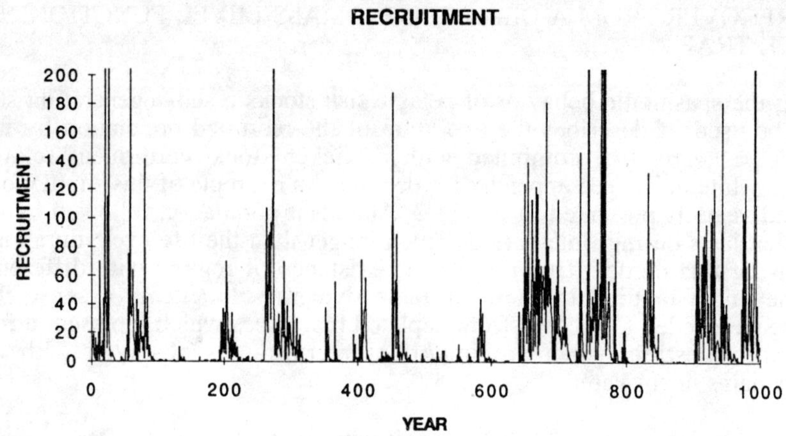

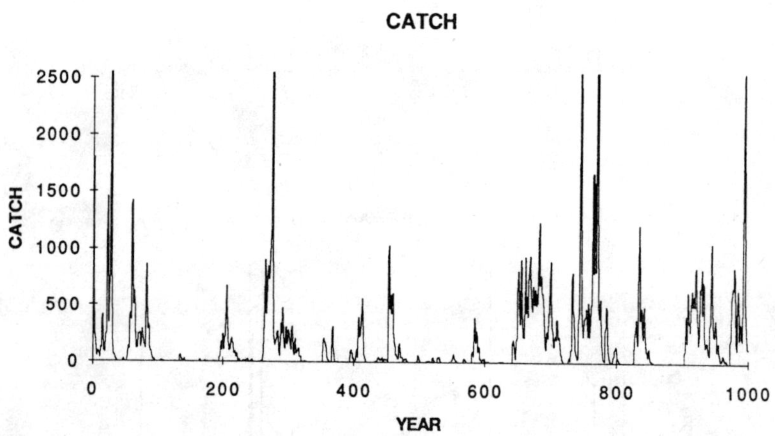

Figure 2. continued.

The underlying mechanisms that could lead to a better understanding of these long term changes in fish stocks are far from being identified, probably because of the complexity of the interactions between the environment and the populations. The theories of natural regulation of population size are not mutually exclusive, and a synthesis of several approaches may be most useful (Krebs 1978). Most studies of change in fish stock abundance have analyzed the quantitative impact of extrinsic factors (Cushing 1982) but some changes in abundance are also the result of changes in the ecological adaptability of the species (Andrewartha and Birch 1954, Myers and Krebs 1974, Mueller et al. 1991). Evolutionary aspects are often considered in terrestrial ecology but are not frequently invoked when analyzing changes in pelagic fish abundance. The purpose of this note is to discuss possible impacts of some qualitative factors on fish abundance.

INTERSPECIFIC COMPETITION AT LOW STOCK LEVEL AND ITS IMPLICATIONS

When two species compete and one becomes rare, they suffer asymmetrical selective pressures (Haldane, 1932). The dominated species has primarily interspecific contacts while the dominant species has essentially only intraspecific contacts. Thus the severity of interspecific competition would depend on fish stock size.

It is particularly difficult to analyze and show such subtleties in marine ecology. We could refer to experimental research in terrestrial ecology (Moore 1952, Pimentel et al. 1965, Mitchell and Arthur 1991) to understand the possible evolutionary implications of competition on population changes. The experiments referred to (Moore 1952, Pimentel et al. 1965) were as follows: two competitive species of flies, of which one is more efficient and thus rapidly becomes more abundant at the beginning of the experiment, were placed together in an experimental cage. After several generations a competitive inversion may lead to the elimination of the more abundant species. This new equilibrium reflects the intervention of a selective process and confirms that there would thus be a modification in the behavior of the dominated species, becoming competitively superior within a few generations.

These evolutionary changes are newly adopted properties which allow a new function and consequently permit the opening of a new adaptive zone (Mayr 1974). Subtle developmental mechanisms may have important consequences on individual survival and may lead to the expansion of a population (Mitchell and Arthur 1991).

The stock biomass of competitive species increased or were abundant after the recent collapses of the Japanese sardine, Pacific sardine and Peruvian and Chilean anchoveta. Respectively the mackerel (*Scomber japonicus*) (Tanaka 1983) and the Japanese anchovy (*Engraulis japonica*) (Kondô 1980, 1991), the anchoveta (*Engraulis mordax*) (Lasker and MacCall 1983) and the sardine (*Sardinops sagax*) and the horse mackerel (*Trachurus murphyi*) (Jordan 1983) have all increased their stock biomass (Figure 3). Consequently the interspecific competition is probably severe for the depleted stocks.

ADAPTIVE POTENTIAL OF PELAGIC SPECIES

Are the coastal pelagic species able to exhibit evolutionary changes within their population ? In other words, do they have a high adaptive potential ?

Two essential characteristics may confer a strong adaptive potential to the pelagic fish populations: biotic potential and demographic plasticity. Pelagic species have a high biotic potential. Every female is able to lay thousands or millions of eggs several times during one year, even though the survival of only two individuals are necessary to maintain the population. The large amount of genetic variation not used in each generation represents an important selective advantage (Mayr 1981). Physical and biological environments fluctuate considerably in the pelagic ecosystem. It is also difficult to predict the particular combination of genes which will be of interest for the next generation. Under these circumstances the probability for certain genotypes to exhibit more optimal fitness under new conditions is high. The species would be able to adapt to a rapid environmental change.

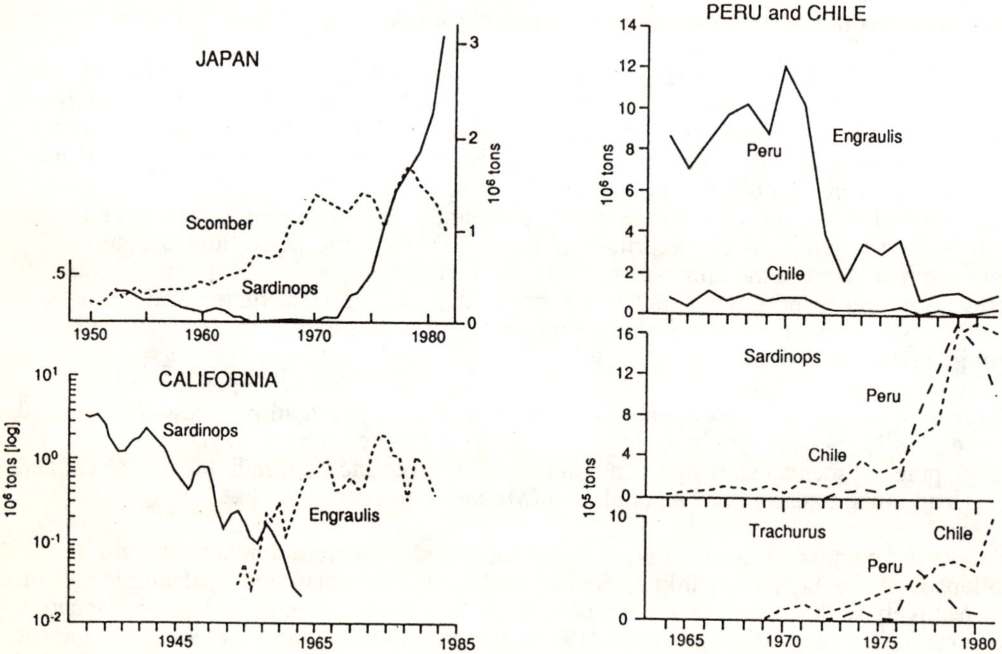

Figure 3. Landings of sardine (*Sardinops*) or anchovy (*Engraulis*) (solid line) and the associated competitive species (dashed line): (a) In Japan, mackerel (*Scomber sp.*) abundance was high after the collapse of the sardine stock; (b) In California, time series of sardine (age 2+) and anchovy (*Engraulis*) spawning biomass reveal that the increase in anchovy abundance lagged the sardine decline. In Peru and Chile decrease in the anchoveta stock level (c) was followed by increases of (d) sardine (*Sardinops*) and (e) mackerel (*Trachurus*) catches.

Variability by itself is part of the species' life strategy. The plasticity of a trait, that is not its mean value but its variance, is also a major adaptive response (Levins 1968). Thus the adaptive mechanisms could be the result of the intervention of the evolution of a trait or the canalization of the variability of this trait (Stearns 1980). Two types of plasticity can be defined (Stearns 1982): continuous and discrete plasticity. Continuous plasticity corresponds to gradual responses to environmental changes, while discrete plasticity, also called developmental switches (Levins 1968), would produce individuals with alternative characteristics that are sometimes better adapted. Fish are phenotypically more variable than other vertebrates (Ryman and Utter 1987) and environmental fluctuations generate large phenotypic variations in pelagic fish (Cury and Fontana 1988). However, as noted by several authors (Ryman and Utter 1987) high levels of phenotypic variation coupled with low heritabilities indicate greater susceptibility to environmental factors.

EVOLUTIONARY CHANGES AND SIZE OF THE POPULATION

Another important factor which is essential to the understanding of the evolutionary mechanism is the population size. These evolutionary changes occur only in small size populations (Mayr 1960). An intuitive approach is to consider that with a large gene pool, the fixation and conservation of evolutionary changes is difficult because of dilution in the population. Abundant populations are inert from an evolutionary point of view (Mayr 1981). The occurrence of an evolutionary change would be favored within small size populations, as in the case of the depleted stocks.

Perturbed or stressed populations appear to be affected by stronger selection than unperturbed populations (Endler 1986). In terms of information a large population will explore all the distribution area but a small one will be able to explore just a portion of this area. Thus, small depleted populations will be confronted to "local surprises" and perhaps to new selective pressures and also will be more subject to changes induced by "error making" (Allen 1988, Allen and McGlade 1989).

Experimental studies provide arguments that density-dependent natural selection at extreme populations densities plays a major role in population dynamics. Population cycles in small rodents were intensively studied. After considering many ecological factors that may affect their dynamics, Myers and Krebs (1974) came to the conclusion that the populations comprise two broad genotypes, one reproductively superior but intolerant of high density, the other adapted to survival under crowded conditions but having a lower reproductive rate. This heterogeneity allows a periodic change in the genetic constitution of the population that is associated with these fluctuations. Recent experimental results on flies (Mueller et al. 1991) confirm earlier observations of fitness trade-offs arising from density-dependent natural selection. At extreme population densities evolution produces alternative life histories due to trade-offs; the trade-offs arise because genotypes with highest fitness at high population densities will not also have high fitness at low density and vice-versa.

CLIMATIC IMPACT AND SIZE OF THE POPULATION

The response of fish stocks to climatic changes have been studied by Skud (1982) who considered competition with other species. This author suggests that the reaction of the stocks to environmental changes may vary with the inversion of dominance. Stock abundance is then controlled by the abundance of the dominant species. The response to environmental factors is weakened. The thesis that climatic factors affect the abundance of a species but do not govern its absolute population density has also been supported (Skud 1982).

SELECTIVE PRESSURES AND EVOLUTIONARY CHANGES AS A POSSIBLE MECHANISM FOR RECOVERY

These ecological and evolutionary considerations show that the response of a population to environmental, intraspecific or interspecific selective pressures vary with abundance level. Evolutionary changes of depleted stocks under strong selective pressures, may occur. They are able to appear very rapidly within a population, which permits the observation on an ecological time scale (Ford 1975, Law 1979, Berry 1979, Mitchell and Arthur, 1991, Mueller et al. 1991).

We hypothesize that a low biomass level of lengthy duration could be due to strong interspecific competitive pressures, a poor population fitness and in a minor way, due to climatic changes. The recovery would be the result of an adaptive process. This implies that after recovery there would be a new relationship between the population and its environment, that is a new adaptive strategy adopted by the species under different interspecific and environmental selective pressures.

To model discontinuous changes in natural phenomena, the topological theory of dynamical systems (Thom 1972), now known as the catastrophe theory (Zeeman 1978), may be used. A model with a one-dimensional section of a cusp catastrophe is proposed to illustrate the regulation of these pelagic ecological systems (Figure 4). The model allows a representation of the discontinuities that exist between stock abundance and the environment by investigating a bistable phenomenon which displays hysteresis. The existence of a regime with distinct equilibrium values has been discussed in several ecological contexts (cf. May 1979). For a prosperous stock, interspecific competition is weak and there is a strong climatic impact on stock size. The collapse appears for a weakened stock (jump phenomena on way 1 on Figure 4). Interspecific competition is strong for a depleted stock which fluctuates only slightly under climatic fluctuations. Furthermore evolutionary changes that may appear under strong selective pressures allow recovery to a prosperous stock level (jump phenomena on way 2 on Figure 4).

CATASTROPHE-TYPE REGULATION AND THE CHANGING NATURE OF POPULATIONS

The adaptations and the mechanist points of view represent the two major approaches to understanding long term changes (Krebs 1978). They are not mutually exclusive but overlap and together they can contribute to a better comprehension of long term regulation in fish stocks. Natural populations are moulded by changing environment (Peterman et al. 1979) and evolutionary changes may occur rapidly in a depleted stock, so that the evolutionary time-scale approaches the ecological time-scale. Periodic changes in the genetic constitution of the population are sometimes associated with numerical fluctuations. The ultimate causes of the recovery of a population perhaps lie in the composition of the population itself. In this perspective, temporary extinctions could represent a process of renewal population adaptability to the environment. During depleted phases, natural selection could restore populations adaptability by readjusting, reorientating and diversifying the populations intrinsic parameters of their life history (Figure 5). Natural populations are evolving under our eyes and evolutionary genetics is a discipline only on the verge of recognition (Berry 1979).

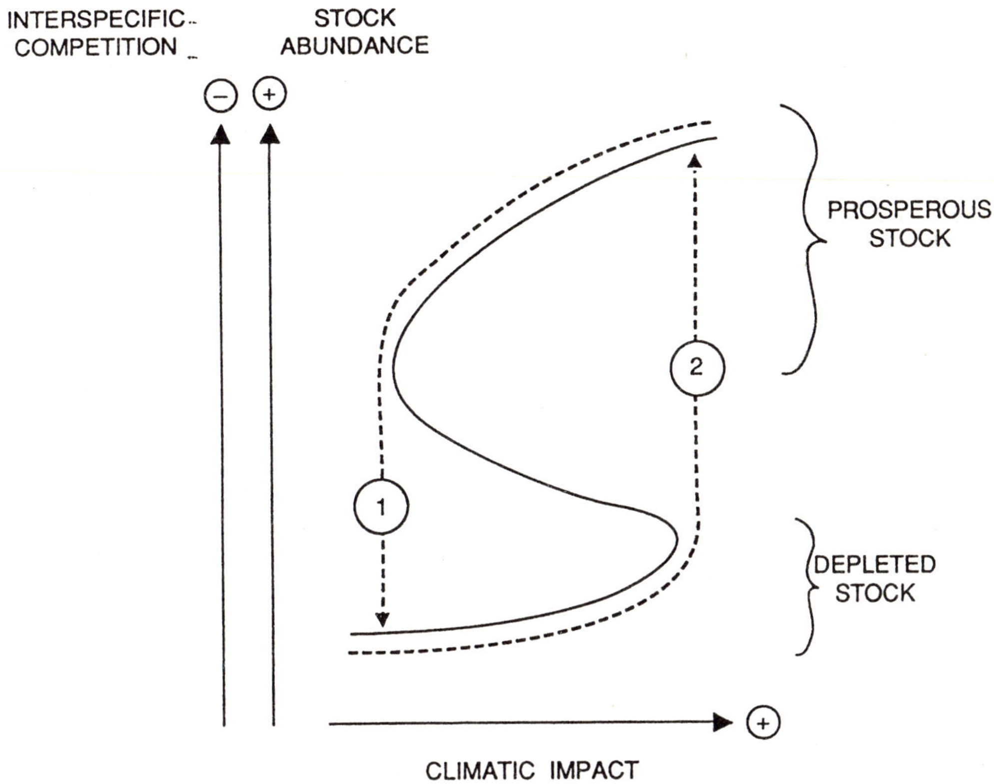

Figure 4. The catastrophe type regulation of fish stock abundance can be expressed with a one dimensional section of a cusp catastrophe. The dashed line represents the possible way on the theoretical topology (solid line). The severity of interspecific competition is correlated with stock abundance: it is weak for a prosperous stock and strong when the stock is depleted (Y-axis). The stock collapses when its level is low under unfavourable climatic impact (way 1). The stock is able to recover to a prosperous level under favourable climatic conditions, when evolutionary changes occur under strong selective pressures (interspecific competition) (way 2).

MANAGEMENT OF UNCERTAINTY USING A "GROPING STRATEGY" ?

Environment, fish population, fishermen and markets are all part of the fisheries. These components interact and change through time. Fisheries are changing sometimes very slowly and sometimes drastically. A fishery is functioning like a kaleidoscope; a small change in one of its components will produce a new landscape of the same components arranged in a new way. Sometimes this new landscape looks like the previous one; figures and colors are quite similar, and sometimes the landscape has nothing to do with another previous one. As seen previously, pelagic fish stocks are good examples of renewable resources where unexpected and drastic changes may appear very rapidly.

Learning from the past is perhaps the most probing way to avoid errors. And we are now in a good situation for learning from the past if managers, as well as scientists, learn from change (Holling 1978). We have good reasons to think that there are a lot of different factors which can drag a fishery system to a collapse or to a recovery. Unexpected events do play an important role when considering year-to-year fluctuations in pelagic fisheries or longer time periods that still are of interest for fisheries management. The question "what should have been done to avoid a pelagic fish collapse (i.e. the Peruvian anchoveta) ?" will receive many different and incompatible (in term of management) answers according to a fishery biologist, a fisherman, a fishery manager or a multinational company (Glantz and Thompson 1981, Glantz 1992). Trying to predict what will happen next year to the stock level is a great source of confusion. Exploring possible changes of a dynamical system and alternatives in management appears to be more creative (Allen and McGlade 1986, 1987). Renewable resources are constantly changing and evolving. Management should try to consider more closely what has been done and better take into account the unexpected events, the surprises and the changes that are inevitable (Brewer 1983, Walters and Holling 1990). And to use a "groping strategy" by developing adaptive strategies will be certainly full of new scientific and management challenges (Cury and Roy 1991).

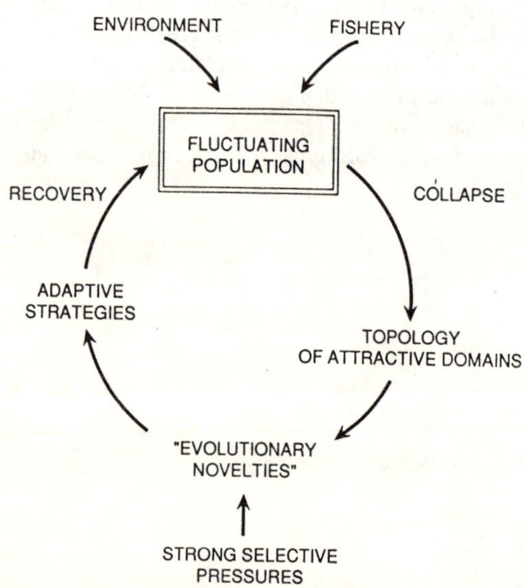

Figure 5. Theoretical scheme of long term regulation in catastrophe-type fish stocks dynamics where density-dependent natural selection plays a major role (adapted from Peterman et al.1979).

ACKNOWLEDGEMENTS

I thank R. Thom, R. Margalef, T.K. Stokes and F. Laloë for their advice. Thanks to Antoinette Nichols who prepared the figures. This is a contribution to the CEOS (Climate and Eastern Ocean Systems) project, a NOAA-ORSTOM-ICLARM cooperative project supported by the NOAA Climate and Global Change Marine Ecosystem Response project and by the TOA department of ORSTOM.

REFERENCES

Allen, P. M. 1988. Evolution: why the whole is greater than the sum of the parts ? Pp. 2-30, in W. Wolff, C.-J. Soeder and F.R. Drepper eds. Ecodynamics. Springer Verlag, Berlin.

Allen, P. M. and J. M. McGlade. 1986. Dynamics of discovery and exploitation: the case of the Scotian shelf fisheries. Canadian Journal of Fisheries and Aquatic Sciences. **43**: 1187-1200.

Allen, P. M. and J. M. McGlade. 1987. The modelling of human systems: a fisheries example. European Journal of Operational Research **30**: 147 - 167.

Allen P. M. and J. M. McGlade. 1989. Optimality, adequacy and the evolution of complexity. Pp. 3-21, in P. L. Christiansen and R. D. Parmentier eds. Structure, coherence and chaos in dynamical systems. Manchester University Press, Manchester.

Andrewartha H. G. and L. C. Birch. 1954. The ecological web. University of Chicago Press, Chicago.

Berry R. J. 1979. Genetical factors in animal population dynamics. Pp 53-80, in R. M. Anderson, B. D. Turner and L. R. Taylor.eds. Population dynamics. Blackwell Scientific Publications, Oxford.

Beverton R. J. H. 1990. Small marine pelagic fish and the threat of fishing; are they endangered ? Journal of Fish Biology **37**: 5-16.

Brewer G. D. 1983. The management challenges of world fisheries. Pp. 195-209, in B.J. Rothchild ed. Global fisheries: Perspectives for the 1980s. Springer Verlag, Berlin.

Crawford R. J. M., L.G. Underhill, L. V. Shannon, D. Lluch-Belda, W. R. Siegfried and C. A. Villacastin-Herrero. 1991. An empirical investigation of trans-oceanic linkages between areas of high abundance of sardine. Pp. 319-322, in T. Kawasaki, S. Tanaka, Y. Toba and A. Taniguchi eds. Long-term variability of pelagic fish populations and their environment. Pergamon Press.

Cury, P. 1988. Pressions sélectives et nouveautés évolutives: une hypothèse pour comprendre certains aspects des fluctuations à long terme des stocks des poissons pélagiques côtiers. Canadian Journal of Fisheries and Aquatic Sciences. **45**: 1099-1107.

Cury, P. and A. Fontana. 1988. Compétition et stratégies démographiques comparées de deux espèces de sardinelles (*Sardinella aurita* et *Sardinella maderensis*) des côtes ouest-africaines. Aquatic Living Resoures **1**: 165-180.

Cury, P. and C. Roy (eds). 1991. Les pêcheries ouest-africaines: variabilité, instabilité et changement. ORSTOM, Paris, France.

Cushing, D. H. 1982. Climate and fisheries. Academic Press.

De Vries, T. J. and W. G. Pearcy 1982. Fish debris in sediments of the upwelling zone off central Peru: a late Quaternary record. Deep-Sea Res. **29**: 87-109.

Endler, J. A. 1986. Natural selection in the wild. Princeton University Press, Princeton.

Ford, E. B. 1975. Ecological genetics. Chapman & Hall, London.

Glantz, M. H. (ed.). 1992. Climate variability, climatic change and fisheries. Cambridge University Press, Cambridge.

Glantz, M. H. and J. D. Thompson 1981. Resource management and environmental uncertainty: lessons from coastal upwelling fisheries. Advances in environmental science and technology **11**: 491p.

Haldane J .B. S. 1932. The causes of evolution. Longman, New York.

Hiramoto, K. 1991. The sardine fishery and ecology in the Joban and Boso waters of central Japan. Pp 117-128T, in Kawasaki, S. Tanaka, Y. Toba and A. Taniguchi eds, Long-term variability of pelagic fish populations and their environment. Pergamon Press.

Holling, C. S. (ed.). 1978. Adaptive environmental assessment and management. Wiley, New York.

Jordan, R. S. 1983. Variabilidad de los recursos pelagicos en el Pacifico sudeste. In G.D. Sharp and J. Csirke eds. FAO Fish Reports. **291**. FAO, Rome.

Kawai, T. and K. Isibasi 1983. Change in abundance and species composition of neritic pelagic fish stocks in connection with larval mortality caused by cannibalism and predatory loss by carnivorous plankton. Pp. 1081-1111, in G.D. Sharp and J. Csirke eds. FAO Fish Reports **291.** FAO, Rome.

Kawasaki, T. 1983. Why do some pelagic fishes have wide fluctuations in their numbers ? Biological basis of fluctuation from the viewpoint of evolutionary ecology. Pp. 1065-1080, in G. D. Sharp and J. Csirke eds.. FAO Fish Reports **291**. FAO, Rome.

Kondô, K. 1980. The recovery of the Japanese sardine. The biological basis of stock-size fluctuations. Rapports et Procès--Verbeaux des Réunions, Conseil International pour l'Exploration de la Mer. **177**: 332-354.

Kondô, K. 1991. Interspecific relation between Japanese sardine and anchovy populations that reflects the essential mutual relation between fluctuation mechanisms of the two species based on 'organism-environment' coupling. Pp. 129-134, in T. Kawasaki, S. Tanaka, Y. Toba and A. Taniguchi eds. Long-term variability of pelagic fish populations and their environment. Pergamon Press.

Krebs, C. J. 1978. Ecology: the experimental analysis of distribution and abundance. Harper and Row, New York.

Lasker, R. and A. MacCall 1983. New ideas on the fluctuations of the clupeoid stocks off California in Pp.110-117 in Proceedings of the Joint Oceanographic Assembly 1982; General Symposia. Canadian National Committee Scientific Committee on Oceanic Research,. Ottawa.

Laurec, A., A. Fonteneau, and C. Champagnat 1980. A study of the stability of some stocks described by self-regenerating stochastic models. Rapports et Procès-Verbeaux des Réunions, Conseil International pour l'Exploration de la Mer **177**: 423-438.

Law, R. 1979. Ecological determinants in the evolution of life histories. Pp 81-103 in R. M. Anderson, B. D. Turner and L. R. Taylor eds. Population dynamics. Blackwell Scientific Publications, Oxford.

Levins, R. 1968. Evolution in changing environments. Princeton University Press, Princeton.

MacCall, A. 1983. Variability of pelagic fish stocks off California. Pp. 101-112 in G. D. Sharp and J. Csirke eds. FAO Fish Reports **291.** FAO, Rome.

May, R .M. 1979. The structure and dynamics of ecological communities. Pp. 385-407 in R. M Anderson, B. D. Turner and L. R. Taylor eds., Population dynamics. Blackwell Scientific Publications, Oxford.

Mayr E, 1960. The emergence of evolutionary novelties.Pp. 349-380, in S. Tax ed. Evolution after Darwin, Volume 7. University. of Chicago Press, Chicago.

Mayr E,. 1974. Populations, espèces et évolution. Hermann. Edition française. .

Mayr E, 1981. La biologie de l'évolution. Hermann. Edition française.

Mitchell, P. and W. Arthur. 1991. Extinction due to evolution of a competitor. Evolution **45**:307-313.

Moore, J.A. 1952. Competition between *Drosophila melanogaster* and *Drosophila simulans*. II. The improvement of competitive ability through selection. Proceedings of the National Academy of Sciences, U.S.A. **38**: 813-817.

Mueller, L. D., Guo P. and F. J. Ayala 1991. Density-dependent natural selection and trade-offs in life history traits. Science **253**: 433-435.

Myers, J. H. and C. Krebs 1974. Population cycles in rodents. Scientific American **230**: 38–46.

Peterman, R. M. , W. C. Clark and C. S. Holling 1979. Pp. 321-341, in: R. M Anderson, B. D. Turner and L. R. Taylor eds., The dynamics of resilience: shifting stability domains in fish and insect systems. Blackwell Scientific Publications, Oxford.

Pimentel, D., E. H. Feinberg, P. W. Wood and J. T. Hayes 1965. Selection, spatial distribution, and the coexistence of competing fly species. American Naturalist **99**: 97-109.

Ryman, N. and F. Utter (eds) 1987. Population genetics and fishery management. University of Washington Press, Seattle.

Shannon, L.V., R. J. M. Crawford and D. C. Duffy 1984. Pelagic fisheries and warm events: a comparative study. South African Journal of Science. **80**: 51-60.

Skud, B. E. 1982. Dominances in fishes: the relation between environment and abundance. Science, **216**:144-149.

Soutar, A. and J. D. Isaacs 1969. History of fish population inferred from fish scales in anaerobic sediments off California. California Cooperation on Oceanic Fisheries Investigations Reports **13**: 63-70.

Stearns, S. C. 1980. A new view of life history evolution. Oikos **35**: 266-281.

Stearns, S. C. 1982. The role of development in the evolution of life histories. In J.T. Bonner. Evolution and development. Dahlem Konferenzen. Springer- Verlag, Berlin.

Steele, H. and E. W. Henderson 1984. Modeling long-term fluctuations in fish stocks. Science **224**: 985-987.

Tanaka, S. 1983. Variation of pelagic fish stocks in waters around Japan. Pp. 17-36, in G. D. Sharp and J. Csirke eds.. FAO Fish Reports **291**. FAO, Rome.

Thom, R. 1972. Stabilité structurelle et morphogénèse. Interéditions. Benjamin, Paris, France.

Tsuboi, M. 1984. Progress Report of synthesis investigations of the Japanese Fisheries, Project in 1983. Issued from Dai-Nihon Suisankai, March 1984.

Walters, C. J. and C. S. Holling 1990. Large scale management experiments and learning by doing. Ecology **71**: 2060-2068.

Watanabe, T. 1983. Stock assessment of common mackerel and Japanese sardine along the Pacific coast of Japan by spawning survey. Pp. 57-81, in G .D. Sharp and J. Csirke eds. FAO Fish Reports **291**. FAO, Rome

Zeeman, E. C. 1978. Catastrophe theory. Addison-Wesley, Reading, Massachusetts.

Zuta, S., Tsukayama I. and R. Villanueva 1983. El ambiente marino y las fluctuaciones de las principales poblaciones pelagica de la costa peruana. Pp. 179-253, in G. D. Sharp and J. Csirke eds. FAO Fish Reports **291**. FAO, Rome

APPENDIX

For a description of this model refer to Laurec et al. (1980).

An annual time scale (index i) for NB age classes is used. The number of individuals at age class i is noted N_i and N_{i+1} for the following year, this number depends on natural mortality rate M_i and fishing mortality rate F_i and it follows an exponential model:

$$N_{i+1} = N_i \, e^{-(Fi+Mi)} \tag{1}$$

Total weight of the cohort is obtained by multiplying the number of individuals (N_i) by their mean weight (W_i). The stock biomass (B) is given by:

$$B = \Sigma_{i=1,NB} N_i * W_i \tag{2}$$

the fecund biomass of age classes is obtained by multiplying individual fecundity (F_{cri}) by number of individuals (N_i) and their mean weight (W_i). Fecundity of the stock (S) is given by:

$$S = \Sigma_{i=1,NB} N_i * W_i * F_{cri} \tag{3}$$

Annual recruitment (R) is calculated using a Ricker stock-recruitment relationship:

$$R = \alpha S \, e^{-\beta S} \tag{4}$$

The large natural variability observed in the recruitment is introduced by multiplying it by a white noise. A stochastic factor (ra) following a log-normal distribution with a mean equal to 1 is utilized. To take into account the large variability observed for low biomass level, the standard error is a function of the fecund biomass and equal to $a+b/S$. Figure 6 presents the form of the stock-recruitment curve and the associated variability for each level of fecund biomass.

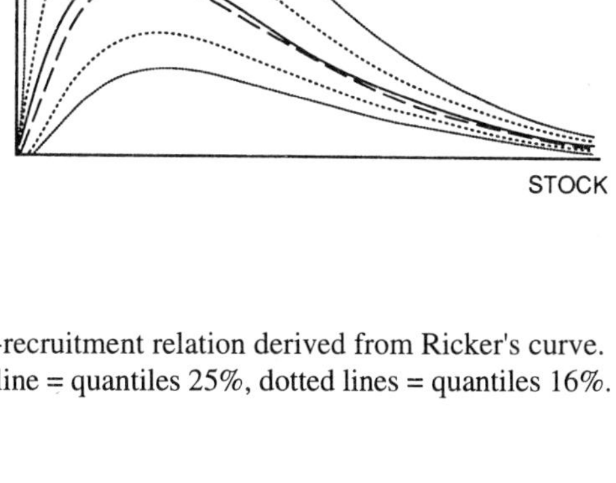

Figure 6. Stochastic stock-recruitment relation derived from Ricker's curve. Solid line = average, dashed line - median, short dashed line = quantiles 25%, dotted lines = quantiles 16%.

Fishing effort (f) is constant. However to take into account the decrease of fishing mortality (F) for depleted biomass, catchability (Q) is described as a function of the biomass:

$$Q = 1 - 0.01 * BV / B \qquad (5)$$

where

BV is the Virgin Biomass (unexploited stock at equilibrium) and B = Stock Biomass.

Figure 7 shows Q as a function of stock biomass (Equation 5). Catchability of the stock really decreases only for very low stock levels. It permits to describe the withdrawal of the fishing activity when the stock is depleted.

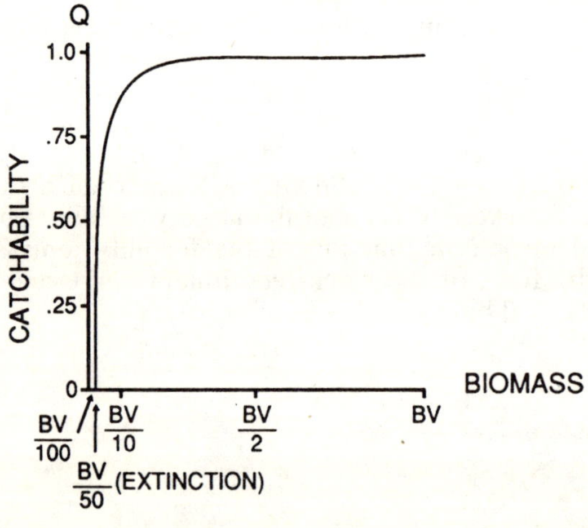

Figure 7. Evolution of catchability as a function of stock biomass.

Total catch (C) is given by:

$$C = Q * f * B = F * B \qquad (6)$$

with:

Q = catchability, f = fishing effort, B = exploited stock biomass.

Parameters values have been chosen close to real values to be more realistic. The fish lives 5 years, demographic and exploitation characteristics are the following:

Natural mortality :	$M = 1.0$
Fishing effort :	$f = 1000$
Coefficients of the stock-recruitment relation	$\alpha = 0.45$ $\beta = 0.0025$
Coefficients of the relation $a+b/S$:	$a = 1.0$ $b = 15.0$

Every year the recruitment (Equation (4) is calculated from the fecund stock biomass of the previous year (3). This recruitment is multiplied by a stochastic factor (*ra*). The total stock biomass for the year is calculated from (1) and (2). Then catch are calculated using (6).

The values of the coefficient per age class that are utilized in this simulation are given in the following table:

Age	1	2	3	4	5
Weight	5	40	80	120	140
Fecundity	0	0.5	0.5	0.5	0.5

At time $t = 0$, the stock is at equilibrium. We postulate that there is a temporary extinction when the stock reaches a value lower or equal to $1/50^{th}$ of the virgin biomass. To make this simulation more realistic a constraint of a minimal value of the recruitment has been fixed to 5‰ of equilibrium recruitment. This constraint allows a possible recovery of the stock when it collapses.

SIZE-SELECTIVE HARVESTING AND AGE-AT-MATURITY I: SOME THEORETICAL IMPLICATIONS FOR MANAGEMENT OF EVOLVING RESOURCES

S P Blythe[1] *and T K Stokes*[2]

[1] Department of Statistics and Modelling Science, University of Strathclyde, Glasgow G1 1XH, UK.
[2] MAFF, Directorate of Fisheries Research, Fisheries Laboratory, Pakefield Rd., Lowestoft, Suffolk NR33 OHT, UK.

ABSTRACT

We show how simple stage- and size-structured models may be used to assess the possible evolutionary consequences of size-selective harvesting regimes. Clonal models are used, and invasibility techniques are applied to show that equilibrium adult biomass is the key quantity and to derive a straightforward fitness measure, R. For a population with an ESS in the character(s) of interest (life-history traits) prior to harvesting, it is possible to use R, and the equilibrium biomass yield per unit time, Y, to explore possible outcomes of harvesting strategies.

INTRODUCTION

The harvesting methods used in the exploitation of natural animal populations as food resources are often size-selective (see e.g. Policansky this volume). Perhaps most commonly, it is larger, and hence normally older, individuals which are preferred. This is explicitly the case for commercial fisheries, where mesh-sizes are set to select particular size ranges. Where harvesting is age-specific, or specific to life history stages, size-selectivity emerges as a consequence rather than as an end in itself.

There has been a growing recognition in recent years that size-selective harvesting will produce selection pressure within the exploited population, perhaps inducing evolutionary shifts in such characters as age-at-maturity, growth rate, and the timing and magnitude of fecundity. In addition to laboratory studies where such trait changes have been induced, confirmatory evidence comes from recent results on long-running experiments manipulating stage-specific mortality, due to predators, in field guppy populations (Reznick et al. 1990).

Law and Grey (1988, 1989) have raised the question of whether evolutionary stable optimal harvesting strategies (ESOHS) are possible. An ESOHS would be characterised by a harvesting regime such that, over long periods of time, there is no selection pressure acting on individuals in the exploited population to change the life-history characteristics expressed in future generations, and such that the harvester cannot increase the sustainable yield by changing the strategy. Blythe and Stokes (1990) looked at a simple stage-structured model for size-selective harvesting, considering some management consequences if the individual growth rate responded to selective pressure. In this paper we use a slightly more general form of the Blythe and Stokes (1990) model, and focus on the age-at-maturity as a responsive trait. The companion paper in this volume (Stokes and Blythe) applies the approaches which we outline here to a number of real exploited fish populations.

As a major object of this study is to consider the possible evolutionary effects of harvesting, we should be careful that in the absence of harvesting there is an evolutionary stable strategy (ESS) for the population, with respect to the parameters of interest. That is, individuals' life-history traits are optimal in the pre-harvesting environment. If this were not so, we would be unable to distinguish between a response due to harvesting, and one due to background selection pressure. We thus start with the pre-harvesting model.

THE MODEL WITHOUT HARVESTING

We consider a model of a clonal population with just two life-history stages (juveniles and adults). Neonates have mass w_0, and grow according to a specific schedule giving mass $w(a)$ at age a. Upon reaching a threshold age τ they become adults, and start reproduction. In this model, juveniles and adults have different food supplies, with no juvenile density dependence but competition among adults. The feeding rate of adults is proportionate (f) to body size, so that the factor mediating competition is total adult biomass. A constant fraction q of adult food intake is assimilated as reproductive tissue, each unit mass of which produce $b = fq/w_0$ neonates of size w_0; this reproduction is assumed instantaneous and continuous. Finally, juveniles and adults have distinct, constant, age-independent *per capita* mortalities δ and μ, respectively.

This is an extremely simplistic stage-structure model; for a discussion of this large and useful class of models, see Metz and Diekmann (1986). In particular the treatment of size-structure here is very definitely a rough approximation only, ignoring maintenance costs and the detailed budgeting of assimilated energy.

The model may be formulated as a small set of delay-differential equations (c.f. Gurney et al. 1986; Blythe and Stokes 1990), but we choose here to use the equivalent integral equation form,

$$V(t) = e^{-\delta\tau} \int_{\tau}^{\infty} w(u)\beta e^{-\mu(u-\tau)} V(t-u) C[V(t-u)] du \tag{1}$$

for the total adult biomass $V(t)$. In the absence of competition, an adult of age u, and hence size $w(u)$, $u \geq \tau$ produces $\beta\, w(u)$ neonates of size w_0 per unit time, a fraction $(e^{-\delta\tau})$ of which survive to maturity. Food intake is reduced by a factor $C[V]$ by competition; we assume $C[0] = 1$, and $C[V]$ declines monotonically with increasing V towards zero as $V \to \infty$. Implicit in our use of Equation (1) is the assumption that t is sufficiently large that initial conditions may be ignored (i.e. it is several times $\tau + 1/\mu$).

There are two equilibrium solutions to Equation (1): a persistent solution

$$V^* = C^{-1}\left[1/R\right] \tag{2}$$

where

$$R = \gamma e^{-\delta\tau}\int_{\tau}^{\infty} w(u)e^{-\mu(u-\tau)}du > 1 \tag{3}$$

with C^{-1} the inverse function of C, and $V^* = 0$ (i.e. extinction), for $R < 1$. R is the average number of offspring an adult would produce, over a lifetime, which survive to maturity, at very low population levels. $C^{-1}[1/R]$ is a monotonically increasing function of $R > 1$.

We may use the invasibility method (Blythe and Stokes 1990; Metz and Diekmann 1986) to illuminate the appropriate fitness measure in this model. Say there are two clonal strains, the resident (subscript r) and the invader (subscript i), which may differ in any function or parameter. Then we have the competition model

$$V_x(t) = \beta_x e^{-\delta_x\tau_x}\int_{\tau}^{\infty} W_x(u)\beta e^{-\mu_x(u-\tau_x)}V_x(t-u)C_x[V_r(t-u)+V_i(t-u)]du$$

for $x = r,i$. The possible equilibria are:

(i) (0,0) where both become extinct,
(ii) $(V_r^*,0)$,
(iii) $(0, V_i^*)$ (in (ii) and (iii) one strain excludes the other, and has an equilibrium biomass given by Equations (2) and (3) with appropriate subscripts), and
(iv) a coexistence curve where $R_i = R_r$ (c.f. Blythe and Stokes 1990).

(i) is uninteresting and (iv) requires a precise mathematical balance of parameter values, so that in general, if $R_r > 1$ and $R_i > 1$, one strain will exclude the other. To see which, set $V_r(t) = V_r^* + \varepsilon_r(t)$ and $V_i(t) = V_i^* + \varepsilon_i(t)$, where the ε_x are small. Then linearising the equation for $V_i(t)$ gives

$$\varepsilon_i(t) = \beta_i e^{-\delta_i\tau_i}C_i\int_{\tau_i}^{\infty} w_i(u)e^{-\mu_i(u-\tau_i)}\varepsilon_i(t-u)du$$

which is a form of Lotka integral. $\varepsilon_i(t)$ will grow (and hence eventually displace the resident strain, because $(V_r^*,0)$ is thus unstable) provided that

$$V^*_i > V^*_r \tag{4}$$

Note that Equation (4) implies $R_i > R_r$ if C_i and C_r are the same function. That being the case, in this paper, where we focus on τ, we may usefully take R as a measure of fitness; in particular, a strain which maximises its R cannot be invaded, so its strategy is an ESS (see Nisbet et al. 1989 for methods of extending this technique to cases where $V_r(t)$ is not at equilibrium. The steady-state invasibility criterion is, however, apparently rather robust in this respect (*op. cit.*).

It is worth commenting at this point on two features of the above analysis. First, we see that maximising fitness does not necessarily maximise population numbers. Functionally, this happens because the population quantity through which competition operates is biomass, not numbers, and because we are dealing with competing clonal strains, which do not of course interbreed. More generally, and we are indebted to Joel Brown for pointing this out at the meeting, the system we have constructed is one where frequency-dependent selection will operate, and under those circumstances the traditional view, that maximising individual fitness maximises population numbers, will rarely hold.

The second point concerns the use of invasibility criteria versus ESS (evolutionary stable strategy) techniques. In the former case, as in this paper, we proceed by searching for conditions under which a new strain might replace an indigenous one, by examining the initial-increase or -decrease rate of the invader, thereby determining whether it will successfully invade or not. This is essentially a matter of calculating the local stability properties of the invader. The ESS approach is usually taken to be a game-theoretic one, which aims at finding non-invadable strategies. It is still something of an open question as to if and how these two approaches may differ in the strategies they find; however, this issue and its inherent ambiguities have been noted in the literature on evolutionary theory, and attempts have been made to provide unified approaches to best strategies (in particular, see Lessard 1989; also, see Brown and Parman this volume for an ESS approach to finding ESOHS in the Blythe and Stokes (1990) model). There is scope for future work in the context of harvesting response models, aimed at determining whether concentrating on the invader or the resident strain yields the more useful information.

A GROWTH RATE EXAMPLE

To look at the growth rate in detail, we choose a particular but fairly general example. First, suppose that $w(a)$ does not depend in any way upon τ, the evolutionary parameter. Then from Equation (3),

$$\frac{\partial R(\tau)}{\partial \tau} = -\gamma e^{-\delta\tau} w() - (\delta - \mu) R(\tau), \tau \geq 0$$

Clearly, if $\delta > \mu$, then regardless of the form of $w(a)$, there can be no ESS for $\tau > 0$. If $\mu > \delta$, an ESS may exist - essentially we require that $w(a)$ increase (less quickly than exponentially) asymptotically as $\tau \to \infty$.. We consider some of the implications of this result in the discussion.

To get an ESS for $\delta > \mu$, we must either invoke reaction norms (essentially a parameter of (say) growth becomes a function of, in the present case, τ), or use stage-specific growth (still less than exponentially fast); either way, $w(a)$ becomes $w(a;\tau)$, and we may in some cases get a $\delta > \mu$ ESS.

We will consider stage-specific growth,

$$w(a,\tau)\begin{cases} u(a) & ,a \le \tau \\ u(\tau)+v(a),a>\tau,v(\tau)=0 \end{cases}$$

In particular, a slightly generalised version of the Blythe and Stokes (1990) model has $u(a) = w_0 + ga$, and $v(a) = 0$ (linear growth of juveniles, no growth of adults). Then

$$R = \frac{\gamma}{\mu} e^{-\delta\tau}(w_0 + g\tau)$$

To find the optimal value of reproductive age τ, we look for a maximum in fitness R. Differentiating R with respect to τ, we find that a maximum occurs for $\tau = \tilde{\tau} = \delta^{-1} - w_0/g$. If neonate mass w_0 is very small, and linear growth rate g reasonably large, then $\tilde{\tau} \approx \delta^{-1}$. A strain with a τ far from $\tilde{\tau}$ will be selected against by one with τ closer to $\tilde{\tau}$. If we regard the conditions prior to the onset of harvesting as being evolutionarily stable in that milieu, then we expect that the pre-harvesting clonal strain can be characterised with a reproductive age τ whose value will be $\tilde{\tau}$ for the prevailing conditions.

Note that the 'slight generalisation' of Blythe and Stokes (1990) model which we employ here, namely to give neonates a non-zero mass w_0 and to incorporate a background death-rate δ, is actually rather an important one! Without these additions (crucially that of background death rates), the model does not admit of a stable strategy in the absence of harvesting, which in the present context is a considerable liability. As a result of these changes, there are significant differences between the behaviour of the model as employed here and that in the 1990 paper.

SIZE SELECTIVE HARVESTING

If $z \le w(\tau)$ is the size at first harvesting, then $T = (z - w_0)/g \le \tau$ is the age at first harvesting, and we replace Equation (1) by

$$V(t) = \gamma e^{-\delta\tau - H(\tau - T)} \int_{\tau}^{\infty} w(u,\tau) e^{-(\mu+H)(u-\tau_s)} V(t-u) C[V(t-u)]du$$

where $H > 0$ is the harvesting rate. As before, we have $V^* = C^{-1}[1/R]$, where now

$$R = \frac{\gamma}{\mu + H} e^{-\delta\tau - H(\tau - T)} (w_0 + g\tau) \tag{5}$$

for the Blythe and Stokes (1990) growth function with $w_0 \geq 0$. Here we may apply the same invasibility techniques as in the previous section to show that maximising R maximises V^* and fitness. The optimal τ for given z and H (with γ, δ and μ unchanged from the pre-harvesting conditions) is obtained as before, by differentiating R with respect to τ and finding a maximum. Denoting the location of this maximum by $\hat{\tau}$, we thus have an optimal reproductive age under harvesting conditions, given by

$$\hat{\tau} = \frac{1}{\delta + H} - \frac{w_0}{g} < \hat{\tau} \; for \; H > 0;$$

with the requirement that $\delta + H < g/w_0$ usually not a problem as w_0 is very small.

At equilibrium, the yield (harvested biomass per unit time) is

$$Y = \gamma H e^{-\delta\tau} V^* C[V^*] \int_{\tau}^{\infty} w(u) S(T,u) du$$

where $S(T,u)$ is survival from age u to age T. This may be written

$$Y = HV^* \left\{ 1 + e^{-(\delta - \mu)\tau} \frac{\int_T^{\tau} w(u) e^{-(\delta + H)u} du}{\int_{\tau}^{\infty} w(u) e^{-(\mu + H)u} du} \right\}$$

$$= HV^* \left\{ 1 + e^{-(\delta - \mu)\tau} \left(\frac{\mu + H}{\delta + H} \right) \left[\frac{Q(\delta,T) - Q(\delta,\tau)}{Q(\delta,T) + \frac{g}{\mu + H}} \right] \right\} \tag{6}$$

In equation (6) we have introduced the notation $Q(x,y)$ to replace the integrals in the Y -equation with evaluable functions, for the particular model under detailed consideration in this paper (where mass-at-age $w(a)$ has a simple form). The first argument of Q (called x here) refers to one of the death rate δ or μ; the second argument (called y here) refers to one of the derived age parameters τ or T. Note that explicit evaluation requires that some functional form for the competition effect (function C) on fecundity be introduced; in the Blythe and Stokes (1990) model, the function $C(V) = \exp(-V/\theta)$ was used. For the modified Blythe and Stokes model used in this paper, with finite w_0, δ and μ, we may write

$$Q(x,y) = e^{-(x+H)}y\left[w_0 - y\left(\frac{1}{x+H} + y\right)\right]$$

(this reduces to the Y equation of Blythe and Stokes (1990) with $\delta = \mu = w_0 = 0$, with b finite).

We may now consider ESOHS. For the models considered here, an ESOHS would be a harvesting strategy, completely defined by the parameters z and H (essentially mesh-size and fishing mortality in a fisheries context), with the following properties: (a) under any harvesting regime, the fish have an optimal age at maturity (called $\hat{\tau}$ above), maximising their individual fitness R; (b) when a population has an age-at-maturity τ given by the optimal value $\hat{\tau}$, the harvester finds that the yield Y is at a maximum, and the fish still have fitness $R > 1$. Given (a) and (b), neither fish nor harvester has any reason to change strategies (the former through character replacement if another τ was better, the latter by explicit policy changes), and we could expect the system to persist in that configuration.

Given this definition of an ESOHS, the procedure for finding them is straightforward in principle: first, identify optimal reproductive age(s) $\hat{\tau}$ for a population under harvesting, by finding the τ which gives the maximum fitness R. Then substitute the equation for $\hat{\tau}$ into the yield equation Y, and look for strategies (z, H) which produce a maximum in Y. By following this procedure, we are saying that the mediators of change in the system as a whole are those aspects of the harvesting strategy which can be altered (z and H), and that the exploited population will *always* tend to undergo evolutionary change as a response to the harvesting, such that its responsive character (reproductive age τ) will tend to the value which maximises individual fitness R, that is the value $\hat{\tau}$. $\hat{\tau}$ is thus a function of the imposed z and H, and should be put explicitly into the Y equation for analysis of any ESOHS.

There are considerable practical difficulties in performing this analysis in any very general way, however: the competition function C enters into the Y equation via the steady-state biomass V^*, and we need to consider variation with respect to both z and H. While some progress may be made, for example making judicious use of algebraic manipulation packages like Mathematica to expand and re-order the equations for the derivatives of Y, it is highly likely that realistic progress in specifying ESOHS will only come from a case-by-case treatment of particular species and the harvesting options available. What should be achievable, however, is a mathematical analysis indicating what exactly are the requirements necessary for the existence of ESOHS. This is an area for future work.

Brown and Parman (this volume) have analysed the Blythe and Stokes (1990) model from an ESS perspective; their results differ from those of the original paper, but are in agreement with the results of this present one, a strong indication that the features missing in the original model, but included here (particularly that a pre-harvesting optimal strategy exists) are non-trivial. they also suggest that there may be circumstances where the ESS and invading approaches will give different results, underlining the need for further work in this area.

DISCUSSION AND CONCLUSIONS

The main theme of this paper is that it is possible to address quite complex problems involving harvesting and evolution, with models that are reasonably simple in form, but which contain a fair amount of biological detail. These models owe much to the 'stage-structure' formalism of delay-differential equations (explicitly so in Blythe and Stokes, 1990), but also to the more general partial differential and integral representations of physiologically structured populations (in particular Metz and Diekmann 1986). The growth models in particular are really quite primitive representations of this important process; but there is no hindrance in principle to using much more realistic methods within essentially the same framework.

The key result is that invasibility criteria may be used to good advantage in the identification of fitness functions (and hence of ESSs). These criteria need not be restricted to cases where an invader enters a population at equilibrium, but can be extended to the full dynamic, non-linear case; we have exploited, however, observations by Nisbet et al., (1989) that for models structurally similar to those discussed here, the equilibrium invasion results are both typical and robust.

Another potentially useful feature is the ready way that fitness, R, and yield, Y, can be examined as functions of responsive character traits. This allows a qualitative assessment of the kinds of changes which can be expected in an exploited population on an evolutionary time scale and also permits the issue of finding ESOHS (or at least ELSOHS, in the sense of the previous section) to be addressed in a simple, essentially graphical manner (for example, see Blythe and Stokes 1990, or Stokes and Blythe this volume). The fact that this ELSOHS represents a *stage-specific* harvesting regime is intriguing, given the stage-specificity of many predators and parasitoids.

The examination of ESOHS is performed using the equilibrium solution(s) of the biological model, and the invasibility criteria are extracted from a linear analysis of stability of steady-states. It is pertinent to consider what restrictions this may place upon conclusions reached. We should recall the difference in time-scales between the action of the harvester (hopefully an ecologically enlightened individual, as posited by Brown and Parman this volume), and the evolutionary response of the exploited population. In fisheries for example, quotas may be set annually, and mesh-sizes are subject to change, but the stock may react considerably more slowly (see Law and Rowell this volume). To the stock, the world is a dynamic and fast-changing place where adaptation proceeds at whatever pace it must, and towards goals largely set by the prevailing harvesting regime; to the harvester, the stock behaviour is perhaps close to static, on a scale of a few years at least; though there will be annual fluctuations arising from recruitment variation and general stochastic effects, the overall stability properties of the system will not change greatly year by year except when the stock is highly stressed. We may thus expect that an analysis based on equilibria and local stability will be adequate unless the stock is near extinction. As in this latter case stochastic effects will in any case become highly significant, and any deterministic analysis prone to an assortment of errors, this restriction is not an onerous one, and we may make use of the equilibrium/ invasibility results for most cases.

It should be clear, however, that any application as a serious management tool is going to require that the model used be as accurate a reflection of the biology of the exploited population as possible. The degree to which these evolutionary conjectures are sensitive to the details of model construction is not yet clear, but there will certainly be cases where it is significant. However, note that we say these models should be accurate, not that they should be immensely detailed. We suspect that in evolutionary management, at least as much as in ecology , the degradation of prediction quality when models become more elaborate than confirmed knowledge can sustain, will be a potent limiting factor (e.g. Costanza and Sklar 1985; Walters 1986).

Finally, we would like briefly to speculate on the future need for the identification and implementation of ESOHS. In many, if not most, parts of the world, the only large-scale exploitation of natural animal populations is by fisheries, and certainly the concept of ESOHS has a significant role to play if these resources are to be exploited for very much longer. So there is undoubtedly a need for considerable advancement of the theory of ESOHS in this area. But we should not forget that the future of a great part of the world ecosystem depends to a large extent upon human management. Management strategies must have evolutionary stable consequences, and it is perhaps not entirely fanciful to suggest that we may soon have to influence some predators and parasites to evolve a "strategy" of exploitation of their resources which is itself stable in evolutionary time, and which reflects our own expectations and requirements.

ACKNOWLEDGEMENTS

We would like to thank Ram Myers, Joel Brown, Robert Smith and Keith Nelson for their useful comments at the Jülich Workshop on the work upon which this paper is based.

REFERENCES

Bennett, A.F., Dao, K.M. and R.E. Lenski 1990. Rapid evolution in response to high-temperature selection. Nature **346**: 79-81.

Blythe, S.P. and T.K. Stokes 1990. Some consequences of size-selective harvesting on fitness and on yield. Journal of Mathematics Applied to Medicine and Biology **7**: 41-53.

Costanza, R. and F.H.Sklar 1985. Articulation, accuracy and effectiveness of mathematical models: a review of freshwater wetland applications. Biosciences **27**:45-68.

Derrida, B. and L. Pelitti 1991. Evolution in a flat fitness landscape. Bulletin of Mathematical Biology **53**: 355-82.

Gurney, W.S.C., Nisbet, R.M. and S.P. Blythe 1986. The systematic formulation of models of stage structured populations. Lecture Notes Biomathematics **68**: 475-94.

Law, R. and D.R. Grey 1988. Maximum sustainable yields and the self-renewal of exploited populations with age-dependent vital rates. In B. Ebenmann and L. Persson, eds. Size-structured populations. Springer-Verlag, Berlin.

Law, R. and D.R. Grey 1989. Life-history evolution and sustainable yields from populations with age-specific cropping. Evolutionary Ecology **3**: 343-59.

Lessard, S. 1989. Resource Allocation in Mendelian Populations; Further in ESS Theory. Pp.207-246, in M.W. Feldman ed. Mathematical Evolutionary Theory, Princeton University Press, New Jersey.

Metz, J.A.J and O. Diekmann. 1986. Physiologically Structured Populations Lecture Notes in Biomathematics **68**.

Myers, R.A. and R.W. Doyle. 1983. Predicting natural mortality rates and reproduction-mortality tradeoffs from fish life history data. Canadian Journal of Fisheries and Aquatic Sciences **40**: 612-20.

Myers, R.A. 1989. Estimating bias in growth caused by size-selective fishing mortality. I.C.E.S. C.M./**D**:8.

Nisbet, R.M., Gurney, W.S.C. and J.A.J. Metz 1989. Stage structured models applied to evolutionary ecology. Lecture Notes. Biomathematics **78**: 428-49.

Reznick, D.A., Brygs, H. and J.A. Endler 1990. Experimentally induced life-history evolution in a natural population. Nature **346**: 357-359.

Smith, R.H. 1991. Genetic and phenotypic aspects of life-history evolution in animals. Advances in Ecological Research **21**: 63-120.

Stearns, S.C. and R.E. Crandall 1984. Plasticity for age and size at sexual maturity; a life history response to some unavoidable stress. Pp. 13-33, in G.W. Potts and R.J. Wootton, eds. Fish Reproduction: strategies and tactics. Academic Press, London.

Walters, C. 1986. Adaptive Management of Renewable Resources. Macmillan Publishing Company, New York.

SIZE-SELECTIVE HARVESTING AND AGE-AT-MATURITY II: REAL POPULATIONS AND MANAGEMENT OPTIONS

T K Stokes[1] *and S P Blythe*[2]

[1]MAFF, Directorate Of Fisheries Research, Fisheries Laboratory, Pakefield Rd, Lowestoft, Suffolk, NR33 OHT, UK.
[2]Department Of Statistics And Modelling Science, University Of Strathclyde, Glasgow, G1 1XH, UK.

ABSTRACT

Parameters for a simple stage-structured harvesting model are derived for five North Sea demersal fish stocks. The parameters are used to generate contour plots of fitness and yield on axes of size-at-maturity Vs size-at-first-capture, and age-at-maturity Vs age-at-first-capture. Such graphs are used qualitatively to assess possible changes in the exploited populations that might result from management actions such as altering mesh size (and hence size-at-capture) and/or harvesting rate.

INTRODUCTION

The most important, commercially exploited North Sea demersal fish stocks are currently managed internationally by the setting of quotas, so-called total allowable catches (TACs). Advice on the appropriate TAC levels is provided by I.C.E.S. (The International Council for the Exploration of the Sea) working groups and other bodies. Traditionally, the advice given is based on single species assessments. In recent years these assessments have mostly been *ad hoc* tuned virtual population analyses (VPAs) using the Laurec-Shepherd method (Laurec and Shepherd 1983; Pope and Shepherd 1985 ; e.g. Anon 1990a, 1990b). Increasingly, however, fisheries managers are using approaches other than quotas (i.e. indirect effort control) to influence fisheries and are seeking advice on biological interactions, fleet interactions and bioeconomics (Hilborn and Walters 1992, Stokes 1993). In particular, the use of so-called technical measures (e.g. mesh size changes, mesh geometry changes, closed seasons or areas) has become common in addition to quotas (Anon 1990c, 1991).

Technical measures are usually aimed at improving the exploitation pattern (at-age) of fishing so as to protect juvenile fish from high fishing mortalities to reduce "growth over fishing" (Pitcher and Hart 1982) and to preserve (or rebuild) the spawning stock biomass (SSB). In mixed demersal fisheries, such as those prosecuted in the North Sea, it is not usually possible to target individual species and compromise mesh sizes are used (see e.g. Beverton and Holt 1957). The following is an example of the type of difficulty that managers have to face.

In the North Sea there are mixed fisheries for cod (*Gadus morhua*), haddock (*Melanogrammus aglefinus*) and whiting (*Merlangius merlangus*). In simplistic terms, cod and whiting are fished in the south-eastern North Sea whilst haddock, whiting and cod are fished in the north-western North Sea. Assessments of all species are conducted, and quotas are set, for the whole area. Cod grow to a large size (> 14kg) and take approximately 4 or more years to reach maturity. Haddock and whiting grow to lesser sizes and take a shorter time to mature (haddock: > 2kg , 3 years; whiting: > 1kg , 3 years). Cod and haddock are valuable and hence are target species. Whiting are much less valuable and are often taken as by-catch in mixed fisheries. Setting mesh size large enough to catch cod such that sufficient numbers survive to maturity would involve the loss of much (or all) of the marketable whiting and haddock. An additional problem is that whiting eat juvenile haddock and cod. Catching only large cod and reducing mortality on whiting might therefore result in increased predation mortality on small cod and haddock which could outweigh the reduction in mortality caused by increasing mesh size.

Traditionally, fisheries managers and politicians have to consider short, medium and long term consequences of their decisions. Clearly, whilst long term conservation and yield are important, so too are short term financial performance, employment levels, production capacity etc. Because of the complex pressures on fish stocks, fishermen and fisheries managers, the field of fisheries management has developed a necessary pragmatism which does not easily admit apparently academic considerations such as (very) long term evolutionary consequences or ESOHS (Evolutionary Stable Optimal harvesting Strategies - Law and Grey 1988, 1989). Nevertheless, it ill behoves fisheries scientists to neglect such possibilities.

In this paper, a simple graphical approach , based on the models presented in Blythe and Stokes (1991 and this volume), is adopted to investigate how management measures such as changing mesh size might influence exploited fish populations in the long term. Parameter estimates for the models are made for five North Sea demersal species - cod, haddock, whiting, plaice (*Pleuronectes pleuronectes*) and sole (*Solea solea*) - and contour plots of yield and fitness are plotted for all species on size-at-maturity vs. size-at-first-capture and age-at-maturity vs. age-at-first-capture. These plots are interpreted and discussed in terms of current fishing and management practice.

METHODS

Parameter estimates for the model proposed in Blythe and Stokes (1991; see also Blythe and Stokes this volume) have been made for five North Sea demersal species and the calculated fitness and yield are plotted for size-at-maturity (w) Vs size-at-first-capture (z) and for age-at-maturity (τ) Vs age-at-first-capture (T).

PARAMETER ESTIMATES

Estimates were derived for all years for which data were available from the relevant I.C.E.S. Working Group reports. Annual fishing mortality rate, H; weight-at-maturity (w) and weight-at-first-capture (z) and age-at-maturity (τ) were taken directly from the I.C.E.S. North Sea Roundfish Working Group Report (Anon. 1990a) and North Sea Flatfish Working Group Report (Anon. 1990b). The age-at-first-capture was taken as the age at which fishing mortality rate first becomes large (i.e. greater than natural mortality) whilst age-at-maturity was taken as the age at which at least 50 per cent of fish were mature. S, the number of individuals, that survive to maturity, produced per unit of body weight by each spawning individual was estimated as

$$S = N_{t+\tau} / (SSB_t \,.\, g)$$

where $N_{t+\tau}$ is the population estimate at time $t+\tau$ (derived by Virtual Population Analysis, VPA, (Gulland 1965; see Pope 1972)); SSB_t is the Spawning Stock Biomass at time t (estimated as

$$\sum_0^{a\,\max} N_{a,t} w_{a,t} m_{a,t}$$

where $m_{a,t}$ is the proportion of the population mature at age a, time t) and g is the proportion of body weight allocated to reproduction (i.e. gonad weight/body weight) over one year. $N_{t+\tau}$ and SSB_t are taken directly from Anon (1990a and 1990b) whilst g is guessed to be 0.1 in all cases. Data permitting accurate estimates of g for particular species are not available; work on cod , however, suggests that 0.1 is not inappropriate (Daan 1975). Hislop (1984), however, points out that the energy allocated to egg production (as a percentage of total energy allocation) is highly variable both within and between species.

Figure 1. shows plots of w and z through time for the five fish stocks. Although annual fluctuations are evident , the only significant trends (all at < 1 per cent) are for a decrease in both w and z for plaice from 1972 to 1989 and for sole for the entire period (increasing) or separately for the periods 1959 to 1971 (increasing) and 1972 to 1989 (decreasing). There are no significant trends in w or z in any period for any of the round fish species. These results are similar to those of Rijnsdorp et al. (1991) but those authors noted decreases in plaice and sole length-at-age from 1973 which were not significant.

Figure 2. shows w/z and $H\tau$ plotted through time for the five species. These plots are shown because when $H\tau$ exceeds w/z, the model of Blythe and Stokes (1991) predicts that selection pressure should act to reduce size-at-age in the long term. It is clear from Figure 2. that for cod, there has not been pressure for reduced size until recent years whilst for the four other species the pressure has existed for a longer duration. This may be surprising given that cod has traditionally been exploited at high H and has a relatively long τ. Cod, however, also has a very high age-at-maturity and hence w/z ratio. It is notable that the two flatfish species exhibit trends in size which match with the cross-over points on the w/z Vs $H\tau$ plots.

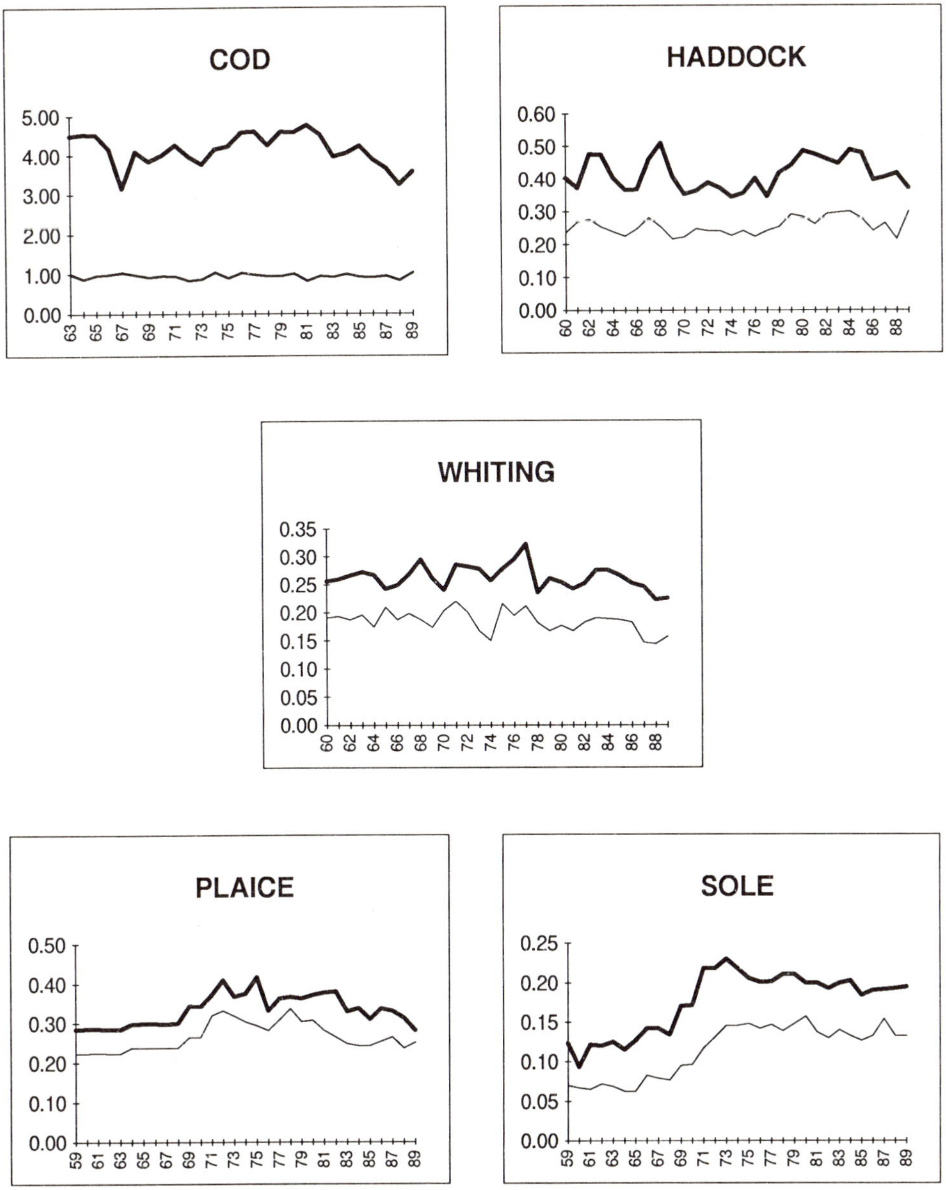

Figure 1. Plots of size-at-maturity, w shown as bold line, and size-at-first-capture, z shown as normal line, through time, for five demersal species.

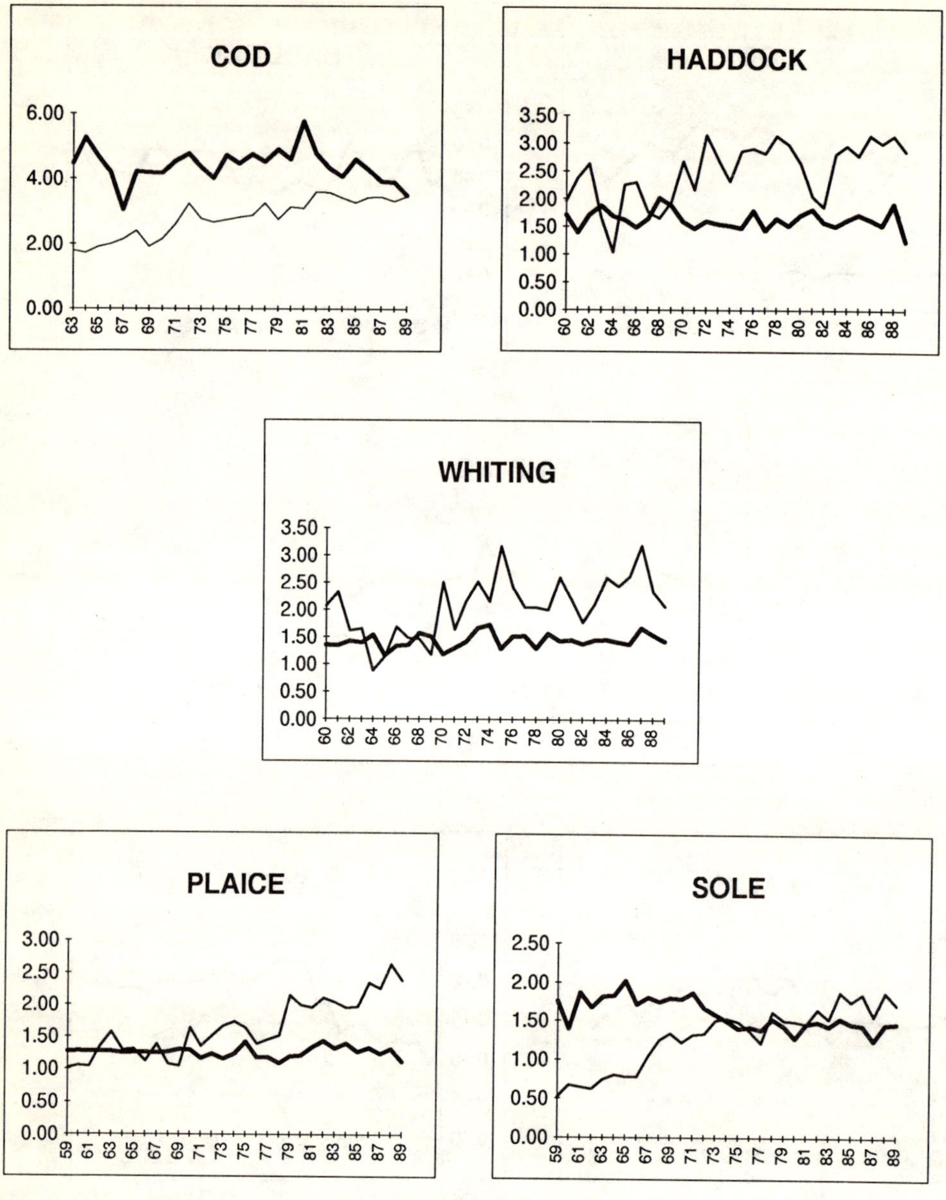

Figure 2. Plots of w/z (bold line) and $H\tau$ (normal line) through time (expressed as year less 1900), for five demersal species.

Table 1. Parameter estimates used for the contour plots shown in Figures 3-7.

	YEARS	H	S	τ	w
COD	63-67	0.56			
	85-89	0.84			
	63-89		9.20	4	4.135
HADDOCK	60-64	0.60			
	85-89	0.99			
	60-89		27.10	3	0.415
WHITING	60-64	0.46			
	85-89	0.85			
	60-89		30.70	3	0.262
PLAICE	59-63	0.30	8.90		
	85-89	0.58	15.40		
	59-89			3	0.334
SOLE	59-63	0.25	9.10		0.120
	85-89	0.58	16.30		0.190
	59-89			3	

Figure 2. shows w/z and $H\tau$ plotted through time for the five species. These plots are shown because when $H\tau$ exceeds w/z, the model of Blythe and Stokes (1991) predicts that selection pressure should act to reduce size-at-age in the long term. It is clear from Figure 2. that for cod, there has not been pressure for reduced size until recent years whilst for the four other species the pressure has existed for a longer duration. This may be surprising given that cod has traditionally been exploited at high H and has a relatively long τ. Cod, however, also has a very high age-at-maturity and hence w/z ratio. It is notable that the two flatfish species exhibit trends in size which match with the cross-over points on the w/z Vs $H\tau$ plots.

Fluctuations in S through time (not shown) are highly variable for all species. Because S has been calculated using SSB (fairly constant) and g (constant), the fluctuations reflect the highly variable recruitment observed in many marine fish species (e.g. Sissenwine et al 1988). Note that sole and haddock are especially variable. The only trends in S are for the whole time period for plaice and sole (at the 2 per cent and 8 per cent levels respectively).

For the purposes of producing contour plots of fitness and yield for the five species, averages of w, τ, H and S have been taken for the first and last five year data periods for each species, or for the entire period. Where the averages for each five year period are not statistically different to the overall average, the overall value is used. The resulting estimates are shown in Table 1. It should be remembered that these estimates are quite crude; the coefficients of variation for w, H and S are in the ranges 0.1 - 0.2, 0.2 - 0.3 and 0.5-1.8 respectively. The numbers-at-age, weights-at-age and fishing mortality rates given in the I.C.E.S. reports do not have associated errors reported. Indeed, the standard *ad hoc* methods (see e.g. Pope and Shepherd 1985) used to infer numbers and fishing mortality rates do not, as a matter of course, produce such estimates. The resulting contour plots, therefore, should be regarded as 'broad brush' and indicative rather than as the final word on the stocks in question.

For future work, it would be appropriate to conduct more detailed parameter estimation with greater emphasis on testing model validity and the sensitivity of conclusions to structural and parameter estimation uncertainty.

CONTOUR PLOTS

Contours of fitness, R, and yield, Y, are calculated for ranges of w Vs z and τ Vs T using equations 3.6 and 4.2 of Blythe and Stokes (1991), i.e.

$$R = S\tau wP / H\tau R = S\tau wP / H\tau,$$

$$Y = HwA^* \left[\left(\frac{\frac{1}{H\tau} + \frac{z}{w}}{P} \right) - \frac{1}{H\tau} \right],$$

where

$$P = e^{-H(\tau - T)},$$

$$A^* = \frac{\theta}{w} \ln\left(\frac{S\tau}{H\tau} wP \right)$$

and θ is an unspecified biomass level (to which yield is scaled) , the magnitude of which characterises the intensity of density dependence.

In the plots with size varying, the time delays τ and T are assumed constant and size variation thus reflects variation in growth rate, g. In the plots with the time delays varying, g is assumed constant. Note that the graph of fitness along the w or τ axis is equivalent to the selection gradients for those traits (see Kirkpatrick this volume). Thus, with careful parameterisation and a known (or assumed) additive genetic covariance function (see Kirkpatrick et al. 1990 and Kirkpatrick this volume), the predicted selection response could be calculated for any z or T.

RESULTS

Contour plots as detailed above are presented in Figures 3 to 7 for cod, haddock, whiting, plaice and sole respectively. In all plots, fitness is shown as solid lines whilst yield is represented by dashed lines. In all cases, the rightmost solid line shows fitness at 0.9 of maximum; subsequent leftward solid lines show fitness as 0.8, 0.7..0.1 of maximum. The innermost dashed contour shows yield at 0.9 of maximum with levels of 08, 0.7..0.1 radiating outwards. The closer together the contours lie, the steeper the yield or fitness surface.

COD, HADDOCK AND WHITING

The contour plots of fitness and yield on w Vs z show a change from the 1960s to the 1980s. Specifically, the fishing intensity has increased such that H, the annual fishing mortality, is now approaching 1 (i.e. culling approximately 65 per cent of the stock each year). Whilst there has not apparently been any selection pressure over recent decades for

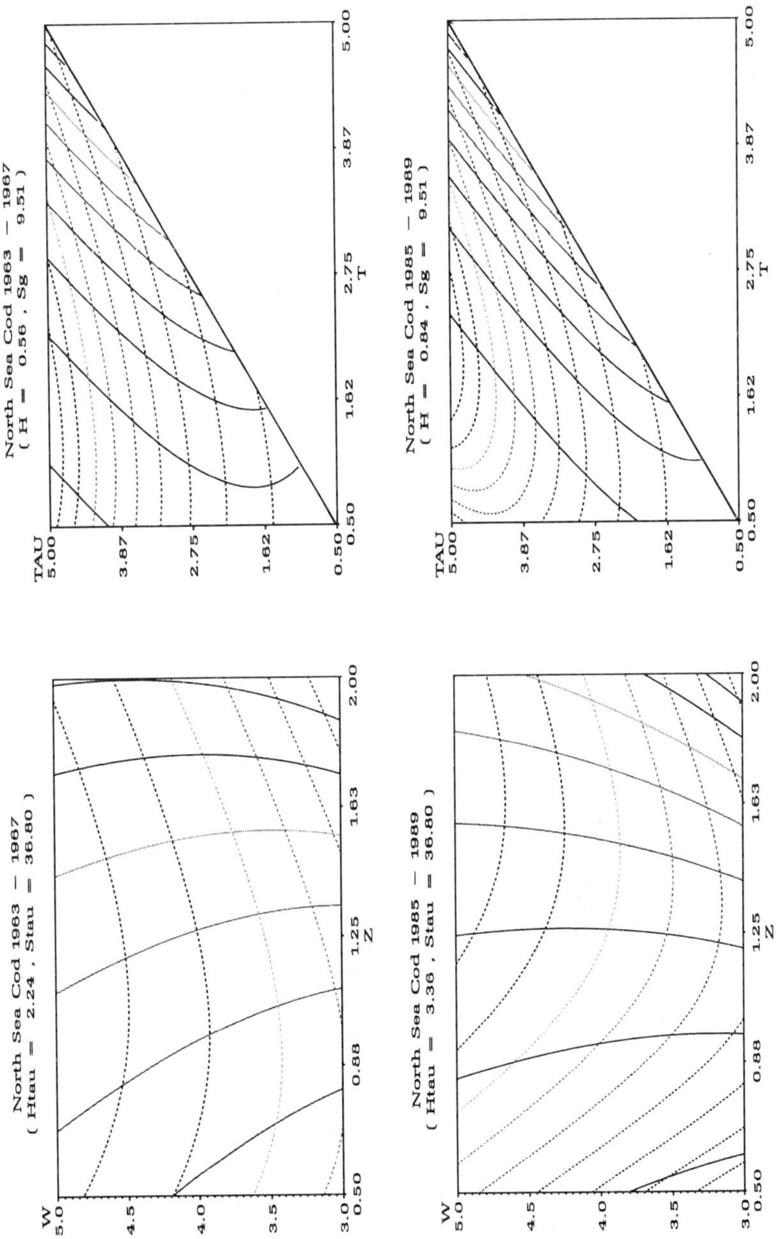

Figure 3. Contour plots of fitness and yield on w vs z and τ vs T for cod, *gadus morhua*. see main text for explanation.

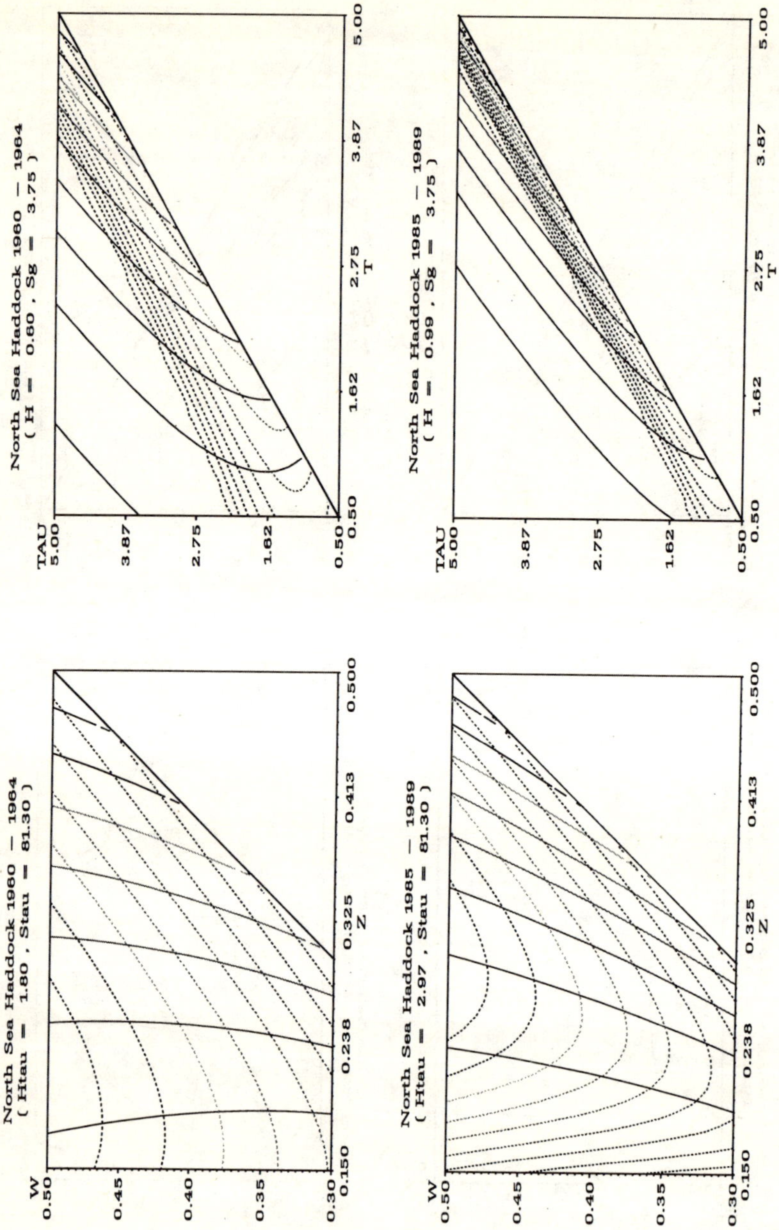

Figure 4. Contour plots of fitness and yield on w vs z and τ vs T for haddock, *Melanogrammus aglefinus*. see main text for explanation.

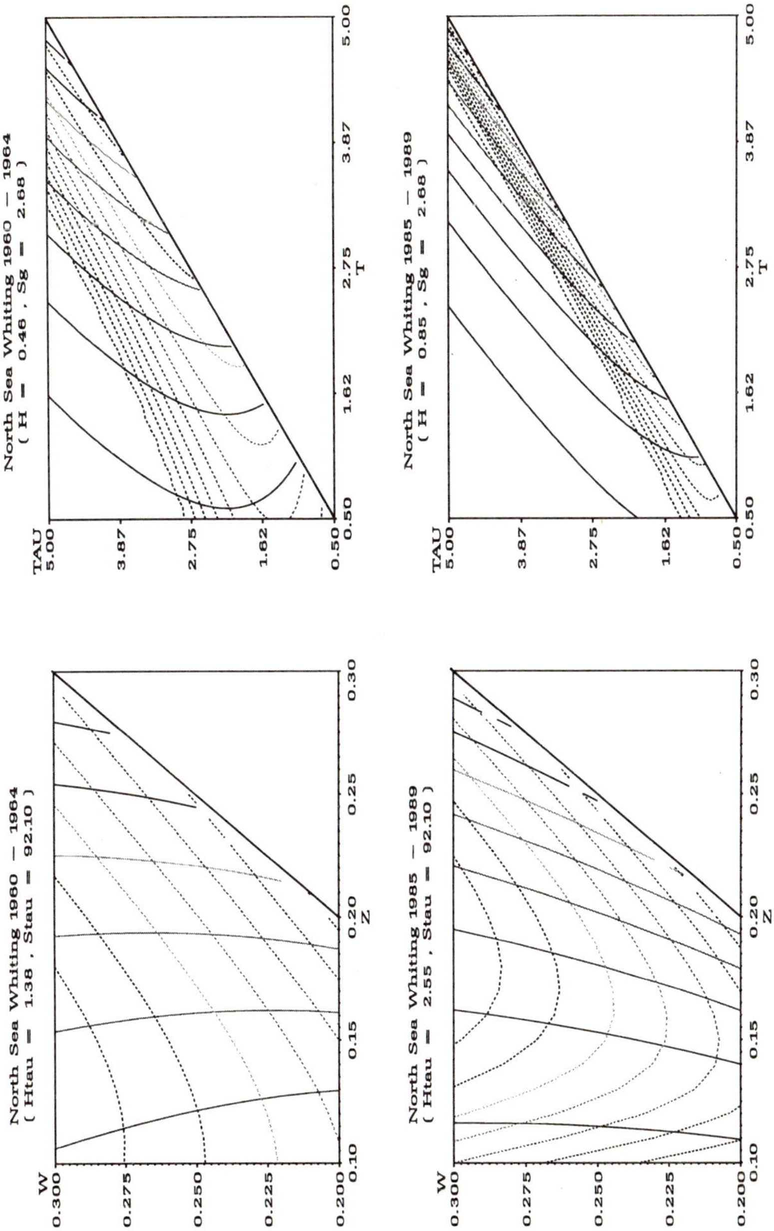

Figure 5. Contour plots of fitness and yield on w vs z and τ vs T for whiting, *Merlangius merlangus*. see main text for explanation.

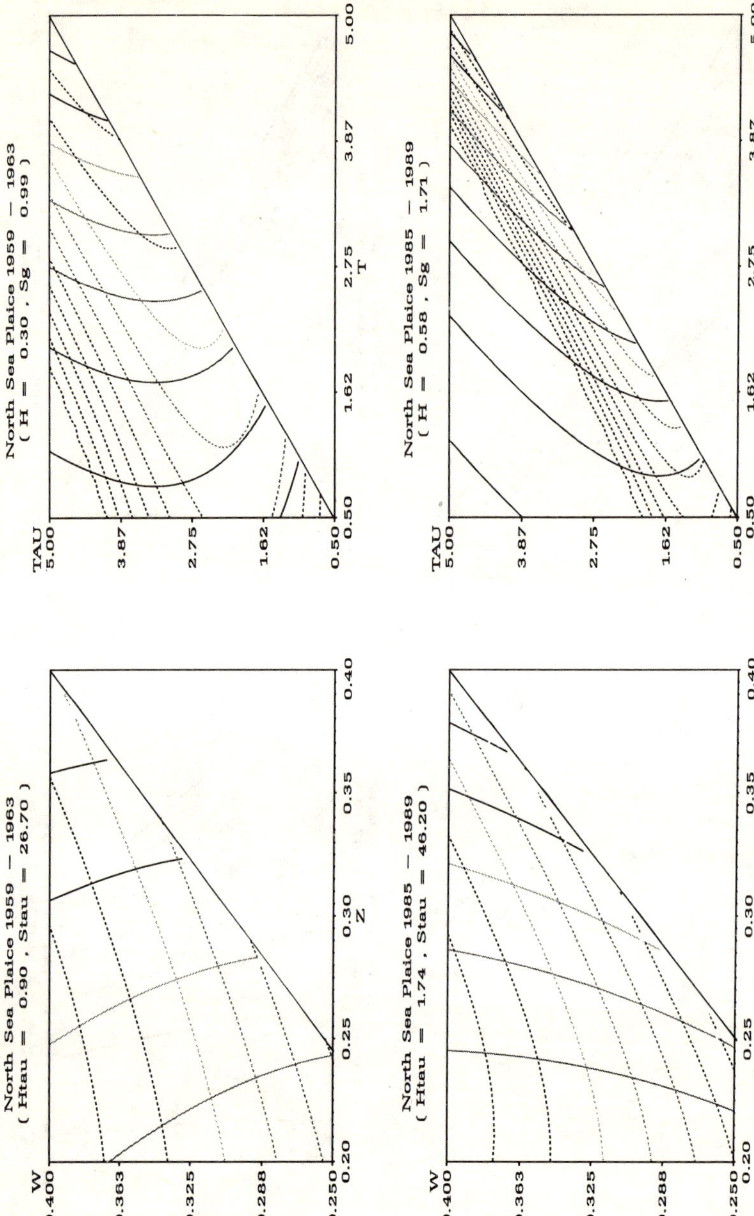

Figure 6. Contour plots of fitness and yield on w vs z and τ vs T for plaice, *Pleuronectes pleuronectes*. see main text for explanation.

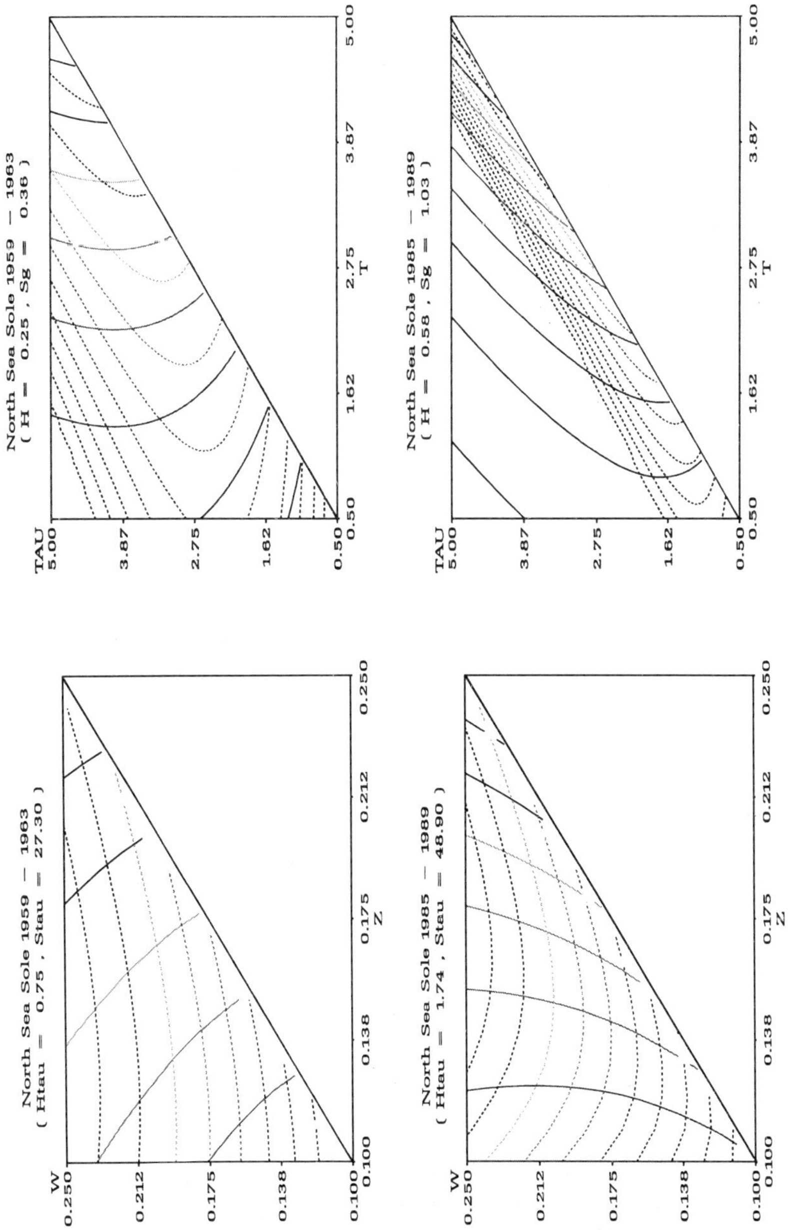

Figure 7. Contour plots of fitness and yield on w vs z and τ vs T for sole, *Solae solea*. see main text for explanation.

reduced w, at the current high H values this is no longer true. A long-term harvesting regime at such high H should be expected to result in reduced w. Law and Rowell (this volume) have calculated the selection differential for size-at-age-1 for cod to be just 1 cm over a 40 year period. Such a small selection differential is to be expected qualitatively from Figure 3. The plot of w/z Vs $H\tau$ (Figure 2.) also indicates that cod is only now reaching a phase when reduced w might be expected in the long-term. Oosthuizen and Daan (1974) suggested that length-at-first-maturity of North Sea cod decreased between the 1930s and the 1970s. Rijnsdorp et al. (1991) argue, however, that there is no evidence to support the hypothesis of a major change.

Cod is currently harvested in mixed fisheries with various combinations of haddock, whiting and plaice. Because all of these other species are smaller than cod, the mesh sizes used are commonly small (e.g. 80mm to 90mm diamond mesh). Cod, therefore, is fully recruited to the fishery at age 2, with a weight of approximately 1kg. Age-at-maturity is approximately 4 years and weight-at-maturity is about 4kg. Figure 3. suggests that for w = 4kg, increasing z will not result in a much increased equilibrium yield. Reducing the harvesting rate, however, would not only increase equilibrium yield but would also reduce the pressure for reduced w.

The w Vs z plots assume that τ and T are constant and that g is varying. It is perhaps more realistic, however, to regard age-at-maturity as a more plastic trait than growth rate *per se* (see Rowell this volume). The plots of fitness and yield on τ Vs T show that there has been pressure for reduced τ from throughout the 1960s up till the present time. This pressure is greater than ever before due to the increased harvesting rates. At any given value of τ, yield is fairly flat with increasing T.

As for cod, the harvesting rate of haddock and whiting has increased markedly from the 1960s to the present time (see Table 1.). Figures 4 and 5. illustrates that the pressure for reduced w (and especially reduced τ) is much increased. Interestingly, the contour plots of w Vs z for the period 1985-1990 indicate that the current z is almost optimal in terms of yield (i.e. for haddock: for w = 0.41, yield is maximum at approximately z = 0.24; actual z = 0.22; for whiting: for w = 0.26, yield is maximum at approximately z = 0.15; actual z = 0.14 (Anon 1990a)). In evolutionary time, however, decreasing w implies that z will need to decrease to maintain yield. It is evident from the plots of τ Vs T, however, that only by increasing T (hence losing yield) and/or reducing H, can fitness be maintained. Note that in the τ Vs T plots, yield goes to zero, and fitness declines dramatically, due to the very high H levels.

Overall, the North Sea haddock stock is apparently close to collapse. This is in accord with recent I.C.E.S. advice (e.g. 1990d). Haddock, though, is susceptible to severe recruitment fluctuations. With such high fishing mortality, the stock is dominated by very young fish; coupled with the very variable recruitment, this results in a very variable SSB and total stock size. The long-term health of the stock can only be guaranteed if the stock age-structure accommodates more older fish - a condition that can only be achieved at reduced harvesting rates.

Figure 5. for whiting is similar to the haddock plots in Figure 4. Whiting and haddock are similar sized fish which mature at the same age and are recruited to the fisheries at the same age. The two species are taken in quite different circumstances and whereas haddock has been under long-term pressure, whiting stocks have been at healthy levels. The contour plots are indicative of a very stressed stock fished at too high a harvesting rate but with a reasonable age-at-first-capture. Recently, it has been suggested that because whiting is a predator on young haddock and cod, that whiting should be selectively fished - essentially as a form of pest control. Based on standard I.C.E.S. approaches to modelling biological interactions (Sparre 1991) this might be sensible. The single species stage-structured model considered here, however, would suggest that such a route could have profound consequences for the stock of North Sea whiting.

PLAICE AND SOLE

Plaice and sole are both caught in directed flatfish fisheries in the North Sea. Sole is caught almost exclusively in such fisheries and is restricted to the Southern North Sea. Plaice, however, is caught both in directed fisheries but also alongside cod and other round fish in mixed demersal operations throughout the North Sea. Plaice grow to about 1kg whilst sole grow to about 0.6kg but both mature and recruit to the fishery at age 3, weighing in excess of 0.2kg. Both stocks are currently regarded as healthy but at higher harvesting rate than is desirable. Figures 6 and 7 show very similar pictures for the two stocks. In brief, harvesting rate has increased about twofold since the 1960s and there is now pressure both for reduced w and τ. It can be seen from Figure 1. that both w and z have decreased for plaice since 1972 and that w and z for sole both increased from 1959 to 1972 but have since decreased.

The τ Vs T plots show that in the period 1959-1963, there was no downward pressure on either w or τ for plaice or sole and that whilst increasing T might be beneficial in terms of fitness and yield, increasing z would have aided fitness but not yield. The situation in the early 1960s thus represented a stable fishery which was presumably also an ESOHS. In more recent years, however, as harvesting rates have increased, the picture has changed such that the stocks are surviving almost entirely due to the significant increase in S over the period.

DISCUSSION

Blythe and Stokes (this volume) point out that the application of approaches outlined in that and this paper, require that the models used be an accurate reflection of the biology of the exploited population. The simple stage-structured model used by Blythe and Stokes (1991, this volume) is not intended to represent the complete biological structure of North Sea fish stocks. It does, however, capture most of the features of models currently used to assess and manage fish stocks and upon which much of fisheries received wisdom is based. It lacks complications such as biological interactions and it does not attempt to consider technical interactions between fishing fleets. Also, due to lack of knowledge, assumptions concerning recruitment are necessarily crude.

Additionally, the parameter estimates presented are not intended to be definitive ; they do, however, permit the creation of a broad picture of the state of the stocks and of the likely consequences of certain management actions. It is clear that changes in mesh size (i.e. changes in T or z) do not always have clear-cut outcomes in the long-term. In the long-term, of course, changes in z do not necessarily relate simply to changes in T as evolutionary pressures may produce changes in growth rate. The question of whether or not such putative changes would be masked by the natural variations in growth rate or by density-dependent changes has not been addressed here and remains contentious (see Horwood this volume).

Nevertheless, even the simple approach taken in this paper allows an easy, qualitative look at how certain management choices might affect fish stocks in the long-term. In the case of plaice and sole, it can be seen that at the current harvesting rates, there is pressure for reduced size-at-age and for reduced age-at-maturity. That both species have been declining in size since the early 1970s is intriguing. The contour plots also reveal in an obvious manner that the current harvesting rates on plaice and sole are far too high and that a return to the rates of the early 1960s would be beneficial in terms of long-term stability and yield.

For the round fish species, the contour plots reveal that there is downward pressure on size-at-age and age-at-maturity for all stocks but that the age-at-first-capture (which relates to mesh size) is about right for whiting and haddock. If the harvesting rate was at levels as experienced in the early 1960s, there would be little pressure for reduced size-at-age or age-at-maturity and the size-at-first-capture of cod would tie-in better with that of haddock and whiting. All species are caught at too high a harvesting rate.

REFERENCES

Anon 1990a. Report of the North Sea Roundfish Working Group. I.C.E.S. C.M./**Assess**:7.

Anon 1990b. Report of the North Sea Flatfish Working Group. I.C.E.S. C.M./**Assess**:6.

Anon 1990c. 16th. Report of the Scientific and Technical Committee for Fisheries. Commission of the European Communities SEC **90** 1327.

Anon 1990d. Report of the I.C.E.S. Advisory Committee on Fisheries Management. I.C.E.S. Co-operative Research Report no. **173**.

Anon 1991. 17th. Report of the Scientific and Technical Committee for Fisheries. Commission of the European Communities SEC **90** 2173.

Beverton, R.J.H. and S.J. Holt. 1957. On the Dynamics of Exploited Fish populations. Fisheries Investigations London. Series II **19**: 1-533.

Blythe, S.P. and T.K. Stokes. 1991. Some consequences of size-selective harvesting on fitness and on yield. IMA Journal of Mathematics Applied in Medicine and Biology **7**: 41-53.

Daan, N. 1975. Consumption and production in North Sea cod (*Gadus morhua*): an assessment of the ecological status of the stock. Netherlands Journal of Sea Research **9**: 24-55.

Gulland, J.A. 1965. Estimation of mortality rates. Annex to Arctic Fisheries Working Group Report. I.C.E.S. C.M./**Assess**:3.

Hilborn, R. and C.J. Walters. 1992. Quantitative Fisheries Stock Assessment: Choice, Dynamics and Uncertainty. Chapman and Hall, New York.

Hislop, J.R.G. 1984. A comparison of the reproductive tactics and strategies of cod, haddock, whiting and Norway pout in the North Sea. In Potts, G.W. and R.J. Wootton eds. Fish reproduction: strategies and tactics. Academic Press, London.

Kirkpatrick, M., D. Lofsvold and M. Bulmer. 1990. Analysis of the inheritance, selection and evolution of growth trajectories. Genetics **124**: 979-993.

Laurec, A. and J.G. Shepherd. 1983. On the analysis of catch and effort data. Journal du Conseil International pour l' Exploration de la Mer **41**: 81-84.

Law, R. and D.R. Grey. 1988. Maximum sustainable yields and the self-renewal of exploited populations with age-dependent vital rates. In B. Ebenmann and L. Persson, eds. Size-structured populations. Springer-Verlag, Berlin .

Law, R. and D.R. Grey. 1989. Life-history evolution and sustainable yields from populations with age-specific cropping. Evolutionary Ecology **3**: 343-359.

Oosthuizen, E. and N. Daan. 1974. Egg fecundity and maturity of North Sea cod, *Gadus morhua*. Netherlands Journal of Sea Research **8**: 378-397.

Pitcher, T.J. and P.J.B. Hart. 1982. Fisheries Ecology. Croom Helm, London.

Pope, J.G. 1972. An investigation into the accuracy of virtual population analysis using cohort analysis. International Commission for the Northwest Atlantic Fisheries Research Bulletin **9**: 65-74.

Pope, J.G. and J.G. Shepherd. 1985. A comparison of the performance of various methods of tuning VPAs using effort data. Journal du Conseil International pour l' Exploration de la Mer **42**: 129-151.

Rijnsdorp, D., N. Daan, F.A. van Beek and H.J.L. Heessen. 1991. Reproductive variability in North Sea plaice, sole and cod. Journal du Conseil International pour l' Exploration de la Mer **47**: 352-375.

Sissenwine, M.P., M.J. Fogarty and W.J. Overholz. 1988. Some fisheries management implications of recruitment variability. In J.A. Gulland ed. Fish Population Dynamics. Wiley, Chichester.

Sparre, P. 1991. Introduction to multispecies virtual population analysis. I.C.E.S. Marine Science Symposia **193**: 12-21.

Stokes, T.K. 1993. An overview of the North Sea multispecies modelling work in I.C.E.S.. South African Journal of Marine Science (in press).

CONSEQUENCES OF SIZE-SELECTIVE HARVESTING AS AN EVOLUTIONARY GAME

Joel S. Brown and Alan O. Parman

Department of Biological Sciences, University of Illinois, Box 4348, Chicago, Illinois, 60680, USA

ABSTRACT

We model the evolution of prey body size in response to size-selective harvesting by humans. The manager (= humans) selects a harvest rate and a threshold size of harvest (below which no smaller prey are harvested) so as to maximize yield. In response to the manager's harvesting regime, the adult size of the prey evolves to an evolutionarily stable strategy (ESS). Over a wide range of parameter values, the prey's ESS is only influenced by the manager's harvest rate and is independent of the manager's threshold size of harvest. Increasing the harvest rate selects for smaller adult prey. We contrast the harvest regime of an ecologically enlightened manager (one who only considers the ecological consequences of harvesting on the prey stock) with that of an evolutionarily enlightened manager who considers the consequences of harvesting on the prey's ESS adult size. The ecologically enlightened manager maintains a higher harvest rate and consequently selects for smaller adults. The evolutionarily enlightened manager uses an evolutionarily stable optimal harvest strategy (ESOHS, sensu Law and Grey 1989) that sacrifices increased yield in ecological time to maintain a larger adult prey size and higher yield in evolutionary time.

INTRODUCTION

The harvesting of commercially valuable species represents a new and often large source of mortality for the harvested population. In response, the harvested species experiences selection to evolve adaptations that either mitigate or accommodate the actions of the new predator (e.g. pest and disease species and the evolution of resistance to pesticides and drugs, respectively). At present, a number of fish stocks have shown striking changes in their size, age, and fertility distributions (Murphy 1967, Borisov 1978, Garrod 1988). Such changes can be ascribed to phenotypic plasticity in response to changing environmental conditions or to evolved changes in phenotypic plasticity and other life-history characteristics (Macdonald and Chinnappa 1989, Trexler and Travis 1990, Trexler et al. 1990). At present, it can be difficult to separate changes due to plasticity and changes due to evolution (but see Edley and Law 1988, Reznick et al. 1990, Hillesheim and Stearns 1991). However, insofar as heritable variation with respect to relevant life history characteristics exists, evolution will or has occurred in exploited populations. If so, the question becomes one of how much change will occur and how quickly will this evolutionary change happen.

Two theoretical works have modelled the evolutionary changes of fish populations in response to size- and age-specific harvesting (Law and Grey 1989, Blythe and Stokes 1990). In both, humans may promote evolutionary change through the intensity of harvesting and through the size-selectivity of fishing gear. Both are strategic models in that the models assume an evolutionarily feasible set of heritable phenotypes (= strategies) without specifying the specific genetics that permit and constrain the strategy set.

Law and Grey (1989) define and apply the concept of an *evolutionarily stable optimal harvesting strategy* (ESOHS) to an age-structured and size-structured model of life history evolution. In response to human exploitation, the fish population evolves to its ESS (evolutionarily stable strategy, Maynard Smith and Price 1973). By definition, the fish's ESS confers resistance to invasion by any rare alternative life history strategy drawn from the set of evolutionarily feasible strategies. Given that the fish evolve to their ESS and given that the fish's ESS changes with fishing intensity, the ESOHS of the fishery is the 'harvest pattern' (Law and Grey 1989) that maximizes yield at equilibrium. As such, the ESOHS takes into consideration both the ecological and evolutionary consequences of human exploitation of commercially valuable species.

Blythe and Stokes (1990) consider a simpler size-structured model of life-history evolution than Law and Grey (1989). In Blythe and Stokes' model, growth and reproduction cannot occur concomitantly. The evolutionary strategy consists of when to switch from growth to reproduction. Rather than using the ESS concept as the outcome of evolution by natural selection, Blythe and Stokes consider evolution to favour the strategy that, when rare, can invade the strategy of the resident population (Nisbet et al. 1989).

Interestingly, the two analyses and models can give strikingly different results. For the case of Arcto-Norwegian cod, Law and Grey found that harvesting selected for slower growth rates, increased reproductive effort, and reproduction at an earlier age and size. Specifically, the average size of adult fish evolves to a smaller size. Blythe and Stokes obtained an opposite result. Human exploitation selected for later reproduction and a larger size at maturation. In fact, above a certain adult size a strategy with a longer growth period could always invade a strategy with a shorter growth period. Their model predicts that ever larger adult fish should evolve in response to harvesting.

The differing results of Law and Grey (1989) and Blythe and Stokes (1990) may be the result of the functional forms of the models under consideration or the result of the different criteria used for determining the consequences of evolution by natural selection. Here, we take the model of Blythe and Stokes and formulate it as an evolutionary game (following Vincent and Brown 1988). We analyse the model according to the evolutionary criterion of Law and Grey (1989). We characterize the ESS of the fish population and assume that the fish evolve to their ESS in response to harvesting. We then consider the equilibrium yield function of the fisheries. We compare the resulting fishing patterns when harvesting is based strictly on ecological considerations with the ESOHS when harvesting is based on both ecological and evolutionary considerations. When analysed as an evolutionary game the model of Blythe and Stokes (1990) generates predictions that accord with those of Law and Grey (1989).

THE MODEL

The following model is drawn directly from Blythe and Stokes (1990) and follows closely models of Gurney et al. (1980) and Blythe et al. (1982). The exploited species consists of juveniles and adults. Juveniles grow linearly at rate g and achieve adult size, u, at age T where $u = gT$. Upon reaching adulthood growth ceases and reproduction commences. Let fecundity be linearly related to adult size, u. Let s be the maximum fecundity rate per unit of adult size. We assume that s is the product of neonate production (per unit adult size) and neonate survivorship. Once a neonate has survived it becomes a juvenile of effectively zero size and zero age. For simplicity (see Blythe and Stokes 1990), we assume that juveniles and adults suffer no mortality other than harvesting by humans. Let z be the threshold harvesting size and let H be the instantaneous harvest rate on all individuals (juveniles and adults) sized greater than or equal to z. For there to be any harvesting at all, z must be less than or equal to u. Given a threshold size z and harvest rate H, we can determine the probability, P, that a juvenile survives to adulthood:

$$P = \exp[-H(u - z) / g] \qquad u \geq z$$

$$P = 1 \qquad u < z \tag{1}$$

where $(u - z)/g$ gives the amount of time that a juvenile maturing at size u is exposed to harvesting. As expected, the probability of surviving to adulthood declines with increasing harvest rate, H, increasing size at maturity, u, and decreasing threshold harvesting size, z.

We introduce density-dependent effects, by assuming that competition for food among adults reduces neonate production below its maximal level. Following a Ricker stock-recruitment relationship, we assume that actual neonate production scales as a negative exponential of adult biomass: $\exp(- uN/R)$ where N is adult population size and R scales the intensity of the density dependence. Throughout we will assume equilibrium population dynamics where N^* denotes the stable adult population size.

At this point we shall deviate from Blythe and Stokes (1990) by explicitly formulating the above model as an evolutionary game. We will let size at adulthood, u, be the evolutionary strategy drawn from the strategy set defined by $0 < u$. We define a fitness generating function (sensu Vincent and Brown 1984, Brown and Vincent 1987) as the per capita growth rate (in units of adults) of an adult using strategy u in a population using strategy u_1. The fitness generating function, G, for this model can be written as:

$$G(u, u_1) = suP(u) \exp(-u_1 N^* / \mathrm{R}) - H \tag{2}$$

The first term on the RHS of (2) gives the expected rate of adult recruitment from the production of neonates by an adult. It is the product of maximal rate of neonate production per unit adult size, parent size, probability of surviving the juvenile period of growth, and the density dependent effect of other adult fish on the maximal rate of neonate production. The second term on the RHS is simply the rate of adult mortality due to harvesting. In this formulation, the strategy of the individual, u, influences its maximal fecundity rate and influences its probability of surviving to adulthood, $P(u)$. The strategies of others, u_1, influences the equilibrium population size, N^*, and the density-dependent depreciation of the maximal fecundity rate.

To determine the equilibrium population size, we must consider how adult population size changes with time as a function of the population's strategy u_1. The population's dynamics can be written as:

$$\frac{dN_t}{dt} = su_1 P(u_1) N_{t-T} \exp(-u_1 N_{t-T} / R) - HN_t \tag{3}$$

where recruitment of adults is determined by the number of adults at time t - T, N_{t-T}, and adult mortality is determined by the number of adults at time t, N_t. Recruitment has a time delay equal to the time it takes for a neonate to achieve adulthood. Equation (3) is identical to equation (2.3) of Blythe and Stokes (1990). Furthermore equation (3) has some obvious similarities to equation (2) with the exception that the strategy of the individual in (2) has been set equal to the strategy of the population in (3). As such, equation (2) permits the individual to make a unilateral change in strategy while holding the strategies of others constant. Equation (3) only permits collective changes in strategy.

At equilibrium $N_{t-T} = N_t = N^*$ and growth rate equals zero; i.e., $dN_t/dt = 0$. Substituting N^* for N_{t-T} and N_t, and setting equation (3) equal to zero yields:

$$N^* = (R/u_1)\log\,[su_1 P(u_1)/H\,] \tag{4}$$

As expected, the equilibrium population size increases with R, increases with maximum fecundity rate, s, and decreases with harvest rate, H (see equation (2.4) of Blythe and Stokes 1990).

THE ESS SIZE AT MATURITY

Equation (2) can be used to find candidate solutions for the ESS value of u. We can apply an ESS maximum principle (Vincent and Brown 1988) that requires u to maximize G when $u = u_1$. For u_1 to be an ESS the fitness of an individual must be maximized when its strategy, u, equals the strategy of everyone else, u_1. To find the candidate solutions for an ESS we take the derivative of equation (2) with respect to u, set u equal to u_1, set the resulting expression equal to zero, and solve for u_1. Consider the derivative of equation (2) with respect to u:

$$\frac{\partial G}{\partial u} = s\exp(-u_1 N^* / R)[u(\partial P / \partial u + P(u))] \tag{5}$$

Substituting u_1 for u and setting equation (5) equal to zero yields:

$$u_1 \frac{\partial P}{\partial u} + P(u_1) = 0 \tag{6}$$

From equation (1) note that when $u > z$, $P/u = -(H/g)P$. Substituting this into equation (6) and rearranging shows that the ESS value for u_1 is given by:

$$u_1^* = g/H \quad \text{when } g/H \geq z$$
$$u_1^* = z \quad \text{when } g/H < z \tag{7}$$

The two ESS conditions given by equation (7) result from the discontinuity of P at $u = z$ (see equation (1)). When $u \geq z$, then equation (6) has the solution that is given by the first part of equation (7). When $u < z$, then equation (6) has no solution, equation (5) is always positive, and fitness will always increase with an increase in u. The second part of equation (7) states that the adults will never evolve to be smaller than the threshold harvest size. At u_1*, the second derivative of G with respect to u is negative; hence, u_1* maximizes fitness (at least locally).

In this model the size of adults should evolve to the values given by equation (7). Both the threshold harvest size, z, and the harvest rate, H, influence the ESS size of adults. Interestingly, under a given set of parameter values either z or H but not both determine the ESS. When the threshold harvest size is small or when harvest rates are low then harvest rates alone influence the ESS. When the threshold harvest size is large or when harvest rates are high then only the threshold harvest size influences the ESS. The ESS adult size tends to decrease with harvest rate, H, and increase with the growth rate of juvenile fish, g (Fig. 1).

YIELD

The yield function that we shall use is drawn directly from Blythe and Stokes and we reproduce their derivation in what follows (see equations 4.1 and 4.2 of Blythe and Stokes 1990).

The yield from exploiting a population with adult size, u_1, has two components. The first includes rate of harvest of juveniles between sizes z and u_1 weighted by their size and the second includes the rate of harvest of adult fish weighted by their size u_1. The first component requires integrating yield across the time period during which juveniles are exposed to harvesting weighted by their size and size-specific population sizes; note that there are successively fewer juveniles as juvenile size increases. Adding these two components gives the rate of biomass harvested, Y:

$$Y(u_1, H, z) = H \int_{z/g}^{T} gxe^{-H(T-x)} B^* \, dx + Hu_1 N^* \tag{8}$$

where

$$B^* = HN^* / P(u_1) \tag{9}$$

gives the equilibrium number of juvenile fish alive at age z/g required to maintain the equilibrium adult population size of N^* given the harvest rate H and the adult size u_1. The numerator of equation (9) is the rate at which adult fish must be replaced to maintain an equilibrium and the denominator of (9) gives the probability of fish aged z/g becoming replacement adults.

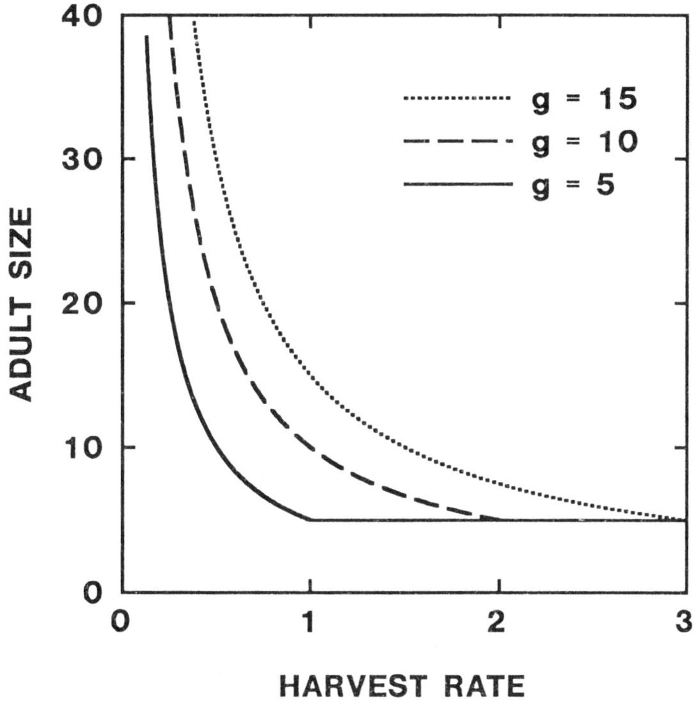

Figure 1. The effect of harvest rate and juvenile growth rate (g) on the ESS adult size of the prey. Adult prey size increases with juvenile growth rate and decreases with harvest rate. As harvest rate increases, the prey ESS will never be smaller than the threshold size of harvest which has been set equal to five for this example. All three curves merge with this lower bound on adult prey size.

The yield function (8) can be rewritten by substituting equation (9) for B^*, substituting equation (4) for N^*, solving the integral, and rearranging the resulting algebraic expression:

$$Y(u_1,H,z) = \frac{R}{u_1}\log\left[\frac{su_1p}{H}\right]\left[\frac{Hz+g}{P} - g\right] \tag{10}$$

where yield is influenced by all of the parameters and variables that we have defined for the model. In particular, it can be shown that for all positive (and hence biologically meaningful) yields, Y increases with an increase in R (a decrease in the density dependent effect of others), an increase in the maximum rate of neonate production, s, and an increase in the juvenile growth rate, g.

In what follows we assume that Y comprises the objective function of the population manager. It is the goal of the manager to set a harvest rate, H, that maximizes yield, Y. There are two ways the manager can go about this. First, the manager can take the adult size of the population as given, and maximize equation (10) with respect to H. Such a manager takes into account the ecological consequences of harvesting on N^* but not the evolutionary consequences of harvesting on u^*. We will refer to such a manager as an ecologically enlightened manager. Second, the manager can include the evolutionary consequences of harvesting into the yield function by first substituting $u_1 = g/H$, from equation (7), into equation (10) and then maximizing yield with respect to H. Such a management strategy will result in the ESOHS of Law and Grey (1989) and we will refer to such a manager as an evolutionarily enlightened manager.

ECOLOGICALLY ENLIGHTENED MANAGER

The objective here is to maximize yield, equation (10), with respect to H where the manager takes adult size, u_1, as a given (later we shall discuss why we are not considering the manager's strategy of varying the minimum size of harvest, z). The first order necessary condition for a maximum requires $\partial Y/\partial H = 0$ at H^*. Taking the derivative of equation (10) with respect to H, setting equal to zero, and rearranging yields:

$$\frac{\partial Y}{\partial H} = [H^* u_1 z + g u_1 - H^* z^2]\left[\log\left(\frac{s u_1 p}{H}\right) - 1\right] \tag{11}$$

where $u_1 \geq z$ and $u_1 = g/H^*$ if $H^* \leq g/z$.

The various components of equation (11) have the following signs:

$$[H^* u_1 z + g u_1 - H^* z^2] > 0$$

$$gP(u_1 - z) - \frac{g^2}{H}(1-P) < 0$$

and, hence, at the optimal harvest rate, H^*, the following must be true:

$$\log\left(\frac{s u_1 P}{H}\right) > 1$$

To determine the ESS value for the population's adult size we need to determine when $\underline{H}^* < \underline{g}/\underline{z}$. This will occur when $\partial \underline{Y}/\partial \underline{H}$ evaluated at $\underline{H} = \underline{g}/\underline{z}$ and $\underline{u}_1 = \underline{z}$ is negative; i.e. yield will increase by decreasing $\underline{H}$ below $\underline{H} = \underline{g}/\underline{z}$. Evaluating $\partial \underline{Y}/\partial \underline{H}$ at $\underline{H} = g/\underline{z}$ and $\underline{u}_1 = \underline{z}$ gives:

$$\frac{\partial Y}{\partial H} = R\left[\log\left(\frac{s z^2}{g}\right) - 1\right] \tag{12}$$

Equation (12) has the following implications:

$$ge/s > z^2 \rightarrow H^* < g/z,\ u_1 = g/H^*, \text{ and } 0 < H^* < s\, u_1/e$$

$$ge/s < z^2 \rightarrow H^* > g/z,\ u_1 = z, \text{ and } H^* = sz/e$$

Increasing juvenile growth rate, g, increases the likelihood that adult prey size is determined by harvest rate, H. Increasing the maximum rate of neonatc production, s, or increasing the threshold size of harvesting, z, increases the likelihood that adult prey size is determined by z. Furthermore, when $u_1 = z$ then the optimal harvest rate is determined solely by the maximum rate of neonate production and the threshold size of harvesting. Increasing either of these will increase the optimal harvest rate, H^*. When $u_1 = g/H^*$ then a numerical solution becomes necessary as there is no explicit solution for equation (11) (see Figure 2).

EVOLUTIONARILY ENLIGHTENED MANAGER

The objective of this manager is to maximize yield, equation (10), with respect to H where the manager knows that u_1 will in time evolve to either $u_1 = z$ or $u_1 = g/H$ according to the conditions of equation (7). When $u_1 = g/H$, the necessary condition for the optimal harvest rate, H', is found by substituting g/H for u_1 in equation (10), taking the derivative of (10) with respect to H, and setting equal to zero. These operations yield:

$$\frac{\partial Y}{\partial H} = [H' gz + g^2(1-P) - H' z)^2]\left[\log(\frac{sgP}{H'^2}) - 1\right] - gzH' P - g^2(1-P) = 0 \tag{13}$$

Although equation (13) has the same number of terms with the same signs, this expression differs from equation (10) of the ecologically enlightened manager. In particular, u_1 does not occur in equation (13).

To determine the ESS value for the population's adult size we evaluate $\partial Y / \partial H$ at $H = g/z$ and $u_1 = z$:

$$\frac{\partial Y}{\partial H} = -R < 0$$

Because this expression is always negative, the optimal harvest rate, H', will always be less than g/z, and consequently the adult prey size will always be greater than z and equal to g/H'. For an evolutionarily enlightened manager (in this model), it always pays to harvest at a rate sufficiently low to evolve fish greater in size than the threshold harvest size, z, rather than evolve fish equal in size to the threshold harvest size (obviously $u_1 < z$ is unsatisfactory as there would be no yield).

In comparing the ecological with the evolutionarily enlightened managers it can be shown that:

$$\left.\frac{\partial Y}{\partial H}\right|_{\text{ECOLOGY}} = \left.\frac{\partial Y}{\partial H}\right|_{\text{EVOLUTION}} + \frac{R}{g^2}\left[\log\frac{sgP}{H^2} - 1\right] + R \tag{14}$$

where $u_1 = g/H$.

Equation (14) gives the result that the ecologically enlightened manager will always choose a higher optimal harvest rate than the evolutionarily enlightened manager; i.e., $H^* > H'$. Consequently, the ecologically enlightened manager will select for a smaller adult prey size than the evolutionarily enlightened manager. The harvest rate of the evolutionarily enlightened manager is the ESOHS for the Blythe and Stokes (1990) model.

A plot of yield versus H can show the relationship between H^* and H' (Figure 2). The surface for the ecologically enlightened manager plots yield as a function of H where u_1 is set to g/H^*. This surface takes on a maximum at H^*. The surface for the evolutionarily enlightened manager plots yield as a function of H where u_1 varies and always equals the larger of g/H or z. At the point where $g/H = z$ the surface has a kink. This surface will take on a maximum at H'. Furthermore the surface of the evolutionarily enlightened manager must intersect that of the ecologically enlightened manager at H^*.

The yield of the evolutionarily enlightened manager at H' is greater than that of the ecologically enlightened manager at H^*. While H' may seem the better management strategy than H^*, this is only true in evolutionary time and, as a result, H^* may be a more stable management strategy than H'. The strategy H^* is stable in the sense that in ecological time there is no benefit to changing strategy. Given an adult prey size of g/H^*, the ecological management strategy maximizes yield. In contrast, the management strategy H' is not stable in the sense that in ecological time there is an incentive to change management strategy. Given an adult prey size of g/H', the evolutionary management strategy does not maximize yield. In fact, increasing the harvest rate above H' will increase yield when $u_1 = g/H'$ (see Figure 2). In ecological time a strategy greater than H' will reap higher yields although in evolutionary time yields will drop as smaller adult prey evolve. In fact, if the managers continue to chase the highest yields in ecological time, the system in evolutionary time will settle down to a harvest rate of H^* and an adult prey size of g/H^*. In this sense, the managers strategy of H^* and the prey's strategy of $u_1 = g/H^*$ represent the ESS in a model of predator-prey coevolution (see Brown and Vincent 1992).

DISCUSSION

We considered a simple life-history model of an organism that grows linearly as a juvenile and then at some size switches from growth to reproduction (see Cohen 1966, Brown and Venable 1986 for a similar model of life history evolution in annual plants). We assumed that the species possesses a heritable strategy that determined when this switch occurs. The species is under selection to evolve an adult size that maximizes the per capita rate of adult recruitment into the population. As a frequency- dependent evolutionary game, the strategy of the individual, the strategies of others, and the population size all influence fitness. We solved the model for the ESS size of adults.

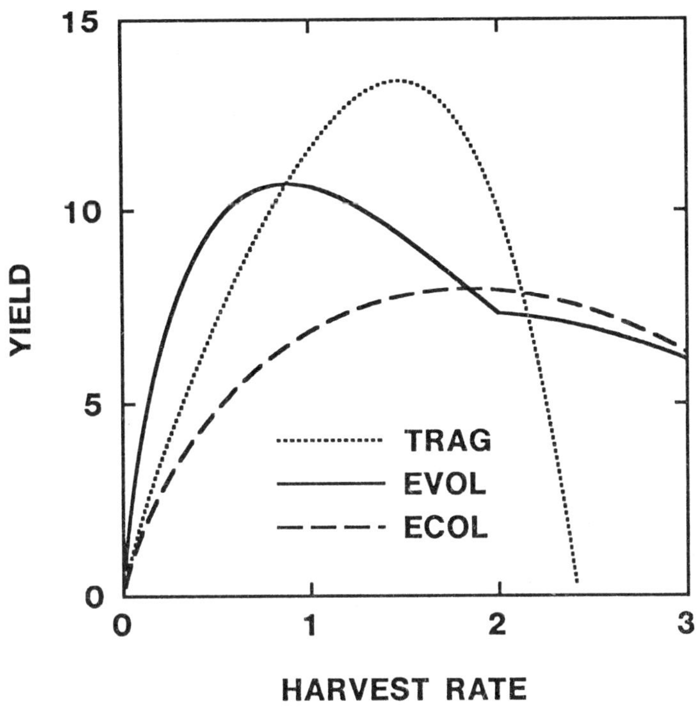

Figure 2. Effect of harvest rate on yield for three management strategies. For this example: $R = 4$ (scaling of density-dependent effects), $s = 1$ (product of neonate production and survivorship), $g = 10$ (juvenile growth rate), and $z = 5$ (threshold size of harvest). The ecologically enlightened manager (ECOL) takes adult prey size as fixed and selects the harvest rate that maximizes yield. Over evolutionary time this results in a harvest rate ($H^* = 1.84$) that maximizes yield for the ESS adult prey size at H^* ($u^* = g/H^* =$ 5.43). The curve labelled ECOL assumes that $u = 5.43$ throughout. The evolutionarily enlightened manager (EVOL) assumes that the adult prey size is always ESS ($u = \max g/H, z$and selects a yield maximizing harvest rate with this in mind ($H' = 0.87$, $u' = 11.4$). The EVOL and ECOL curve must intersect at (H^*, u^*). While H' generates a higher yield than H^*, the strategy of the evolutionarily enlightened strategy is susceptible, in ecological time, to the tragedy of the commons (TRAG). The curve labelled TRAG shows how yield varies with harvest rate when prey adult size is fixed at $u' = 11.4$. In ecological time, yields can be temporally increased above H' by selecting a yield maximizing harvest rate of $H = 1.47$. However, at this harvest rate the prey will evolve to be smaller and yields will drop. While $H^* = 1.84$, $u^* = 5.43$ and $H' = 0.87$, $u' = 11.4$ are stable in evolutionary time, $H = 1.47$, $u = 11.4$ is not stable in evolutionary time.

The model also includes a predator in the form of harvesting by humans. The human's harvest rate and threshold size of harvest both influence the prey's fitness and influence the ESS adult size of the prey. In addition, harvest rate and threshold size of harvest influence the rate of prey biomass harvested. We assume that the human managers desire to maximize this yield. In this way, the prey and the manager form a model of predator-prey coevolution in which the prey evolve to their ESS adult size and the manager selects a harvest rate that maximizes yield; adult prey size influences the manager's selection of harvest rate and the harvest rate influences the prey's ESS.

From the prey's perspective, increasing the harvest rate either has no effect on the prey's ESS or selects for decreased adult size. Increasing the threshold size of harvest either has no effect on the prey's ESS or selects for an increase in adult size. This is because either the harvest rate or the threshold size of harvest determine the prey's ESS, but never both. This result runs slightly counter to the conventional wisdom that size-selective harvesting determines the direction of selection and the harvest rate determines the strength of selection. Interestingly, over a wide range of size thresholds, the harvest rate alone selects for smaller adult size (also see Edley and Law 1988). In summary the prey ESS either equals the threshold size of harvest (when the threshold or harvest rate are high) or it equals the ratio of the juvenile growth rate and the harvest rate (when the threshold or harvest rate are low).

In determining the manager's optimal harvest rate we considered two scenarios. In the first, an ecologically enlightened manager considers the consequences of harvest rate on the prey's population size but takes the prey's adult size as a given. In the second, an evolutionarily enlightened manager also considers the effect of the harvest rate on the prey's adult size (the ESOHS of Law and Grey 1989). The first harvest rate maximizes yield in ecological time and the second maximizes yield in evolutionary time. In accord with Law and Grey, we find that the evolutionarily enlightened manager selects a lower harvest rate than the ecologically enlightened manager. The evolutionarily enlightened manager sacrifices yield in ecological time to preserve yield in evolutionary time. By scaling back harvest rates, such a manager simultaneously conserves ecological and evolutionary resources. While still an open question, we feel that this result will be general across many models of human exploitation of prey species. Another open question regards the relevance of conserving evolutionary resources. If the predicted evolutionary changes require millennia they may be irrelevant to the planning horizons of most managers (and politicians!).

DeAngelis et al. (1985) consider a predation model where larger prey are less harvestable. This introduces additional conflicting selective pressures. Like our model, increasing predation pressure may select for smaller adults. But in contrast, predation may also select for larger adult size as an anti-predator adaptation. Selection becomes disruptive as adults of intermediate size are disfavoured.

Our results accord well with those of Law and Grey (1989) who consider a different model but use a similar game-theoretic modelling approach. Our results differ completely from those of Blythe and Stokes (1990) who used the same model but a different criterion for modelling evolutionary outcomes. Blythe and Stokes conclude that harvesting may favour increased adult size and that the model lacks an ESS adult prey size. We feel our results differ for two reasons. First, Blythe and Stokes do not consider the full effect of changing adult body size on fitness as adult body size appears both explicitly and implicitly in their fitness function (see equations 3.1 through 3.2, Blythe and Stokes 1990). Second, under some circumstances, the use of the ability to invade, rather than resistance to invasion, as the criterion for evolutionary stability may generate divergent results.

In the present model it is tempting to ask what value for the threshold size of harvest will maximize yield. We have not presented the analysis of varying threshold size because the models structure yields nonsense results. In the model the only source of mortality to juveniles and adults is harvesting. In the absence of harvesting the prey evolve to be as large as possible and the equilibrium population size goes to infinite. The lack of any natural sources of mortality also determines the optimal value for the threshold size of harvest. If the manager can jointly co-ordinate harvest rate and threshold size, then the manager should let the threshold size go to infinite and the harvest rate go to zero. In so doing the prey evolve to an infinite adult size. This lack of realism can be remedied by introducing a background rate of mortality for juveniles and adults.

Ubiquitous evidence exists for the influence of predators on the evolution of their prey (Edley and Law 1988, Reznick et al. 1990, Parejko and Dodson 1991). Usually, the predators select for characteristics in their prey that make them less accessible or favourable to the predators (cats and similar predators that colonize islands with flightless species of birds find themselves temporarily in nirvana). When faced with a novel predator there are generally two outcomes. The prey population, in ecological time, may crash and become extinct (at least locally). Or, the prey population may persist and through successive generations evolve in response to predation risk and mortality (the evolution of drilling by carnivorous gastropods and the subsequent extinctions and evolutionary responses of their prey provide a global example, Vermeij 1977).

The outcome of human exploitation of natural prey populations should follow closely evolutionary and ecological principles of predator-prey interactions (Favro et al. 1982, Nelson and Soulé 1987, Vincent and Brown 1987). Some prey populations will be exploited to extinction. Those that persist even in the face of heavy harvesting (e.g. North Atlantic cod, Garrod 1988) should evolve. Here we advocate formulating ecological management models into evolutionary games as a way of providing insights into the ecological and evolutionary consequences of harvesting.

ACKNOWLEDGEMENTS

We thank Richard Law, Jacqueline McGlade and Kevin Stokes for organizing a stimulating and enjoyable workshop. Presentations by and discussions with Steven Blythe, David Grey, R. Law, and K. Stokes inspired the ideas presented herein. We received helpful comments from R. Law and S. Blythe. This work was supported by National Science Foundation grant No. BSR 9106785.

REFERENCES

Blythe, S. P., R. M. Nisbet and W. S. C. Gurney 1982. Instability and complex dynamic behaviour in models with long time delay. Theoretical Population Biology **2**: 147-76.

Blythe, S. P. and T. K. Stokes 1990. Some consequences of size-selective harvesting on fitness and on yield. IMA Journal of Mathematics Applied in Medicine & Biology **7**: 41-53.

Borisov, V. M. 1978. The selective effect of fishing on the population structure of species with a long life cycle. Journal of Ichthyology **18**: 896-904.

Brown, J. S. and D. L. Venable. 1986. Evolutionary ecology of seed-bank annuals in temporally varying environments. American Naturalist **127**: 31-47.

Brown, J. S. and T. L. Vincent. 1987. A theory for the evolutionary game. Theoretical Population Biology **31**: 140-166.

Brown, J. S. and T. L. Vincent. 1992. Organization of predator-prey communities as an evolutionary game. Evolution (in press).

Cohen, D. 1966. Optimizing reproduction in a randomly varying environment. Journal of Theoretical Biology **12**: 119-129.

DeAngelis, D. L., J. A. Kitchell and W. M. Post. 1985. The influence of naticid predation on evolutionary strategies of bivalve prey: conclusions from a model. American Naturalist **126**: 817-842.

Edley, M. T. and R. Law. 1988. Evolution of life histories and yields in experimental populations of *Daphnia magna*. Biological Journal of the Linnean Society **34**: 309-326.

Favro, L. D., K. P. Kuo, J. F. McDonald, D. D. Favro and A. D. Kuo. 1982. A multilocus genetic model applied to the effects of selective fishing on the growth rate of trout. Journal of the Fisheries Research Board of Canada **39**: 1540-1543.

Garrod, G. J. 1988. North atlantic cod: fisheries and management to 1986. Pp.185 -218, in J.A. Gulland ed., Fish Population Dynamics (Second edition). Wiley, London.

Gurney, W. S. C., S. P. Blythe and R. M. Nisbet. 1980. Nicholson's blowflies revisited. Nature **287**: 17-21.

Hillesheim, E. and S. C. Stearns. 1991. The responses of *Drosophila melanogaster* to artificial selection on body weight and its phenotypic plasticity in two larval food environments. Evolution **45**: 1909-1923.

Law, R. and D. R. Grey. 1989. Evolution of yields from populations with age-specific cropping. Evolutionary Ecology **3**: 343-359.

Macdonald, S. E. and C. C. Chinnappa. 1989. Population differentiation for phenotypic plasticity in the *Stellaria longipes* complex. American Journal of Botany **76**: 1627-1637.

Maynard Smith, J. and G. R. Price. 1973. The logic of animal conflict. Nature **246**: 15-18.

Murphy, G. I. 1967. Vital statistics of the pacific sardine (*Sardinops caerulea*) and the population consequences. Ecology **48**: 731-736.

Nelson, K. and M. Soulé. 1987. Genetical conservation of exploited fishes. Pp. 345-368, in N. Ryman and F. Utter eds., Population Genetics and Fishery Management. University of Washington Press, Seattle.

Nisbet, R. M., W. S. C. Gurney and J. A. J. Metz. 1989. Stage structure models applied in evolutionary ecology. Lecture Notes in Biomathematics **18**: 428-449.

Parejko, K. and S. I. Dodson. 1991. The evolutionary ecology of an antipredator reaction norm: *Daphnia pulex* and *Chaoborus americanus*. Evolution **45**: 1665-1674.

Reznick, D. A., H. Bryga and J. A. Endler. 1990. Experimentally induced life-history evolution in a natural population. Nature **346**: 357-359.

Trexler, J. C. and J. Travis. 1990. Phenotypic plasticity in the sailfin molly, *Poecilia latipinna* (Pisces:Poeciliidae). I. Field experiments. Evolution **44**: 143-156.

Trexler, J. C., J. Travis and M. Trexler. 1990. Phenotypic plasticity in the sailfin molly, *Poecilia latipinna* (Pisces: Poeciliidae). II. Laboratory experiment. Evolution **44**: 157-167.

Vincent, T. L. and J. S. Brown. 1984. Stability in an evolutionary game. Theoretical Population Biology **26**: 408-427.

Vincent, T. L. and J. S. Brown. 1987. An evolutionary response to harvesting. Lecture Notes in Biomathematics **72**: 80-95.

Vincent, T. L. and J. S. Brown. 1988. The evolution of ESS theory. Annual Review of Ecology and Systematics **19**: 423-443.

Vermeij, G. J. 1977. The mesozoic marine revolution: evidence from snails, predators and grazers. Paleobiology **3**: 245-258.

LIST OF ADDRESSES OF PARTICIPANTS

Dr. S P. Blythe
Biometrics Unit
Center for Applied Mathematics
Cornell University
Ithaca, New York
USA
Tel: 44-742-768555 Ext. 4297
Fax: 44-742-739826

Dr. J. A. Brown
Ocean Sciences Centre
Memorial University
St. John's AIC 557
CANADA

Dr. J. S. Brown
Dept. of Biological Sciences
University of Illinois at Chicago
Box 4348, Illinois 60680
USA
Tel: 1-312-996-4289
Fax: 1-312-413-2435

Dr. R. W. Doyle
Dept. of Biology
Dalhousie University
Halifax, Nova Scotia B3H 4J1
CANADA
Tel: 1-902-494-3515
Fax: 1-902-494-3736

Professor W. M. Getz
College of Natural Resources
Agricultural Experiment Station
University of California
Berkeley CA 94720
USA
Tel: 1-415-642-8745
Fax: 1-415-642-4612

Dr. D.R. Grey
Dept. of Probability & Statistics
The University
Sheffield S3 7RH
UK

Dr. J. Hutchings
Dept. of Zoology
University of Edinburgh
West Mains Road
Edinburgh EH9 3JT
UK
Tel: 44-31-650-5440
Fax: 44-31-667-3210

Dr. V. Kaitala
Systems Analysis Laboratory
Helsinki University of Technology
Otakaari 1
02150 Espoo
FINLAND
Tel: 358-0-4513096
Fax: 358-0-4513056

Dr. M. Kirkpatrick
Dept. of Zoology
University of Texas
Austin
Texas TX 78712
USA
Tel: 1-512-471-7131
Fax: 1-512-471-9651

Dr. R. Law
Dept. of Biology
University of York
York YO1 5DD
UK
Tel: 44-904-432820
Fax: 44-904-432860

Professor J. M. McGlade
Biological Sciences
University of Warwick
Coventry CV4 7AL
UK

Tel: 44-203-524620
Fax: 44-203-524619

Dr. R. A. Myers
Science Branch
Dept. of Fisheries & Oceans
P.O. Box 5667, St. John's
Newfoundland A1C 5X1
CANADA
Tel: 1-709-772-2341
Fax: 1-709-772-2156

Dr. K. Nelson
Bodega Marine Laboratory
P.O. Box 247, Bodega Bay
CA 94923
USA
Tel: 1-707-875-221
Fax: 1-707-875-2089

Dr. D. Policansky
National Research Council
Board of Environmental Studies & Toxicology
2101, Constitution Avenue
Washington D.C. 20418
USA
Tel: 1-202-334-2234
Fax: 1-202-334-2752

Professor D. A. Rand
Mathematics Institute
University of Warwick
Coventry CV4 7AL
UK
Tel: 44-203-523599
Fax: 44-203-524619

Dr. D. Reznick
Dept. of Biology
University of California
Riverside
CA 92521
USA
Tel: 1-714-787-5903
Fax: 1-714-787-4286

Dr. A. Rijnsdorp
Rijksinstitut voor Visserijonderzoek
Postbus 68
1970 AB IJmuiden
The Netherlands
Tel: 31-2550-64646
Fax: 31-2550-64644

Dr. C. Rowell
Dept. of Zoology
University of Cambridge
Cambridge CB2 3EJ
UK
Tel: 44-223-336600
Fax: 44-223-336676

Dr. A. Sharov
VNIRO
17, V. Krasnoselskaya
Moscow B-140
107140 RUSSIA
Tel: 284-9454
Fax: 95-265-9187

Professor R. H. Smith
School of Biological Sciences
University of Leicester
Leicester LE1 7RH
UK
Tel: 44-533-523381
Fax: 44-533-522791

Dr. T. K. Stokes
Directorate of Fisheries Research
Pakefield Road
Lowestoft NR33 0HT
UK
Tel: 44-502-524360
Fax: 44-502-513865